Introduction to Spectral Analysis

Introduction to Spectral Analysis

Petre Stoica
Upsala University

Randolph L. Moses
Ohio State University

Prentice Hall
Upper Saddle River, New Jersey 07458

Library of Congress Cataloging-in-Publication Data

Stoica, Petre.
Introduction to spectral analysis/ Petre Stoica and Randolph L. Moses.
p. cm.
Includes bibliographical references and index..
ISBN:0-13-258419-0
1. Spectral theory (Mathematics) I. Moses, Randolph L.
II. Title
QA320.S86 1997 96-35123
519.5'4--dc20 CIP

Acquisitions editor: Tom Robbins
Editor in Chief: Marcia Horton
Production editor: Ann Marie Longobardo
Copy editor: Sharyn Vitrano
Director of production and manufacturing: David W. Riccardi
Managing editor: Bayani Mendoza de Leon
Cover designer: Karen Salzbach
Manufacturing buyer: Donna Sullivan
Editorial assistant: Nancy Garcia

A Pearson Education Company
Upper Saddle River, NJ 07458

Printed in the United States of America

10 9 8 7 6 5 4 3

ISBN 0-13-258419-0

Prentice-Hall International (UK) Limited,London
Prentice-Hall of Australia Pty. Limited, Sydney
Prentice-Hall Canada Inc., Toronto
Prentice-Hall Hispanoamericana, S.A., Mexico
Prentice-Hall of India Private Limited, New Delhi
Prentice-Hall of Japan, Inc., Tokyo
Pearson Education Asia Pte. Ltd., Singapore
Editora Prentice-Hall do Brasil, Ltda., Rio de Janeiro

The MathWorks, Inc.
24 PrimePark Way
Natick, MA 01760
Tel: (508) 647-7000
Fax: (508)647-7001
E-mail: info@mathworks.com
WWW:http://www.mathworks.com

CONTENTS

LIST OF EXERCISES

Chapter 3

Chapter 4

Chapter 5

Chapter 6

PREFACE

Spectral analysis considers the problem of determining the spectral content (*i.e.*, the distribution of power over frequency) of a time series from a finite set of measurements, by means of either nonparametric or parametric techniques. The history of spectral analysis as an established discipline started about one century ago with the work by Schuster on detecting cyclic behavior in time series. An interesting historical perspective on the developments in this field can be found in [MARPLE 1987]. This reference notes that the word "spectrum" was apparently introduced by Newton in relation to his studies of the decomposition of white light into a band of light colors, when passed through a glass prism (as illustrated on the front cover). This word appears to be a variant of the Latin word "specter" which means "ghostly apparition". The contemporary English word that has the same meaning as the original Latin word is "spectre". Despite these roots of the word "spectrum", we hope the student will be a "vivid presence" in the course that has just started!

This text is designed to be used with a first course in spectral analysis that would typically be offered as a senior or first–year graduate course on the subject. The book should also be useful for self-study, as it is largely self-contained. The text is concise by design, so that it gets to the main points quickly and should hence be appealing to those who would like a fast appraisal on the classical and modern approaches of spectral analysis.

In order to keep the book concise, without sacrificing the rigor of presentation or skipping over essential aspects, we have limited or avoided coverage of some advanced topics in spectral estimation. For example, we do not include a detailed quantitative analysis of the statistical properties of the various estimators. A complete coverage of these statistical properties would greatly increase the length of the book, and even then provide only asymptotic (in data length or signal-to-noise ratio) analytical results. Also, we do not consider the order estimation problem for parametric techniques. On the other hand, we address the aforementioned topics in a number of ways. We include algorithms that have near–optimum statistical performance, and provide qualitative discussions on accuracy that are based on the quantitative arguments in the research literature. We also include computer exercises that illustrate many of the main points on estimator accuracy and order estimation via numerical examples.

In keeping with our theme of a concise introductory text, we also have omitted the following topics: spectral analysis based on higher–order statistics, time-frequency analysis and scale representations, and multidimensional spectral analysis. While these topics are receiving considerable attention in the current research literature, we consider them "too advanced" for a concise first course in spectral analysis.

At the end of each chapter we have included text complements, analytical exercises and computer problems. The complements present material which we consider to be more advanced or perhaps not in the mainstream of the text material. The analytical exercises are more–or–less ordered from least to most difficult; this ordering also approximately follows the chronological presentation of material in the chapters. The more difficult exercises sometimes explore advanced topics in spectral analysis and provide results which are not available in the main text. The computer problems are designed to illustrate the main points of the text and to provide the reader with first–hand information on the behavior and performance of the various spectral analysis techniques considered. The computer exercises also illustrate the relative performance of the methods and explore other issues as well, such as statistical accuracy, order estimation, resolution properties, and the like, that are not analytically developed in the book. We have used MATLAB[1] to minimize the programming chore and to encourage the reader to "play" with other examples. We provide a set of MATLAB functions for data generation and spectral estimation that forms a basis for a comprehensive set of spectral estimation tools; these functions are available at the text web site `www.prenhall.com/~stoica`.

Supplementary material may also be obtained from the text web site. We have prepared a set of overhead transparencies which can be used as a teaching aid for a spectral analysis course. We believe that these transparencies are useful not only to course instructors but also to other readers, because they summarize the principal methods and results in the text. For the readers who study the topic on their own, it should be a useful exercise to refer to the main points addressed in the transparencies after completing the reading of each chapter.

A Solutions Manual, which contains solutions to all analytical and computer exercises in the text, is available to course instructors, and can be obtained by writing to Prentice-Hall.

The text is organized as follows. Chapter 1 introduces the spectral analysis problem, motivates the definition of power spectral density functions, and reviews some important properties of autocorrelation sequences and spectral density functions. Chapters 2 and 5 consider nonparametric spectral estimation. Chapter 2 presents classical techniques, including the periodogram, the correlogram, and their modified versions to reduce variance. We include an analysis of bias and variance of these techniques, and relate them to one another. Chapter 5 considers the more recent filter bank version of nonparametric techniques, including both data-independent

[1] MATLAB® is a registered trademark of The Mathworks, Inc.

filter design techniques and data-dependent filter design techniques (*e.g.*, the Capon estimator). Chapters 3 and 4 consider parametric techniques; Chapter 3 focuses on continuous spectral models (Autoregressive Moving Average (ARMA) models and their AR and MA special cases), while Chapter 4 focuses on discrete spectral models (sinusoids in noise). We have placed the filter bank methods in Chapter 5, after Chapters 3 and 4, mainly because the Capon estimator has interpretations as both an averaged AR spectral estimator and as a matched filter for line spectral models, and we need the background of Chapters 3 and 4 to develop these interpretations. The data-independent filter bank techniques in Sections 5.1–5.4 can equally well be covered directly following Chapter 2, if desired.

Chapter 6 considers the closely-related problem of spatial spectral estimation in the context of array signal processing. Both nonparametric (beamforming) and parametric methods are considered, and tied into the temporal spectral estimation techniques considered in Chapters 2, 4 and 5.

The Bibliography contains both modern and classical references (ordered both alphabetically and by subject). We include many of the historical references as well, for those interested in tracing the early developments of spectral analysis. However, spectral analysis is a topic with contributions from many diverse fields, including electrical and mechanical engineering, geophysics, mathematical statistics, and econometrics to name a few. As such, any attempt to accurately document the historical development of spectral analysis is doomed to failure. The bibliography reflects our own perspectives, biases, and limitations; while there is no doubt that the list is incomplete, we hope that it gives the reader an appreciation of the breadth and diversity of contributions to the spectral analysis field.

The background needed for this text includes a basic knowledge of linear algebra, discrete-time linear systems, and introductory discrete-time stochastic processes (or time series). A basic understanding of estimation theory is helpful, though not required. Appendix A contains most of the needed background results on matrices and linear algebra, and Appendix B gives a tutorial introduction to the Cramér-Rao bound. Regarding the other topics, we have included concise definitions and descriptions of the required concepts and results in the text where needed. Thus, we have tried to make the text as self-contained as possible.

We would like to thank the many people and organizations that contributed to this book. Torsten Söderström provided the initial stimulus for preparation of lecture notes that led to the book. We are indebted to Hung-Chih Chiang, Peter Händel, Ari Kangas, Erlendur Karlsson and Lee Swindlehurst for careful proofreading and comments, and for many ideas on and early drafts of the computer problems and figures in the text. We would like to thank Jian Li and Lee Potter for adopting a former version of the text in their spectral estimation classes, and for their valuable feedback. We wish to thank the Systems and Control Group at Uppsala University for providing the fertile environment in which much of this book was written. We are very grateful to Jackie Buckner, who patiently and skillfully typed many of our notes. We also wish to thank Wallace Anderson, Alfred Hero, Ralph Hippenstiel, Louis Scharf, and Douglas Williams, who reviewed the manuscript and provided us with numerous useful comments and suggestions. It was a pleasure to work with the excellent staff at Prentice Hall, and we are particularly appreciative of Tom Robbins, Phyllis Morgan, Nancy Garcia, and Ann-Marie Longobardo for their cheerful help and professional expertise.

Many of the topics described in this book are outgrowths of our research programs in statistical signal and array processing, and we wish to thank the sponsors of this research: the Swedish Research Council for Engineering Sciences, the Swedish National Board for Technical Development, The Swedish Institute, the U.S. Army Research Laboratory, the U.S. Air Force Wright Laboratory, and the U.S. Advanced Research Projects Administration.

Finally, we are indebted to Anca and Liz for their continuing support and understanding throughout this project.

Petre Stoica
Uppsala University

Randy Moses
The Ohio State University

NOTATIONAL CONVENTIONS AND ABBREVIATIONS

Symbols:

Dn	the nth definition in Appendix A or B
Rn	the nth result in Appendix A or B
$\|x\|$	the Euclidean norm of a vector x.
$*$	convolution operator
$(\cdot)^T$	transpose of a vector or matrix
$(\cdot)^c$	conjugate of a vector or matrix
$(\cdot)^*$	conjugate transpose of a vector or matrix; also used for scalars in lieu of $(\cdot)^c$
A_{ij}	the (i, j)th element of the matrix A
a_i	the ith element of the vector a
$\hat{x}$	an estimate of the quantity x
$A > 0\ (\geq 0)$	A is positive definite (positive semidefinite); (p. 269)
$\arg\max_x f(x)$	the value of x that maximizes $f(x)$
$\arg\min_x f(x)$	the value of x that minimizes $f(x)$
$\text{cov}\{x, y\}$	the covariance between x and y
$\lvert x\rvert$	the modulus of the (possibly complex) scalar x
$\lvert A\rvert$	the determinant of the square matrix A
$\text{diag}(a)$	the square matrix whose diagonal elements are the elements of the vector a

$\delta_{k,l}$	Kronecker delta: $\delta_{k,l} = 1$ if $k = l$ and $\delta_{k,l} = 0$ otherwise
$\delta(t - t_0)$	Dirac delta: $\delta(t - t_0) = 0$ for $t \neq t_0$; $\int_{-\infty}^{\infty} \delta(t - t_0) dt = 1$
$E\{x\}$	the expected value of x (p. 5)
f	(discrete-time) frequency: $f = \omega/2\pi$, in cycles per sampling interval (p. 8)
$\phi(\omega)$	a power spectral density function (p. 6)
$\text{Im}\{x\}$	the imaginary part of x
$O(x)$	on the order of x (p. 34)
$p(x)$	probability density function
$\text{Pr}\{A\}$	the probability of event A
q^{-1}	unit delay operator: $q^{-1}x(t) = x(t - 1)$ (p. 10)
$r(k)$	an autocovariance sequence (p. 5)
$\text{Re}\{x\}$	the real part of x
t	discrete-time index
$\text{tr}(A)$	the trace of the matrix A (p. 259)
$\text{var}\{x\}$	the variance of x
$w(k)$, $W(\omega)$	a window sequence and its Fourier transform
$w_B(k)$, $W_B(\omega)$	the Bartlett (or triangular) window sequence and its Fourier transform (p. 30)
$w_R(k)$, $W_R(\omega)$	the rectangular (or Dirichlet) window sequence and its Fourier transform (p. 32)
ω	radian (angular) frequency, in radians/sampling interval (p. 3)

Abbreviations:

ACS	autocovariance sequence (p. 5)
AR	autoregressive (p. 87)
ARMA	autoregressive moving-average (p. 87)
BSP	beamspace processing (p. 250)
BT	Blackman-Tukey (p. 38)
CM	Capon method (p. 196)
CRB	Cramér-Rao bound (p. 285)
DFT	discrete Fourier transform (p. 27)
DGA	Delsarte-Genin algorithm (p. 94)
DOA	direction of arrival (p. 222)
DTFT	discrete-time Fourier transform (p. 3)
ESP	elementspace processing (p. 250)
ESPRIT	estimation of signal parameters by rotational invariance techniques (p. 163)
EVD	eigenvalue decomposition (p. 258)
FB	forward-backward (p. 165)
FBA	filter bank approach (p. 182)
FFT	fast Fourier transform (p. 27)
FIR	finite impulse response (p. 18)
flop	floating point operation (p. 27)
GS	Gohberg-Semencul (formula) (p. 123)
HOYW	high–order Yule–Walker (p. 151)
LDA	Levinson–Durbin algorithm (p. 94)
LS	least squares (p. 279)
MA	moving-average (p. 87)
MFD	matrix fraction description (p. 134)
ML	maximum likelihood (p. 287)
MLE	maximum likelihood estimate (p. 287)
MSE	mean squared error (p. 29)
MUSIC	multiple signal classification (or characterization) (p. 155)
MYW	modified Yule–Walker (p. 97)
NLS	nonlinear least squares (p. 140)

PARCOR	partial correlation (p. 97)
PSD	power spectral density (p. 5)
RFB	refined filter bank (p. 186)
QRD	Q-R decomposition (p. 281)
SNR	signal-to-noise ratio (p. 79)
SVD	singular value decomposition (p. 264)
TLS	total least squares (p. 281)
ULA	uniform linear array (p. 230)
YW	Yule–Walker (p. 89)

Introduction to Spectral Analysis

Chapter 1

BASIC CONCEPTS

1.1 Introduction

The essence of the spectral estimation problem is captured by the following informal formulation.

From a finite record of a stationary data sequence, estimate how the total power is distributed over frequency.	(1.1.1)

Spectral analysis finds applications in many diverse fields. In *vibration monitoring*, the spectral content of measured signals give information on the wear and other characteristics of mechanical parts under study. In *economics*, *meteorology*, *astronomy* and several other fields, the spectral analysis may reveal "hidden periodicities" in the studied data, which are to be associated with cyclic behavior or recurring processes. In *speech analysis*, spectral models of voice signals are useful in better understanding the speech production process, and — in addition — can be used for both speech synthesis (or compression) and speech recognition. In *radar and sonar systems*, the spectral contents of the received signals provide information on the location of the sources (or targets) situated in the field of view. In *medicine*, spectral analysis of various signals measured from a patient, such as electrocardiogram (ECG) or electroencephalogram (EEG) signals, can provide useful material for diagnosis. In *seismology*, the spectral analysis of the signals recorded prior to and during a seismic event (such as a volcano eruption or an earthquake) gives useful information on the ground movement associated with such events and may help in predicting them. Seismic spectral estimation is also used to predict subsurface geologic structure in gas and oil exploration. In *control systems*, there is a resurging interest in spectral analysis methods as a means of characterizing the dynamical behavior of a given system, and ultimately synthesizing a controller for that system. The previous and other applications of spectral analysis are reviewed in [KAY 1988; MARPLE 1987; BLOOMFIELD 1976; BRACEWELL 1986; HAYKIN 1991; HAYKIN 1995; KOOPMANS 1974; PRIESTLEY 1989; PERCIVAL AND WALDEN 1993; PORAT 1994; SCHARF 1991; THERRIEN 1992; PROAKIS, RADER, LING, AND NIKIAS 1992]. The textbook [MARPLE 1987] also contains a well-written historical perspective on spectral estimation which is worth reading. Many of the classical articles on spectral analysis, both application–driven and theoretical, are reprinted in [CHILDERS 1978; KESLER 1986]; these excellent collections of reprints are well worth consulting.

There are *two broad approaches* to spectral analysis. One of these derives its basic idea directly from definition (1.1.1): the studied signal is applied to a bandpass filter with a narrow bandwidth, which is swept through the frequency band of interest, and the filter output power divided by the filter bandwidth is used as a measure of the spectral contents of the input to the filter. This is essentially what the *classical* (or *nonparametric*) *methods* of spectral analysis do. These methods are described in Chapters 2 and 5 of this text (the fact that the methods of Chapter 2 can be given the above filter bank interpretation is made clear in Chapter 5). The second approach to spectral estimation, called the *parametric approach*, is to postulate a model for the data, which provides a means of parameterizing the spectrum, and to thereby reduce the spectral estimation problem to that of estimating the parameters in the assumed model. The parametric approach to spectral analysis is treated in Chapters 3, 4 and 6. Parametric methods may offer more accurate spectral estimates than the nonparametric ones in the cases where the data indeed satisfy the model assumed by the former methods. However, in the more likely case that the data do not satisfy the assumed models, the nonparametric methods may outperform the parametric ones owing to the sensitivity of the latter to model misspecifications. This observation has motivated renewed interest in the nonparametric approach to spectral estimation.

Many real–world signals can be characterized as being *random* (from the observer's viewpoint). Briefly speaking, this means that the variation of such a signal outside the observed interval cannot be determined exactly but only specified in statistical terms of averages. In this text, we will be concerned with estimating the spectral characteristics of random signals. In spite of this fact, we find it useful to start the discussion by considering the spectral analysis of deterministic signals (which we do in the first section of this chapter). Throughout this work, we consider *discrete signals* (or *data sequences*). Such signals are most commonly obtained by the temporal or spatial sampling of a continuous (in time or space) signal. The main motivation for focusing on discrete signals lies in the fact that spectral analysis is most often performed by a digital computer or by digital circuitry. Chapters 2 to 5 of this text deal with *discrete–time signals*, while Chapter 6 considers the case of *discrete–space data sequences.*

In the interest of notational simplicity, the discrete–time variable t, as used in this text, is assumed to be measured in units of sampling interval. A similar convention is adopted for spatial signals, whenever the sampling is uniform. Accordingly, the *units of frequency* are cycles per sampling interval.

The signals dealt with in the text are *complex–valued.* Complex–valued data may appear in signal processing and spectral estimation applications, for instance, as a result of a "complex demodulation" process (this is explained in detail in Chapter 6). It should be noted that the treatment of complex–valued signals is not always more general or more difficult than the analysis of corresponding real–valued signals. A typical example which illustrates this claim is the case of sinusoidal signals considered in Chapter 4. A real–valued sinusoidal signal, $\alpha\cos(\omega t+\varphi)$, can be rewritten as a linear combination of two complex–valued sinusoidal signals, $\alpha_1 e^{i(\omega_1 t+\varphi_1)}+\alpha_2 e^{i(\omega_2 t+\varphi_2)}$, whose parameters are constrained as follows: $\alpha_1=\alpha_2=\alpha/2$, $\varphi_1=-\varphi_2=\varphi$ and $\omega_1=-\omega_2=\omega$. Here $i=\sqrt{-1}$. The fact that we need to consider *two constrained* complex sine waves to treat the case of *one unconstrained* real sine wave shows that the real–valued case of sinusoidal signals can actually be considered to be more complicated than the complex–valued case! Fortunately, it appears that the latter case is encountered more frequently in applications, where often both the *in–phase and quadrature* components of the studied signal are available. (For more details and explanations on this aspect, see Chapter 6's introductory section.)

1.2 Energy Spectral Density of Deterministic Signals

Let $\{y(t); t=0,\pm1,\pm2,\ldots\}$ denote a *deterministic* discrete–time data sequence. Most commonly, $\{y(t)\}$ is obtained by sampling a continuous–time signal. For notational convenience, the time index t is expressed in units of sampling interval; that is, $y(t)=y_c(t\cdot T_s)$, where $y_c(\cdot)$ is the continuous time signal and T_s is the sampling time interval.

Assume that $\{y(t)\}$ has *finite energy*, which means that

$$\sum_{t=-\infty}^{\infty}|y(t)|^2<\infty \tag{1.2.1}$$

Then the sequence $\{y(t)\}$ possesses a *discrete–time Fourier transform* (DTFT) defined as

$$\boxed{Y(\omega)=\sum_{t=-\infty}^{\infty}y(t)e^{-i\omega t} \qquad \text{(DTFT)}} \tag{1.2.2}$$

In this text we use the symbol $Y(\omega)$, in lieu of the more cumbersome $Y(e^{i\omega})$, to denote the DTFT. This notational convention is commented on a bit later, following equation (1.4.6). The corresponding inverse DTFT is then

$$\boxed{y(t)=\frac{1}{2\pi}\int_{-\pi}^{\pi}Y(\omega)e^{i\omega t}d\omega \qquad \text{(Inverse DTFT)}} \tag{1.2.3}$$

which can be verified by substituting (1.2.3) into (1.2.2). The (angular) *frequency* ω is measured in radians per sampling interval. The conversion from ω to the *physical frequency variable* $\bar{\omega}=\omega/T_s$ [rad/sec] can be done in a straightforward manner, as described in Exercise 1.1.

Let

$$\boxed{S(\omega)=|Y(\omega)|^2 \qquad \text{(Energy Spectral Density)}} \tag{1.2.4}$$

A straightforward calculation gives

$$\frac{1}{2\pi}\int_{-\pi}^{\pi} S(\omega)d\omega = \frac{1}{2\pi}\int_{-\pi}^{\pi}\sum_{t=-\infty}^{\infty}\sum_{s=-\infty}^{\infty} y(t)y^*(s)e^{-i\omega(t-s)}d\omega$$
$$= \sum_{t=-\infty}^{\infty}\sum_{s=-\infty}^{\infty} y(t)y^*(s)\left[\frac{1}{2\pi}\int_{-\pi}^{\pi} e^{-i\omega(t-s)}d\omega\right]$$
$$= \sum_{t=-\infty}^{\infty} |y(t)|^2 \qquad (1.2.5)$$

To obtain the last equality in (1.2.5) we have used the fact that $\frac{1}{2\pi}\int_{-\pi}^{\pi} e^{-i\omega(t-s)}d\omega = \delta_{t,s}$ (the Kronecker delta). The symbol $(\cdot)^*$ will be used in this text to denote the complex–conjugate of a scalar variable or the conjugate transpose of a vector or matrix. Equation (1.2.5) can be restated as

$$\sum_{t=-\infty}^{\infty} |y(t)|^2 = \frac{1}{2\pi}\int_{-\pi}^{\pi} S(\omega)d\omega \qquad (1.2.6)$$

This equality is called *Parseval's theorem.* It shows that $S(\omega)$ represents the distribution of sequence energy as a function of frequency. For this reason, $S(\omega)$ is called the ***energy spectral density.***

The previous interpretation of $S(\omega)$ also comes up in the following way. Equation (1.2.3) represents the sequence $\{y(t)\}$ as a weighted "sum" (actually, an integral) of orthonormal sequences $\{\frac{1}{\sqrt{2\pi}}e^{i\omega t}\}$ $(\omega \in [-\pi, \pi])$, with weighting $\frac{1}{\sqrt{2\pi}}Y(\omega)$. Hence, $\frac{1}{\sqrt{2\pi}}|Y(\omega)|$ "measures" the "length" of the projection of $\{y(t)\}$ on each of these basis sequences. In loose terms, therefore, $\frac{1}{\sqrt{2\pi}}|Y(\omega)|$ shows how much (or how little) of the sequence $\{y(t)\}$ can be "explained" by the orthonormal sequence $\{\frac{1}{\sqrt{2\pi}}e^{i\omega t}\}$ for some given value of ω.

Define

$$\rho(k) = \sum_{t=-\infty}^{\infty} y(t)y^*(t-k) \qquad (1.2.7)$$

It is readily verified that

$$\sum_{k=-\infty}^{\infty} \rho(k)e^{-i\omega k} = \sum_{k=-\infty}^{\infty}\sum_{t=-\infty}^{\infty} y(t)y^*(t-k)e^{-i\omega t}e^{i\omega(t-k)}$$
$$= \left[\sum_{t=-\infty}^{\infty} y(t)e^{-i\omega t}\right]\left[\sum_{s=-\infty}^{\infty} y(s)e^{-i\omega s}\right]^*$$
$$= S(\omega) \qquad (1.2.8)$$

which shows that $S(\omega)$ can be obtained as the DTFT of the "autocorrelation" (1.2.7) of the finite–energy sequence $\{y(t)\}$.

The above definitions can be extended in a rather straightforward manner to the case of random signals treated throughout the remaining text. In fact, the only purpose for discussing the deterministic case in this section was to provide some motivation for the analogous definitions in the random case. As such, the discussion in this section has been kept brief. More insights into the meaning and properties of the previous definitions are provided by the detailed treatment of the random case in the following sections.

1.3 Power Spectral Density of Random Signals

Most of the signals encountered in applications are such that their variation in the future cannot be known exactly. It is only possible to make probabilistic statements about that variation. The mathematical device to describe such a signal is that of a *random sequence* which consists of an ensemble of possible realizations, each of which has some associated probability to occur. Of course, from the whole ensemble of realizations, the experimenter can usually observe only one realization of the signal, and then it might be thought that the deterministic definitions of the previous section could be carried over unchanged to the present case. However, this is not possible because the realizations of a random signal, viewed as discrete–time sequences, do not have finite energy, and hence do not possess DTFTs. A random signal usually has finite *average* power and, therefore, can be characterized by an average power spectral density. For simplicity reasons, in what follows we will use the name *power spectral density* (PSD) for that quantity.

The discrete–time signal $\{y(t); t = 0, \pm 1, \pm 2, \ldots\}$ is assumed to be a sequence of random variables with *zero mean*:

$$E\{y(t)\} = 0 \qquad \text{for all } t \tag{1.3.1}$$

Hereafter, $E\{\cdot\}$ denotes the expectation operator (which averages over the ensemble of realizations). The *autocovariance sequence* (ACS) or *covariance function* of $y(t)$ is defined as

$$\boxed{r(k) = E\{y(t)y^*(t-k)\}} \tag{1.3.2}$$

and it is assumed to depend only on the lag between the two samples averaged. The two assumptions (1.3.1) and (1.3.2) imply that $\{y(t)\}$ is a *second–order stationary sequence*. When it is required to distinguish between the autocovariance sequences of several signals, a lower index will be used to indicate the signal associated with a given covariance lag, such as $r_y(k)$.

The autocovariance sequence $r(k)$ enjoys some simple but useful properties:

$$\boxed{r(k) = r^*(-k)} \tag{1.3.3}$$

and

$$\boxed{r(0) \geq |r(k)| \qquad \text{for all } k} \tag{1.3.4}$$

The equality (1.3.3) directly follows from definition (1.3.2) and the stationarity assumption, while (1.3.4) is a consequence of the fact that the *covariance matrix* of $\{y(t)\}$, defined as follows

$$R_m = \begin{bmatrix} r(0) & r^*(1) & \dots & r^*(m-1) \\ r(1) & r(0) & \ddots & \vdots \\ \vdots & \ddots & \ddots & r^*(1) \\ r(m-1) & \dots & r(1) & r(0) \end{bmatrix}$$

$$= E\left\{ \begin{bmatrix} y^*(t-1) \\ \vdots \\ y^*(t-m) \end{bmatrix} [y(t-1) \dots y(t-m)] \right\} \tag{1.3.5}$$

is positive semidefinite for all m. Recall that a Hermitian matrix M is positive semidefinite if $a^*Ma \geq 0$ for every vector a (see Section A.5 for details). Since

$$a^* R_m a = a^* E\left\{ \begin{bmatrix} y^*(t-1) \\ \vdots \\ y^*(t-m) \end{bmatrix} [y(t-1) \dots y(t-m)] \right\} a$$

$$= E\{z^*(t)z(t)\} = E\{|z(t)|^2\} \geq 0 \tag{1.3.6}$$

where

$$z(t) = [y(t-1) \dots y(t-m)]a$$

we see that R_m is indeed positive semidefinite for every m. Hence, (1.3.4) follows from the properties of positive semidefinite matrices (see Definition D11 in Appendix A and Exercise 1.5).

1.3.1 First Definition of Power Spectral Density

The PSD is defined as the DTFT of the covariance sequence:

$$\boxed{\phi(\omega) = \sum_{k=-\infty}^{\infty} r(k) e^{-i\omega k} \qquad \text{(Power Spectral Density)}} \tag{1.3.7}$$

Note that the previous definition (1.3.7) of $\phi(\omega)$ is similar to the definition (1.2.8) in the deterministic case. The inverse transform, which recovers $\{r(k)\}$ from given $\phi(\omega)$, is

$$\boxed{r(k) = \frac{1}{2\pi} \int_{-\pi}^{\pi} \phi(\omega) e^{i\omega k} d\omega} \tag{1.3.8}$$

We readily verify that

$$\frac{1}{2\pi}\int_{-\pi}^{\pi}\phi(\omega)e^{i\omega k}d\omega=\sum_{p=-\infty}^{\infty}r(p)\left[\frac{1}{2\pi}\int_{-\pi}^{\pi}e^{i\omega(k-p)}d\omega\right]=r(k)$$

which proves that (1.3.8) is the inverse transform for (1.3.7). Note that to obtain the first equality above, the order of integration and summation has been inverted, which is possible under weak conditions (such as under the requirement that $\phi(\omega)$ is square integrable; see Chapter 4 in [PRIESTLEY 1989] for a detailed discussion on this aspect).

From (1.3.8), we obtain

$$r(0)=\frac{1}{2\pi}\int_{-\pi}^{\pi}\phi(\omega)d\omega \tag{1.3.9}$$

Since $r(0)=E\left\{|y(t)|^2\right\}$ measures the (average) power of $\{y(t)\}$, the equality (1.3.9) shows that $\phi(\omega)$ can indeed be named PSD, as it represents the distribution of the (average) signal power over frequencies. Put another way, it follows from (1.3.9) that $\phi(\omega)d\omega/2\pi$ is the infinitesimal power in the band $(\omega-d\omega/2,\,\omega+d\omega/2)$, and the total power in the signal is obtained by integrating these infinitesimal contributions. Additional motivation for calling $\phi(\omega)$ a PSD is provided by the second definition of $\phi(\omega)$, given next, which resembles the usual definition (1.2.2), (1.2.4) in the deterministic case.

1.3.2 Second Definition of Power Spectral Density

The second definition of $\phi(\omega)$ is:

$$\boxed{\phi(\omega)=\lim_{N\to\infty}E\left\{\frac{1}{N}\left|\sum_{t=1}^{N}y(t)e^{-i\omega t}\right|^2\right\}} \tag{1.3.10}$$

This definition is equivalent to (1.3.7) under the mild assumption that the covariance sequence $\{r(k)\}$ decays sufficiently rapidly, so that

$$\lim_{N\to\infty}\frac{1}{N}\sum_{k=-N}^{N}|k||r(k)|=0 \tag{1.3.11}$$

The equivalence of (1.3.7) and (1.3.10) can be verified as follows:

$$\begin{aligned}\lim_{N\to\infty}E\left\{\frac{1}{N}\left|\sum_{t=1}^{N}y(t)e^{-i\omega t}\right|^2\right\}&=\lim_{N\to\infty}\frac{1}{N}\sum_{t=1}^{N}\sum_{s=1}^{N}E\left\{y(t)y^*(s)\right\}e^{-i\omega(t-s)}\\&=\lim_{N\to\infty}\frac{1}{N}\sum_{\tau=-(N-1)}^{N-1}(N-|\tau|)r(\tau)e^{-i\omega\tau}\end{aligned}$$

$$
\begin{aligned}
&= \sum_{\tau=-\infty}^{\infty} r(\tau)e^{-i\omega\tau} - \lim_{N\to\infty} \frac{1}{N} \sum_{\tau=-(N-1)}^{N-1} |\tau| r(\tau) e^{-i\omega\tau} \\
&= \phi(\omega)
\end{aligned}
$$

The second equality is proven in Exercise 1.6, and we used (1.3.11) in the last equality.

The above definition of $\phi(\omega)$ resembles the definition (1.2.4) of energy spectral density in the deterministic case. The main difference between (1.2.4) and (1.3.10) consists of the appearance of the expectation operator in (1.3.10) and the normalization by $1/N$; the fact that the "discrete–time" variable in (1.3.10) runs over positive integers only is just for convenience and does not constitute an essential difference, compared to (1.2.2). In spite of these differences, the analogy between the deterministic formula (1.2.4) and (1.3.10) provides further motivation for calling $\phi(\omega)$ a PSD. The alternative definition (1.3.10) will also be quite useful when discussing the problem of estimating the PSD by nonparametric techniques in Chapters 2 and 5.

We can see from either of these definitions that $\phi(\omega)$ is a *periodic function*, with the period equal to 2π. Hence, $\phi(\omega)$ is completely described by its variation in the interval

$$
\boxed{\omega \in [-\pi, \pi] \qquad \text{(radians per sampling interval)}} \tag{1.3.12}
$$

Alternatively, the PSD can be viewed as a function of the frequency

$$
\boxed{f = \frac{\omega}{2\pi} \qquad \text{(cycles per sampling interval)}} \tag{1.3.13}
$$

which, according to (1.3.12), can be considered to take values in the interval

$$
\boxed{f \in [-1/2, 1/2]} \tag{1.3.14}
$$

We will generally write the PSD as a function of ω whenever possible, since this will simplify the notation.

As already mentioned, the discrete–time sequence $\{y(t)\}$ is most commonly derived by sampling a continuous–time signal. To avoid aliasing effects which might be incurred by the sampling process, the continuous–time signal should be (at least, approximately) bandlimited in the frequency domain. To ensure this, it may be necessary to low–pass filter the continuous–time signal before sampling. Let F_0 denote the largest ("significant") frequency component in the spectrum of the (possibly filtered) continuous signal, and let F_s be the *sampling frequency*. Then it follows from Shannon's sampling theorem that the continuous–time signal can be "exactly" reconstructed from its samples $\{y(t)\}$, provided that

$$
F_s \geq 2F_0 \tag{1.3.15}
$$

In particular, "no" frequency aliasing will occur when (1.3.15) holds (see, *e.g.*, [OPPENHEIM AND SCHAFER 1989]). Since the frequency variable, F, associated with the continuous–time signal, is related to f by the equation

$$F = f \cdot F_s \tag{1.3.16}$$

it follows that the interval of F corresponding to (1.3.14) is

$$\boxed{F \in \left[-\frac{F_s}{2}, \frac{F_s}{2}\right] \qquad \text{(cycles/sec)}} \tag{1.3.17}$$

which is quite natural in view of (1.3.15).

1.4 Properties of Power Spectral Densities

Since $\phi(\omega)$ is a power density, it should be real–valued and nonnegative. That this is indeed the case is readily seen from definition (1.3.10) of $\phi(\omega)$. Hence,

$$\boxed{\phi(\omega) \geq 0 \qquad \text{for all } \omega} \tag{1.4.1}$$

From (1.3.3) and (1.3.7), we obtain

$$\phi(\omega) = r(0) + 2\sum_{k=1}^{\infty} \mathrm{Re}\{r(k)e^{-i\omega k}\}$$

where $\mathrm{Re}\{\cdot\}$ denotes the real part of the bracketed quantity. If $y(t)$, and hence $r(k)$, is real valued then it follows that

$$\phi(\omega) = r(0) + 2\sum_{k=1}^{\infty} r(k)\cos(\omega k) \tag{1.4.2}$$

which shows that $\phi(\omega)$ is an even function in such a case. In the case of complex–valued signals, however, $\phi(\omega)$ is not necessarily symmetric about the $\omega = 0$ axis. Thus:

$$\boxed{\begin{array}{l} \text{For real–valued signals:} \\ \qquad \phi(\omega) = \phi(-\omega), \;\; \omega \in [-\pi, \pi] \\ \text{For complex–valued signals:} \\ \qquad \text{in general } \phi(\omega) \neq \phi(-\omega), \;\; \omega \in [-\pi, \pi] \end{array}} \tag{1.4.3}$$

Remark: The reader might wonder why we did not define the ACS as

$$c(k) = E\{y(t)y^*(t+k)\}$$

Comparing with the ACS $\{r(k)\}$ used in this text, as defined in (1.3.2), we obtain $c(k) = r(-k)$. Consequently, the PSD associated with $\{c(k)\}$ is related to the PSD corresponding to $\{r(k)\}$ (see (1.3.7)) via:

$$\psi(\omega) \triangleq \sum_{k=-\infty}^{\infty} c(k)e^{-i\omega k} = \sum_{k=-\infty}^{\infty} r(k)e^{i\omega k} = \phi(-\omega)$$

It may seem arbitrary as to which definition of the ACS (and corresponding definition of PSD) we choose. In fact, from a mathematical standpoint we can use either definition of the ACS, but the ACS definition $r(k)$ is preferred from a practical standpoint as we now explain.

First, we should stress that the PSD describes the spectral content of the ACS, as seen from equation (1.3.7). The PSD $\phi(\omega)$ is sometimes perceived as showing the (infinitesimal) power at frequency ω in the signal itself, but that is not strictly true. If the PSD represented the power in the signal itself, then we should have had $\psi(\omega) = \phi(\omega)$ because the signal's spectral content should not depend on the ACS definition. However, as shown above, in the general complex case $\psi(\omega) = \phi(-\omega) \neq \phi(\omega)$, which means that the signal power interpretation of the PSD is not (always) correct. Indeed, the PSD $\phi(\omega)$ "measures" the *power at frequency ω in the signal's ACS.*

On the other hand, our motivation for considering spectral analysis is to characterize the *average power* at frequency ω *in the signal*, as given by the second definition of the PSD in equation (1.3.10). If $c(k)$ is used as the ACS, its corresponding second definition of the PSD is

$$\psi(\omega) = \lim_{N\to\infty} E\left\{\frac{1}{N}\left|\sum_{t=1}^{N} y(t)e^{+i\omega t}\right|^2\right\}$$

which is the average power of $y(t)$ at frequency $-\omega$. Clearly, the second PSD definition corresponding to $r(k)$ aligns with this average power motivation, while the one for $c(k)$ does not; it is for this reason that we use the definition $r(k)$ for the ACS. ■

Next, we present a useful result which concerns the *transfer of PSD through an asymptotically stable linear system.* Let

$$H(q) = \sum_{k=-\infty}^{\infty} h_k q^{-k} \tag{1.4.4}$$

denote an asymptotically stable linear time–invariant system. The operator q^{-1} is the unit delay operator, defined by $q^{-1}y(t) = y(t-1)$. Also, let $e(t)$ be the stationary input to the system and $y(t)$ the corresponding output, as shown in Figure 1.1. Then $\{y(t)\}$ and $\{e(t)\}$ are related via the convolution sum

$$y(t) = H(q)e(t) = \sum_{k=-\infty}^{\infty} h_k e(t-k) \tag{1.4.5}$$

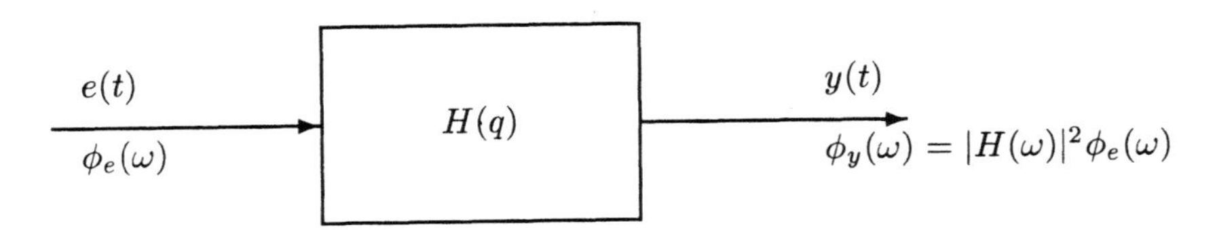

Figure 1.1. Relationship between the PSDs of the input and output of a linear system.

The transfer function of this filter is

$$H(\omega) = \sum_{k=-\infty}^{\infty} h_k e^{-i\omega k} \tag{1.4.6}$$

Throughout the text, we will follow the convention of writing $H(q)$ for the convolution operator of a linear system, $H(z)$ for its corresponding Z-transform, and $H(\omega)$ for its transfer function. We obtain $H(z)$ from $H(q)$ by the substitution $z = q$. For the transfer function, however, we obtain $H(\omega)$ from $H(z)$ by the substitution $z = e^{i\omega}$:

$$\boxed{H(\omega) = H(z)\big|_{z=e^{i\omega}}}$$

While we recognize the slight abuse of notation in writing $H(\omega)$ instead of $H(e^{i\omega})$, we prefer the simplicity of notation it affords.

From (1.4.5), we obtain

$$\begin{aligned} r_y(k) &= \sum_{p=-\infty}^{\infty} \sum_{m=-\infty}^{\infty} h_p h_m^* E\{e(t-p)e^*(t-m-k)\} \\ &= \sum_{p=-\infty}^{\infty} \sum_{m=-\infty}^{\infty} h_p h_m^* r_e(m+k-p) \end{aligned} \tag{1.4.7}$$

Inserting (1.4.7) in (1.3.7) gives

$$\begin{aligned} \phi_y(\omega) &= \sum_{k=-\infty}^{\infty} \sum_{p=-\infty}^{\infty} \sum_{m=-\infty}^{\infty} h_p h_m^* r_e(m+k-p) e^{-i\omega(k+m-p)} e^{i\omega m} e^{-i\omega p} \\ &= \left[\sum_{p=-\infty}^{\infty} h_p e^{-i\omega p}\right] \left[\sum_{m=-\infty}^{\infty} h_m^* e^{i\omega m}\right] \left[\sum_{\tau=-\infty}^{\infty} r_e(\tau) e^{-i\omega\tau}\right] \\ &= |H(\omega)|^2 \phi_e(\omega) \end{aligned} \tag{1.4.8}$$

From (1.4.8), we get the following important formula

$$\phi_y(\omega) = |H(\omega)|^2 \phi_e(\omega) \tag{1.4.9}$$

that will be much used in the following chapters.

Finally, we derive a property that will be of use in Chapter 5. Let the signals $y(t)$ and $x(t)$ be related by

$$y(t) = e^{i\omega_0 t} x(t) \tag{1.4.10}$$

for some given ω_0. Then, it holds that

$$\phi_y(\omega) = \phi_x(\omega - \omega_0) \tag{1.4.11}$$

In other words, multiplication by $e^{i\omega_0 t}$ of a temporal sequence shifts its spectral density by the angular frequency ω_0. Owing to this interpretation, the process of constructing $y(t)$ as in (1.4.10) is called *complex (de)modulation.* The proof of (1.4.11) is immediate: since (1.4.10) implies that

$$r_y(k) = e^{i\omega_0 k} r_x(k) \tag{1.4.12}$$

we obtain

$$\phi_y(\omega) = \sum_{k=-\infty}^{\infty} r_x(k) e^{-i(\omega-\omega_0)k} = \phi_x(\omega - \omega_0) \tag{1.4.13}$$

which is the desired result.

1.5 The Spectral Estimation Problem

The spectral estimation problem can now be stated more formally as follows.

> From a finite–length record $\{y(1), \ldots, y(N)\}$ of a second–order stationary random process, determine an estimate $\hat{\phi}(\omega)$ of its power spectral density $\phi(\omega)$, for $\omega \in [-\pi, \pi]$ (1.5.1)

It would, of course, be desirable that $\hat{\phi}(\omega)$ is as close to $\phi(\omega)$ as possible. As we shall see, the main limitation on the quality of most PSD estimates is due to the quite small number of data samples usually available for processing. Note that N will be used throughout this text to denote the number of points of the available data sequence. In some applications, N is small since the cost of obtaining large amounts of data is prohibitive. Most commonly, the value of N is limited by the fact that the signal under study can be considered second–order stationary only over short observation intervals.

As already mentioned in the introductory part of this chapter, there are two main approaches to the PSD estimation problem. The *nonparametric approach*, discussed in Chapters 2 and 5, proceeds to estimate the PSD by relying essentially only on the basic definitions (1.3.7) and (1.3.10) and on some properties that directly follow from these definitions. In particular, these methods do not make any assumption on the functional form of $\phi(\omega)$. This is in contrast with the *parametric approach*, discussed in Chapters 3, 4 and 6. That approach makes assumptions on the signal under study, which lead to a parameterized functional form of the PSD, and then proceeds by estimating the parameters in the PSD model. The parametric approach can thus be used only when there is enough information about the studied signal, that allows formulation of a model. Otherwise the nonparametric approach should be used. Interestingly enough, the nonparametric methods are close competitors to the parametric ones, even when the model form assumed by the latter is a reasonable description of reality.

1.6 Complements

1.6.1 Coherency Spectrum

Let

$$\boxed{C_{yu}(\omega) = \frac{\phi_{yu}(\omega)}{[\phi_{yy}(\omega)\phi_{uu}(\omega)]^{1/2}}} \tag{1.6.1}$$

denote the so-called *complex-coherency* of the stationary signals $y(t)$ and $u(t)$. In the definition above, $\phi_{yu}(\omega)$ is the cross-spectrum of the two signals, and $\phi_{yy}(\omega)$ and $\phi_{uu}(\omega)$ are their respective PSDs. (We implicitly assume in (1.6.1) that $\phi_{yy}(\omega)$ and $\phi_{uu}(\omega)$ are strictly positive for all ω.) Also, let

$$\epsilon(t) = y(t) - \sum_{k=-\infty}^{\infty} h_k u(t-k) \tag{1.6.2}$$

denote the residues of the least squares problem in Exercise 1.11. Hence, $\{h_k\}$ in equation (1.6.2) satisfy

$$\sum_{k=-\infty}^{\infty} h_k e^{-i\omega k} \triangleq H(\omega) = \phi_{yu}(\omega)/\phi_{uu}(\omega).$$

Show that

$$E\left\{|\epsilon(t)|^2\right\} = \frac{1}{2\pi}\int_{-\pi}^{\pi} (1 - |C_{yu}(\omega)|^2)\phi_{yy}(\omega)\, d\omega \tag{1.6.3}$$

where $|C_{yu}(\omega)|$ is the so-called *coherency spectrum*.

Deduce from (1.6.3) that the coherency spectrum shows the extent to which $y(t)$ and $u(t)$ are linearly related to one another, hence providing a motivation for the name given to $|C_{yu}(\omega)|$. Also deduce from (1.6.3) that $|C_{yu}(\omega)| \leq 1$ with equality, for all ω values, if and only if $y(t)$ and $u(t)$ are related as in equation (1.7.11). Show that $|C_{yu}(\omega)|$ is invariant to linear filtering of $u(t)$ and $y(t)$ (possibly by different filters); that is, if $\tilde{u} = g * u$ and $\tilde{y} = f * y$ where f and g are linear filters, then $|C_{\tilde{y}\tilde{u}}(\omega)| = |C_{yu}(\omega)|$.

Solution: Let $x(t) = \sum_{k=-\infty}^{\infty} h_k u(t-k)$. It can be shown that $u(t-k)$ and $\epsilon(t)$ are uncorrelated with one another for all k. (The reader is required to verify this claim; see also Exercise 1.11). Hence $x(t)$ and $\epsilon(t)$ are also uncorrelated with each other. As

$$y(t) = \epsilon(t) + x(t), \tag{1.6.4}$$

it then follows that

$$\phi_{yy}(\omega) = \phi_{\epsilon\epsilon}(\omega) + \phi_{xx}(\omega) \tag{1.6.5}$$

By using the fact that $\phi_{xx}(\omega) = |H(\omega)|^2 \phi_{uu}(\omega)$, we can write

$$\begin{aligned}
E\left\{|\epsilon(t)|^2\right\} &= \frac{1}{2\pi} \int_{-\pi}^{\pi} \phi_{\epsilon\epsilon}(\omega)\, d\omega \\
&= \frac{1}{2\pi} \int_{-\pi}^{\pi} \left[1 - |H(\omega)|^2 \frac{\phi_{uu}(\omega)}{\phi_{yy}(\omega)}\right] \phi_{yy}(\omega)\, d\omega \\
&= \frac{1}{2\pi} \int_{-\pi}^{\pi} \left[1 - \frac{|\phi_{yu}(\omega)|^2}{\phi_{uu}(\omega)\phi_{yy}(\omega)}\right] \phi_{yy}(\omega)\, d\omega \\
&= \frac{1}{2\pi} \int_{-\pi}^{\pi} \left[1 - |C_{yu}(\omega)|^2\right] \phi_{yy}(\omega)\, d\omega
\end{aligned}$$

which is (1.6.3).

Since the left-hand side in (1.6.3) is nonnegative and the PSD function $\phi_{yy}(\omega)$ is arbitrary, we must have $|C_{yu}(\omega)| \leq 1$ for all ω. It can also be seen from (1.6.3) that the closer $|C_{yu}(\omega)|$ is to one, the smaller the residual variance. In particular, if $|C_{yu}(\omega)| \equiv 1$ then $\epsilon(t) \equiv 0$ (in the mean square sense) and hence $y(t)$ and $u(t)$ must be linearly related as in (1.7.11). Owing to the previous interpretation, $C_{yu}(\omega)$ is sometimes called *the correlation coefficient in the frequency domain.*

Next, consider the filtered signals

$$\tilde{y}(t) = \sum_{k=-\infty}^{\infty} f_k y(t-k)$$

and

$$\tilde{u}(t) = \sum_{k=-\infty}^{\infty} g_k u(t-k)$$

where the filters $\{f_k\}$ and $\{g_k\}$ are assumed to be stable. As

$$\begin{aligned}
r_{\tilde{y}\tilde{u}}(p) &\triangleq E\left\{\tilde{y}(t)\tilde{u}^*(t-p)\right\} \\
&= \sum_{k=-\infty}^{\infty} \sum_{j=-\infty}^{\infty} f_k g_j^* E\left\{y(t-k)u^*(t-j-p)\right\}
\end{aligned}$$

$$= \sum_{k=-\infty}^{\infty} \sum_{j=-\infty}^{\infty} f_k g_j^* r_{yu}(j+p-k),$$

it follows that

$$\phi_{\tilde{y}\tilde{u}}(\omega) = \sum_{p=-\infty}^{\infty} \sum_{k=-\infty}^{\infty} \sum_{j=-\infty}^{\infty} f_k e^{-i\omega k} \, g_j^* e^{i\omega j} \, r_{yu}(j+p-k) e^{-i\omega(j+p-k)}$$

$$= \left(\sum_{k=-\infty}^{\infty} f_k e^{-i\omega k} \right) \left(\sum_{j=-\infty}^{\infty} g_j e^{-i\omega j} \right)^* \left(\sum_{s=-\infty}^{\infty} r_{yu}(s) e^{-i\omega s} \right)$$

$$= F(\omega) G^*(\omega) \phi_{yu}(\omega)$$

Hence

$$|C_{\tilde{y}\tilde{u}}(\omega)| = \frac{|F(\omega)| \, |G(\omega)| \, |\phi_{yu}(\omega)|}{|F(\omega)| \phi_{yy}^{1/2}(\omega) |G(\omega)| \phi_{uu}^{1/2}(\omega)} = |C_{yu}(\omega)|$$

which is the desired result. Observe that the latter result is similar to the invariance of the modulus of the correlation coefficient in the time domain,

$$\frac{|r_{yu}(k)|}{[r_{yy}(0) r_{uu}(0)]^{1/2}},$$

to a scaling of the two signals: $\tilde{y}(t) = f \cdot y(t)$ and $\tilde{u}(t) = g \cdot u(t)$.

1.7 Exercises

Exercise 1.1: Scaling of the Frequency Axis

In this text, the time variable t has been expressed in units of the sampling interval T_s (say). Consequently, the frequency is measured in cycles per sampling interval. Assume we want the frequency units to be expressed in radians per second or in Hertz (Hz = cycles per second). Then we have to introduce the scaled frequency variables

$$\bar{\omega} = \omega / T_s \quad \bar{\omega} \in [-\pi/T_s, \ \pi/T_s] \text{ rad/sec} \tag{1.7.1}$$

and $\bar{f} = \bar{\omega}/2\pi$ (in Hz). It might be thought that the PSD in the new frequency variable is obtained by inserting $\omega = \bar{\omega} T_s$ into $\phi(\omega)$, but this is *wrong*. Show that the PSD, *as expressed in units of power per Hz*, is in fact given by:

$$\bar{\phi}(\bar{\omega}) = T_s \phi(\bar{\omega} T_s) \triangleq T_s \sum_{k=-\infty}^{\infty} r(k) e^{-i\bar{\omega} T_s k}, \qquad |\bar{\omega}| \le \pi / T_s \tag{1.7.2}$$

(See [Marple 1987] for more details on this scaling aspect.)

Exercise 1.2: Time–Frequency Distributions

Let $y(t)$ denote a discrete–time signal, and let $Y(\omega)$ be its discrete–time Fourier transform. As explained in Section 1.2, $Y(\omega)$ shows how the energy in the *whole sequence* $\{y(t)\}_{t=-\infty}^{\infty}$ is distributed over frequency.

Assume that we want to determine how the energy of the signal is distributed in *time and frequency*. If $D(t,\omega)$ is a function that characterizes the time–frequency distribution, then it should satisfy the so–called *marginal properties*:

$$\sum_{t=-\infty}^{\infty} D(t,\omega) = |Y(\omega)|^2 \tag{1.7.3}$$

and

$$\frac{1}{2\pi}\int_{-\pi}^{\pi} D(t,\omega)d\omega = |y(t)|^2 \tag{1.7.4}$$

Use intuitive arguments to explain why the previous conditions are desirable properties of a time–frequency distribution. Next, show that the so–called Rihaczek distribution,

$$\boxed{D(t,\omega) = y(t)Y^*(\omega)e^{-i\omega t}} \tag{1.7.5}$$

satisfies conditions (1.7.3) and (1.7.4). (For treatments of the time–frequency distributions, the reader is referred to [THERRIEN 1992] and [COHEN 1995]).

Exercise 1.3: Two Useful Z–Transform Properties

(a) Let h_k be an absolutely summable sequence, and let $H(z) = \sum_{k=-\infty}^{\infty} h_k z^{-k}$ be its Z–transform. Find the Z–transforms of the following two sequences:

(i) h_{-k}

(ii) $g_k = \sum_{m=-\infty}^{\infty} h_m h_{m-k}^*$

(b) Show that if z_i is a zero of $A(z) = 1 + a_1 z^{-1} + \cdots + a_n z^{-n}$, then $(1/z_i^*)$ is a zero of $A^*(1/z^*)$ (where $A^*(1/z^*) = [A(1/z^*)]^*$).

Exercise 1.4: A Simple ACS Example

Let $y(t)$ be the output of a linear system as in Figure 1.1 with filter $H(q) = (1 + b_1 q^{-1})/(1 + a_1 q^{-1})$, and whose input is zero mean white noise with variance σ^2 (the ACS of such an input is $\sigma^2 \delta_{k,0}$).

(a) Find $r(k)$ and $\phi(\omega)$ analytically in terms of a_1, b_1, and σ^2.

(b) Verify that $r(-k) = r^*(k)$, and that $|r(k)| \leq r(0)$. Also verify that when a_1 and b_1 are real, $r(k)$ can be written as a function of $|k|$.

Exercise 1.5: Alternative Proof that $|r(k)| \le r(0)$

We stated in the text that (1.3.4) follows from (1.3.6). Provide a proof of that statement. Also, find an alternative, simple proof of (1.3.4) by using (1.3.8).

Exercise 1.6: A Double Summation Formula

A result often used in the study of discrete–time random signals is the following summation formula:

$$\sum_{t=1}^{N}\sum_{s=1}^{N} f(t-s) = \sum_{\tau=-N+1}^{N-1} (N-|\tau|) f(\tau) \tag{1.7.6}$$

where $f(\cdot)$ is an arbitrary function. Provide a proof of the above formula.

Exercise 1.7: Is a Truncated Autocovariance Sequence (ACS) a Valid ACS?

Suppose that $\{r(k)\}_{k=-\infty}^{\infty}$ is a valid ACS; thus, $\sum_{k=-\infty}^{\infty} r(k)e^{-i\omega k} \ge 0$ for all ω. Is it possible that for some integer p the partial (or truncated) sum

$$\sum_{k=-p}^{p} r(k)e^{-i\omega k}$$

is negative for some ω? Justify your answer.

Exercise 1.8: When Is a Sequence an Autocovariance Sequence?

We showed in Section 1.3 that if $\{r(k)\}_{k=-\infty}^{\infty}$ is an ACS, then $R_m \ge 0$ for $m = 0, 1, 2, \ldots$. We also implied that the first definition of the PSD in (1.3.7) satisfies $\phi(\omega) \ge 0$ for all ω; however, we did not prove this by using (1.3.7) solely. Show that

$$\phi(\omega) = \sum_{k=-\infty}^{\infty} r(k)e^{-i\omega k} \ge 0 \text{ for all } \omega$$

if and only if

$$a^* R_m a \ge 0 \quad \text{for every } m \text{ and for every vector } a$$

Exercise 1.9: Spectral Density of the Sum of Two Correlated Signals

Let $y(t)$ be the output to the system shown:

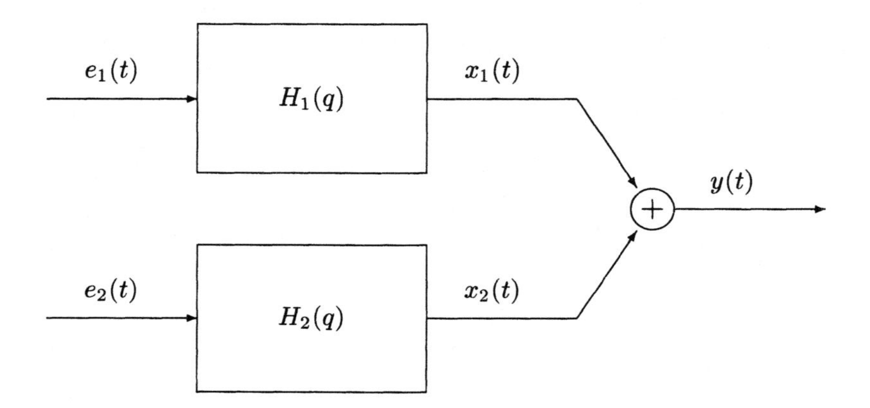

Assume $H_1(q)$ and $H_2(q)$ are linear, asymptotically stable systems. The inputs $e_1(t)$ and $e_2(t)$ are each zero mean white noise, with

$$E\left\{\begin{bmatrix} e_1(t) \\ e_2(t) \end{bmatrix}\begin{bmatrix} e_1^*(s) & e_2^*(s) \end{bmatrix}\right\} = \begin{bmatrix} \sigma_1^2 & \rho\sigma_1\sigma_2 \\ \rho\sigma_1\sigma_2 & \sigma_2^2 \end{bmatrix}\delta_{t,s}$$

(a) Find the PSD of $y(t)$.

(b) Show that for $\rho = 0$, $\phi_y(\omega) = \phi_{x_1}(\omega) + \phi_{x_2}(\omega)$.

(c) Show that for $\rho = \pm 1$ and $\sigma_1^2 = \sigma_2^2 = \sigma^2$, $\phi_y(\omega) = \sigma^2|H_1(\omega) + H_2(\omega)|^2$.

Exercise 1.10: Least Squares Spectral Approximation

Assume we are given an ACS $\{r(k)\}_{k=-\infty}^{\infty}$ or, equivalently, a PSD function $\phi(\omega)$ as in equation (1.3.7). We wish to find a finite–impulse response (FIR) filter as in Figure 1.1, where $H(\omega) = h_0 + h_1e^{-i\omega} + \ldots + h_me^{-im\omega}$, whose input $e(t)$ is zero mean, unit variance white noise, and such that the output sequence $y(t)$ has PSD $\phi_y(\omega)$ "close to" $\phi(\omega)$. Specifically, we wish to find $h = [h_0 \ldots h_m]^T$ so that the approximation error

$$\epsilon = \frac{1}{2\pi}\int_{-\pi}^{\pi} [\phi(\omega) - \phi_y(\omega)]^2\, d\omega \tag{1.7.7}$$

is minimum.

(a) Show that ϵ is a quartic (fourth–order) function of h, and thus no simple closed–form solution h exists to minimize (1.7.7).

(b) We attempt to reparameterize the minimization problem as follows. We note that $r_y(k) \equiv 0$ for $|k| > m$; thus,

$$\phi_y(\omega) = \sum_{k=-m}^{m} r_y(k)e^{-i\omega k} \tag{1.7.8}$$

Equation (1.7.8) and the fact that $r_y(-k) = r_y^*(k)$ mean that $\phi_y(\omega)$ is a function of $g = [r_y(0) \ldots r_y(m)]^T$. Show that the minimization problem in (1.7.7) is quadratic in g; it thus admits a closed-form solution. Show that the vector g that minimizes ϵ in equation (1.7.7) gives

$$r_y(k) = \begin{cases} r(k), & |k| \leq m \\ 0, & \text{otherwise} \end{cases} \tag{1.7.9}$$

(c) Can you identify any problems with the "solution" (1.7.9)?

Exercise 1.11: Linear Filtering and the Cross-Spectrum

For two stationary signals $y(t)$ and $u(t)$, with (cross)covariance sequence $r_{yu}(k) = E\{y(t)u^*(t-k)\}$, the *cross-spectrum* is defined as:

$$\boxed{\phi_{yu}(\omega) = \sum_{k=-\infty}^{\infty} r_{yu}(k)e^{-i\omega k}} \tag{1.7.10}$$

Let $y(t)$ be the output of a linear filter with input $u(t)$,

$$y(t) = \sum_{k=-\infty}^{\infty} h_k u(t-k) \tag{1.7.11}$$

Show that the input PSD, $\phi_{uu}(\omega)$, the filter transfer function

$$H(\omega) = \sum_{k=-\infty}^{\infty} h_k e^{-i\omega k}$$

and $\phi_{yu}(\omega)$ are related through the so-called *Wiener–Hopf equation*:

$$\boxed{\phi_{yu}(\omega) = H(\omega)\phi_{uu}(\omega)} \tag{1.7.12}$$

Next, consider the following least squares (LS) problem,

$$\min_{\{h_k\}} E\left\{ \left| y(t) - \sum_{k=-\infty}^{\infty} h_k u(t-k) \right|^2 \right\} \tag{1.7.13}$$

where now $y(t)$ and $u(t)$ are no longer necessarily related through equation (1.7.11). Show that the filter minimizing the above LS criterion is still given by the Wiener–Hopf equation, by minimizing the expectation in (1.7.13) with respect to the real and imaginary parts of h_k (assume that $\phi_{uu}(\omega) > 0$ for all ω).

Computer Exercises

Exercise C1.12: Computer Generation of Autocovariance Sequences

Autocovariance sequences are two–sided sequences. In this exercise we develop computer techniques for generating two–sided ACSs.

Let $y(t)$ be the output of the linear system in Figure 1.1 with filter $H(q) = (1 + b_1 q^{-1})/(1 + a_1 q^{-1})$, and whose input is zero mean white noise with variance σ^2.

(a) Find $r(k)$ analytically in terms of a_1, b_1, and σ^2 (see also Exercise 1.4).

(b) Plot $r(k)$ for $-20 \leq k \leq 20$ and for various values of a_1 and b_1. Notice that the tails of $r(k)$ decay at a rate dictated by $|a_1|$.

(c) When $a_1 \simeq b_1$ and $\sigma^2 = 1$, then $r(k) \simeq \delta_{k,0}$. Verify this for $a_1 = -0.95$, $b_1 = -0.9$, and for $a_1 = -0.75$, $b_1 = -0.7$.

(d) A quick way to generate (approximately) $r(k)$ on the computer is to use the fact that $r(k) = \sigma^2 h(k) * h^*(-k)$ where $h(k)$ is the impulse response of the filter in Figure 1.1 (see equation (1.4.7)) and $*$ denotes convolution. Consider the case where

$$H(q) = \frac{1 + b_1 q^{-1} + \cdots + b_m q^{-m}}{1 + a_1 q^{-1} + \cdots + a_n q^{-n}}.$$

Write a MATLAB function **genacs.m** whose inputs are M, σ^2, a and b, where a and b are the vectors of denominator and numerator coefficients, respectively, and whose output is a vector of ACS coefficients from 0 to M. Your function should make use of the MATLAB function **filter** to generate $\{h_k\}_{k=0}^{M}$, and **conv** to compute $r(k) = \sigma^2 h(k) * h^*(-k)$ using the truncated impulse response sequence.

(e) Test your function using $\sigma^2 = 1$, $a_1 = -0.9$ and $b_1 = 0.8$. Try $M = 20$ and $M = 150$; why is the result more accurate for larger M? Suggest a "rule of thumb" about a good choice of M in relation to the poles of the filter.

The above method is a "quick and simple" way to compute an approximation to the ACS, but it is sometimes not very accurate because the impulse response is truncated. Methods for computing the exact ACS from σ^2, a and b are discussed in Exercise 3.2 and also in [KINKEL, PERL, SCHARF, AND STUBBERUD 1979; DEMEURE AND MULLIS 1989].

Exercise C1.13: DTFT Computations using Two–Sided Sequences

In this exercise we consider the DTFT of two–sided sequences (including autocovariance sequences), and in doing so illustrate some basic properties of autocovariance sequences.

(a) We first consider how to use the DTFT to determine $\phi(\omega)$ from $r(k)$ on a computer. We are given an ACS:

$$r(k) = \begin{cases} \frac{M-|k|}{M}, & |k| \leq M \\ 0, & \text{otherwise} \end{cases} \tag{1.7.14}$$

Generate $r(k)$ for $M = 10$. Now, in MATLAB form a vector $\mathbf{x}$ of length $L = 256$ as:

$$\mathtt{x} = [r(0), r(1), \ldots, r(M), 0 \ldots, 0, r(-M), \ldots, r(-1)]$$

Verify that `xf=fft(x)` gives $\phi(\omega_k)$ for $\omega_k = 2\pi k/L$. (Note that the elements of `xf` should be nonnegative and real.). Explain why this particular choice of `x` is needed, citing appropriate circular shift and zero padding properties of the DTFT.

Note that `xf` often contains a very small imaginary part due to computer roundoff error; replacing `xf` by `real(xf)` truncates this imaginary component and leads to more expected results when plotting.

A word of caution — do not truncate the imaginary part unless you are sure it is negligible; the command `zf=real(fft(z))` when

$$\mathtt{z} = [r(-M), \ldots, r(-1), r(0), r(1), \ldots, r(M), 0 \ldots, 0]$$

gives erroneous "spectral" values; try it and explain why it doesn't work.

(b) Alternatively, since we can readily derive the analytical expression for $\phi(\omega)$, we can instead work backwards. Form a vector

$$\mathtt{yf} = [\phi(0), \phi(2\pi/L), \phi(4\pi/L), \ldots, \phi((L-1)2\pi/L)]$$

and find `y=ifft(yf)`. Verify that `y` closely approximates the ACS.

(c) Consider the ACS $r(k)$ in Exercise C1.12; let $a_1 = -0.9$ and $b_1 = 0$, and set $\sigma^2 = 1$. Form a vector `x` as above, with $M = 10$, and find `xf`. Why is `xf` not a good approximation of $\phi(\omega_k)$ in this case? Repeat the experiment for $M = 127$ and $L = 256$; is the approximation better for this case? Why?

We can again work backwards from the analytical expression for $\phi(\omega)$. Form a vector

$$\mathtt{yf} = [\phi(0), \phi(2\pi/L), \phi(4\pi/L), \ldots, \phi((L-1)2\pi/L)]$$

and find `y=ifft(yf)`. Verify that `y` closely approximates the ACS for large L (say, $L = 256$), but poorly approximates the ACS for small L (say, $L = 20$). By citing properties of inverse DTFTs of infinite, two–sided sequences, explain how the elements of `y` relate to the ACS $r(k)$, and why the approximation is poor for small L. Based on this explanation, give a "rule of thumb" on the choice of L for good approximations of the ACS.

(d) We have seen above that the `fft` command results in spectral estimates from 0 to 2π instead of the more common range of $-\pi$ to π. The MATLAB command `fftshift` can be used to exchange the first and second halves of the `fft` output to correspond to a frequency range of $-\pi$ to π. Similarly, `fftshift` can be used on the output of the `ifft` operation to "center" the zero lag of an ACS. Experiment with `fftshift` to achieve both of these results. What frequency vector `w` is needed so that the command `plot(w, fftshift(fft(x)))` gives the spectral values at the proper frequencies? Similarly, what time vector `t` is needed to get a proper plot of the ACS with `stem(t,fftshift(ifft(xf)))`? Do the results depend on whether the vectors are even or odd in length?

Exercise C1.14: Relationship between the PSD and the Eigenvalues of the ACS Matrix

An interesting property of the ACS matrix R in equation (1.3.5) is that for large dimensions m, its eigenvalues are close to the values of the PSD $\phi(\omega_k)$ for $\omega_k = 2\pi k/m$, $k = 0, 1, \ldots, m-1$ (see, *e.g.*, [GRAY 1972]). We verify this property here.

Consider the ACS in Exercise C1.12, with the values $a_1 = -0.9$, $b_1 = 0.8$, and $\sigma^2 = 1$.

(a) Compute a vector `phi` which contains the values of $\phi(\omega_k)$ for $\omega_k = 2\pi k/m$, with $m = 256$ and $k = 0, 1, \ldots, m-1$. Plot a histogram of these values with `hist(phi)`. Also useful is the cumulative distribution of the values of `phi` (plotted on a logarithmic scale), which can be found with the command `semilogy( (1/m:1/m:1), sort(phi) )`.

(b) Compute the eigenvalues of R in equation (1.3.5) for various values of m. Plot the histogram of the eigenvalues, and their cumulative distribution. Verify that as m increases, the cumulative distribution of the eigenvalues approaches the cumulative distribution of the $\phi(\omega)$ values. Similarly, the histograms also approach the histogram for $\phi(\omega)$, but it is easier to see this result using cumulative distributions than using histograms.

Chapter 2

NONPARAMETRIC METHODS

2.1 Introduction

The nonparametric methods of spectral estimation rely entirely on the definitions (1.3.7) and (1.3.10) of PSD to provide spectral estimates. These methods constitute the "classical means" for PSD estimation [JENKINS AND WATTS 1968]. The present chapter reviews the main nonparametric methods, their properties and the Fast Fourier Transform (FFT) algorithm for their implementation. A related discussion is to be found in Chapter 5, where the nonparametric approach to PSD estimation is given a filter bank interpretation.

We first introduce two common spectral estimators, the *periodogram* and the *correlogram*, derived directly from (1.3.10) and (1.3.7), respectively. These methods are then shown to be equivalent under weak conditions. The periodogram and correlogram methods provide reasonably high resolution for sufficiently long data lengths, but are poor spectral estimators because their variance is high and does not decrease with increasing data length. (In Chapter 5 we provide an interpretation of the periodogram and correlogram methods as a power estimate based on a *single* sample of a filtered version of the signal under study; it is thus not surprising that the periodogram or correlogram variance is large).

The high variance of the periodogram and correlogram methods motivates the development of modified methods that have lower variance, at a cost of reduced resolution. Several modified methods have been introduced, and we present some of the most popular ones. We show them all to be more–or–less equivalent in their properties and performance for large data lengths.

2.2 Periodogram and Correlogram Methods

2.2.1 Periodogram

The periodogram method relies on the definition (1.3.10) of the PSD. Neglecting the expectation and the limit operation in (1.3.10), which cannot be performed when the only available information on the signal consists of the samples $\{y(t)\}_{t=1}^{N}$, we get

$$\hat{\phi}_p(\omega) = \frac{1}{N}\left|\sum_{t=1}^{N} y(t)e^{-i\omega t}\right|^2 \qquad \text{(Periodogram)} \tag{2.2.1}$$

One of the first uses of the *periodogram spectral estimator*, (2.2.1), has been in determining possible "hidden periodicities" in time series, which may be seen as a motivation for the name of this method [SCHUSTER 1900].

2.2.2 Correlogram

The correlation–based definition (1.3.7) of the PSD leads to the *correlogram spectral estimator* [BLACKMAN AND TUKEY 1959]:

$$\hat{\phi}_c(\omega) = \sum_{k=-(N-1)}^{N-1} \hat{r}(k)e^{-i\omega k} \qquad \text{(Correlogram)} \tag{2.2.2}$$

where $\hat{r}(k)$ denotes an estimate of the covariance lag $r(k)$, obtained from the available sample $\{y(1), \ldots, y(N)\}$. When no assumption is made on the signal under study, except for the stationarity assumption, there are two standard ways to obtain the sample covariances required in (2.2.2):

$$\hat{r}(k) = \frac{1}{N-k}\sum_{t=k+1}^{N} y(t)y^*(t-k), \qquad 0 \le k \le N-1 \tag{2.2.3}$$

and

$$\hat{r}(k) = \frac{1}{N}\sum_{t=k+1}^{N} y(t)y^*(t-k) \qquad 0 \le k \le N-1 \tag{2.2.4}$$

The sample covariances for negative lags are then constructed using the property (1.3.3) of the covariance function:

$$\hat{r}(-k) = \hat{r}^*(k), \qquad k = 0, \ldots, N-1 \tag{2.2.5}$$

The estimator (2.2.3) is called the standard unbiased ACS estimate, and (2.2.4) is called the standard biased ACS estimate. The biased ACS estimate is most commonly used, for the following reasons:

- For most stationary signals, the covariance function decays rather rapidly, so that $r(k)$ is quite small for large lags k. Comparing the definitions (2.2.3) and (2.2.4), it can be seen that $\hat{r}(k)$ in (2.2.4) will be small for large k (provided N

is reasonably large), whereas $\hat{r}(k)$ in (2.2.3) may take large and erratic values for large k, as it is obtained by averaging only a few products in such a case (in particular, only one product for $k = N - 1$!). This observation implies that (2.2.4) is likely to be a more accurate estimator of $r(k)$, than (2.2.3), for medium and large values of k (compared to N). For small values of k, the two estimators in (2.2.3) and (2.2.4) can be expected to behave in a similar manner.

- The sequence $\{\hat{r}(k), k = 0, \pm 1, \pm 2, \ldots\}$ obtained with (2.2.4) is guaranteed to be positive semidefinite (as it should, see equation (1.3.5) and the related discussion), while this is not the case for (2.2.3). This fact is especially important for PSD estimation, since a sample covariance sequence that is not positive definite, when inserted in (2.2.2), may lead to negative spectral estimates, and this is undesirable in most applications.

When the sample covariances (2.2.4) are inserted in (2.2.2), it can be shown that the so–obtained spectral estimate is identical to (2.2.1). In other words, we have the following result.

$\hat{\phi}_c(\omega)$ evaluated using the standard biased ACS estimates coincides with $\hat{\phi}_p(\omega)$	(2.2.6)

A simple proof of (2.2.6) runs as follows. Consider the signal

$$x(t) = \frac{1}{\sqrt{N}} \sum_{k=1}^{N} y(k) e(t-k) \tag{2.2.7}$$

where $\{y(k)\}$ are considered to be fixed (nonrandom) constants and $e(t)$ is a white noise of unit variance: $E\{e(t)e^*(s)\} = \delta_{t,s}$ ($= 1$ if $t = s$; and $= 0$ otherwise). Hence $x(t)$ is the output of a filter with the following transfer function:

$$Y(\omega) = \frac{1}{\sqrt{N}} \sum_{k=1}^{N} y(k) e^{-i\omega k}$$

Since the PSD of the input to the filter is given by $\phi_e(\omega) = 1$, it follows from (1.4.5) that

$$\phi_x(\omega) = |Y(\omega)|^2 = \hat{\phi}_p(\omega) \tag{2.2.8}$$

On the other hand, a straightforward calculation gives (for $k \geq 0$):

$$\begin{aligned} r_x(k) &= E\{x(t)x^*(t-k)\} \\ &= \frac{1}{N} \sum_{p=1}^{N} \sum_{s=1}^{N} y(p) y^*(s) E\{e(t-p)e^*(t-k-s)\} \end{aligned}$$

$$
\begin{aligned}
&= \frac{1}{N}\sum_{p=1}^{N}\sum_{s=1}^{N} y(p)y^*(s)\delta_{p,k+s} = \frac{1}{N}\sum_{p=k+1}^{N} y(p)y^*(p-k) \\
&= \begin{cases} \hat{r}(k) \text{ given by (2.2.4)}, & k = 0,\ldots,N-1 \\ 0, & k \geq N \end{cases}
\end{aligned} \tag{2.2.9}
$$

Inserting (2.2.9) in the definition (1.3.7) of PSD, the following alternative expression for $\phi_x(\omega)$ is obtained:

$$
\phi_x(\omega) = \sum_{k=-(N-1)}^{N-1} \hat{r}(k)e^{-i\omega k} = \hat{\phi}_c(\omega) \tag{2.2.10}
$$

Comparing (2.2.8) and (2.2.10) concludes the proof of the claim (2.2.6).

The equivalence of the periodogram and correlogram spectral estimators can be used to derive their properties simultaneously. These two methods are shown in Section 2.4 to provide *poor estimates* of the PSD. There are two reasons for this, and both can be explained intuitively using $\hat{\phi}_c(\omega)$.

- The estimation errors in $\hat{r}(k)$ are on the order of $1/\sqrt{N}$ for large N (see Exercise 2.4), at least for $|k|$ not too close to N. Because $\hat{\phi}_c(\omega) = \hat{\phi}_p(\omega)$ is a sum that involves $(2N-1)$ such covariance estimates, the difference between the true and estimated spectra will be a sum of "many small" errors. Hence there is no guarantee that the total error will die out as N increases. The spectrum estimation error is even larger than what is suggested by the above discussion, because errors in $\{\hat{r}(k)\}$, for $|k|$ close to N, are typically of an order larger than $1/\sqrt{N}$. The consequence is that the variance of $\hat{\phi}_c(\omega)$ does not go to zero as N increases.

- In addition, if $r(k)$ converges slowly to zero, then the periodogram estimates will be biased. Indeed, for lags $|k| \simeq N$, $\hat{r}(k)$ will be a poor estimate of $r(k)$ since $\hat{r}(k)$ is the sum of only a few lag products that are divided by N (see equation (2.2.4)). Thus, $\hat{r}(k)$ will be much closer to zero than $r(k)$ is; in fact, $E\{\hat{r}(k)\} = [(N-|k|)/N]r(k)$, and the bias is significant for $|k| \simeq N$ if $r(k)$ is not close to zero in this region. If $r(k)$ decays rapidly to zero, the bias will be small and will not contribute significantly to the total error in $\hat{\phi}_c(\omega)$; however, the nonzero variance discussed above will still be present.

Both the bias and the variance of the periodogram are discussed more quantitatively in Section 2.4.

Another intuitive explanation for the poor statistical accuracy of the periodogram and correlogram methods is given in Chapter 5, where it is shown, roughly speaking, that these methods can be viewed as procedures attempting to estimate the variance of a data sequence from a *single* sample.

In spite of their poor quality as spectral estimators, the periodogram and correlogram methods form the basis for the improved nonparametric spectral estimation methods, to be discussed later in this chapter. As such, computation of these two basic estimators is relevant to many other nonparametric estimators derived from them. The next section addresses this computational task.

2.3 Periodogram Computation via FFT

In practice it is not possible to evaluate $\hat{\phi}_p(\omega)$ (or $\hat{\phi}_c(\omega)$) over a continuum of frequencies. Hence, the frequency variable must be sampled for the purpose of computing $\hat{\phi}_p(\omega)$. The following frequency sampling scheme is most commonly used:

$$\boxed{\omega = \frac{2\pi}{N}k, \qquad k = 0, \ldots, N-1} \tag{2.3.1}$$

Define

$$W = e^{-i\frac{2\pi}{N}} \tag{2.3.2}$$

Then, evaluation of $\hat{\phi}_p(\omega)$ (or $\hat{\phi}_c(\omega)$) at the frequency samples in (2.3.1) basically reduces to the computation of the following Discrete Fourier Transform (DFT):

$$Y(k) = \sum_{t=1}^{N} y(t)W^{tk}, \qquad k = 0, \ldots, N-1 \tag{2.3.3}$$

A direct evaluation of (2.3.3) would require about N^2 complex multiplications and additions, which might be a prohibitive burden for large values of N. Any procedure that computes (2.3.3) in less than N^2 flops (1 flop = 1 complex multiplication plus 1 complex addition) is called a Fast Fourier Transform (FFT) algorithm. In recent years, there has been significant interest in developing more and more computationally efficient FFT algorithms. In the following, we review one of the first FFT procedures — the so–called radix–2 FFT — which, while not being the most computationally efficient of all, is easy to program in a computer and yet quite computationally efficient [Cooley and Tukey 1965; Proakis, Rader, Ling, and Nikias 1992].

2.3.1 Radix–2 FFT

Assume that N is a power of 2,

$$\boxed{N = 2^m} \tag{2.3.4}$$

If this is not the case, then we can resort to *zero padding*, as described in the next subsection. By splitting the sum in (2.3.3) into two parts, we get

$$Y(k) = \sum_{t=1}^{N/2} y(t)W^{tk} + \sum_{t=N/2+1}^{N} y(t)W^{tk}$$

$$= \sum_{t=1}^{N/2} [y(t) + y(t+N/2)W^{\frac{Nk}{2}}]W^{tk} \tag{2.3.5}$$

Next, note that

$$W^{\frac{Nk}{2}} = \begin{cases} 1, & \text{for even } k \\ -1, & \text{for odd } k \end{cases} \tag{2.3.6}$$

Using this simple observation in (2.3.5), we obtain:

For $k = 2p = 0, 2, \ldots$

$$Y(2p) = \sum_{t=1}^{\bar{N}} [y(t) + y(t+\bar{N})]\bar{W}^{tp} \tag{2.3.7}$$

For $k = 2p + 1 = 1, 3, \ldots$

$$Y(2p+1) = \sum_{t=1}^{\bar{N}} \{[y(t) - y(t+\bar{N})]W^t\}\bar{W}^{tp} \tag{2.3.8}$$

where $\bar{N} = N/2$ and $\bar{W} = W^2 = e^{-i2\pi/\bar{N}}$.

The above two equations are the core of the radix–2 FFT algorithm. Both of these equations represent DFTs for sequences of length equal to $\bar{N}$. Computation of the sequences transformed in (2.3.7) and (2.3.8) requires roughly $\bar{N}$ flops. Hence, the computation of an N–point transform has been reduced to the evaluation of two $N/2$–point transforms plus a sequence computation requiring about $N/2$ flops. This reduction process is continued until $\bar{N} = 1$ (which is made possible by requiring N to be a power of 2).

In order to evaluate the number of flops required by a radix–2 FFT, let c_k denote the computational cost (expressed in flops) of a 2^k–point radix–2 FFT. According to the discussion in the previous paragraph, c_k satisfies the following recursion:

$$c_k = 2^k/2 + 2c_{k-1} = 2^{k-1} + 2c_{k-1} \tag{2.3.9}$$

with initial condition $c_1 = 1$ (the number of flops required by a 1–point transform). By iterating (2.3.9), we obtain the solution

$$c_k = k2^{k-1} = \frac{1}{2}k2^k \tag{2.3.10}$$

from which it follows that $c_m = \frac{1}{2}m2^m = \frac{1}{2}N\log_2 N$; thus

$$\text{An } N\text{–point radix–2 FFT requires about } \tfrac{1}{2}N\log_2 N \text{ flops} \tag{2.3.11}$$

As a comparison, the number of complex operations required to carry out an N–point ***split–radix*** FFT, which at present appears to be the most practical algorithm for general–purpose computers when N is a power of 2, is about $\frac{1}{3}N\log_2 N$ (see [PROAKIS, RADER, LING, AND NIKIAS 1992]).

2.3.2 Zero Padding

In some applications, N is not a power of 2 and hence the previously described radix–2 FFT algorithm cannot be applied directly to the original data sequence. However, this is easily remedied since we may increase the length of the given sequence by means of zero padding $\{y(1), \ldots, y(N), 0, 0, \ldots\}$ until the length of the so–obtained sequence is, say, L (which is generally chosen as a power of 2).

Zero padding is also useful when the frequency sampling (2.3.1) is considered to be too sparse to provide a good representation of the continuous–frequency estimated spectrum, for example $\hat{\phi}_p(\omega)$. Applying the FFT algorithm to the data sequence padded with zeroes, which gives

$$\hat{\phi}_p(\omega) \text{ at frequencies } \omega_k = \frac{2\pi k}{L}, \qquad 0 \leq k \leq L-1$$

may reveal finer details in the spectrum, which were not visible without zero padding.

Since the *continuous–frequency* spectral estimate, $\hat{\phi}_p(\omega)$, is the same for both the original data sequence and the sequence padded with zeroes, zero padding cannot of course improve the spectral resolution of the periodogram methods. See [Oppenheim and Schafer 1989; Porat 1997] for further discussion.

2.4 Properties of the Periodogram Method

The analysis of the statistical properties of $\hat{\phi}_p(\omega)$ (or $\hat{\phi}_c(\omega)$) is important in that it shows the poor quality of the periodogram as an estimator of the PSD and, in addition, provides some insight into how we can modify the periodogram so as to obtain better spectral estimators. We split the analysis in two parts: bias analysis and variance analysis (see also [Priestley 1989]).

The bias and variance of an estimator are two measures often used to characterize its performance. A primary motivation is that the total squared error of the estimate is the sum of the bias squared and the variance. To see this, let a denote any quantity to be estimated, and let $\hat{a}$ be an estimate of a. Then the mean squared error (MSE) of the estimate is:

$$\begin{aligned} \text{MSE} \triangleq E\left\{|\hat{a}-a|^2\right\} &= E\left\{|\hat{a}-E\{\hat{a}\}+E\{\hat{a}\}-a|^2\right\} \\ &= E\left\{|\hat{a}-E\{\hat{a}\}|^2\right\}+|E\{\hat{a}\}-a|^2 \\ &\quad +2\operatorname{Re}\left[E\{\hat{a}-E\{\hat{a}\}\}\cdot(E\{\hat{a}\}-a)\right] \\ &= \operatorname{var}\{\hat{a}\}+|\operatorname{bias}\{\hat{a}\}|^2 \end{aligned} \tag{2.4.1}$$

By separately considering the bias and variance components of the MSE, we gain some additional insight into the source of error and in ways to reduce the error.

2.4.1 Bias Analysis of the Periodogram

By using the result (2.2.6), we obtain

$$E\left\{\hat{\phi}_p(\omega)\right\} = E\left\{\hat{\phi}_c(\omega)\right\} = \sum_{k=-(N-1)}^{N-1} E\left\{\hat{r}(k)\right\} e^{-i\omega k} \tag{2.4.2}$$

For $\hat{r}(k)$ defined in (2.2.4)

$$E\left\{\hat{r}(k)\right\} = \left(1 - \frac{k}{N}\right) r(k), \qquad k \geq 0 \tag{2.4.3}$$

and

$$E\left\{\hat{r}(-k)\right\} = E\left\{\hat{r}^*(k)\right\} = \left(1 - \frac{k}{N}\right) r(-k), \qquad -k \leq 0 \tag{2.4.4}$$

Hence

$$E\left\{\hat{\phi}_p(\omega)\right\} = \sum_{k=-(N-1)}^{N-1} \left(1 - \frac{|k|}{N}\right) r(k) e^{-i\omega k} \tag{2.4.5}$$

Define

$$w_B(k) = \begin{cases} 1 - \frac{|k|}{N}, & k = 0, \pm 1, \ldots, \pm(N-1) \\ 0, & \text{otherwise} \end{cases} \tag{2.4.6}$$

The above sequence is called the *triangular window*, or the *Bartlett window*. By using $w_B(k)$, we can write (2.4.5) as a DTFT:

$$E\left\{\hat{\phi}_p(\omega)\right\} = \sum_{k=-\infty}^{\infty} [w_B(k) r(k)] e^{-i\omega k} \tag{2.4.7}$$

The DTFT of the product of two sequences is equal to the convolution of their respective DTFTs. Hence, (2.4.7) leads to

$$\boxed{E\left\{\hat{\phi}_p(\omega)\right\} = \frac{1}{2\pi} \int_{-\pi}^{\pi} \phi(\psi) W_B(\omega - \psi) d\psi} \tag{2.4.8}$$

where $W_B(\omega)$ is the DTFT of the triangular window. For completeness, we include a direct proof of (2.4.8). Inserting (1.3.8) in (2.4.7), we get

$$E\left\{\hat{\phi}_p(\omega)\right\} = \sum_{k=-\infty}^{\infty} w_B(k) \left[\frac{1}{2\pi} \int_{-\pi}^{\pi} \phi(\psi) e^{i\psi k} d\psi\right] e^{-i\omega k} \tag{2.4.9}$$

$$= \frac{1}{2\pi} \int_{-\pi}^{\pi} \phi(\psi) \left[\sum_{k=-\infty}^{\infty} w_B(k) e^{-ik(\omega - \psi)}\right] d\psi \tag{2.4.10}$$

$$= \frac{1}{2\pi} \int_{-\pi}^{\pi} \phi(\psi) W_B(\omega - \psi) d\psi \tag{2.4.11}$$

which is (2.4.8).

We can find an explicit expression for $W_B(\omega)$ as follows. A straightforward calculation gives

$$W_B(\omega) = \sum_{k=-(N-1)}^{N-1} \frac{N-|k|}{N} e^{-i\omega k} \tag{2.4.12}$$

$$= \frac{1}{N}\sum_{t=1}^{N}\sum_{s=1}^{N} e^{-i\omega(t-s)} = \frac{1}{N}\left|\sum_{t=1}^{N} e^{i\omega t}\right|^2 \tag{2.4.13}$$

$$= \frac{1}{N}\left|\frac{e^{i\omega N}-1}{e^{i\omega}-1}\right|^2 = \frac{1}{N}\left|\frac{e^{i\omega N/2}-e^{-i\omega N/2}}{e^{i\omega/2}-e^{-i\omega/2}}\right|^2 \tag{2.4.14}$$

or, in final form,

$$W_B(\omega) = \frac{1}{N}\left[\frac{\sin(\omega N/2)}{\sin(\omega/2)}\right]^2 \tag{2.4.15}$$

$W_B(\omega)$ is sometimes referred to as the *Fejer kernel.* As an illustration, $W_B(\omega)$ is displayed as a function of ω, for $N = 25$, in Figure 2.1.

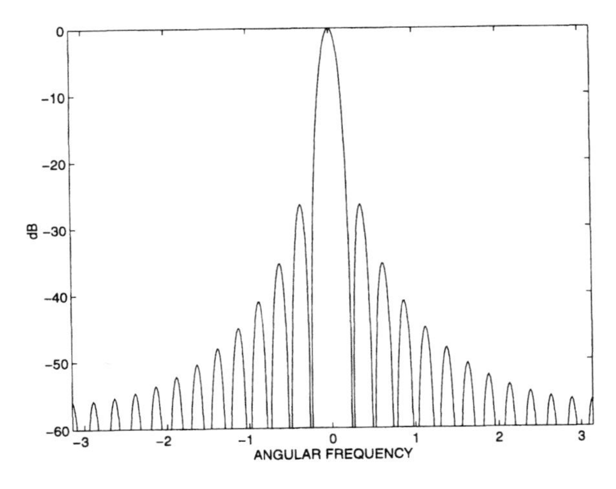

Figure 2.1. $W_B(\omega)/W_B(0)$, for $N = 25$.

The convolution formula (2.4.8) is the key equation to understanding the behavior of the mean estimated spectrum $E\left\{\hat{\phi}_p(\omega)\right\}$. In order to facilitate the interpretation of this equation, the reader may think of it as representing a dynamical system with "input" $\phi(\omega)$, "weighting function" $W_B(\omega)$ and "output" $E\left\{\hat{\phi}_p(\omega)\right\}$. Note that a similar equation would be obtained if the covariance estimator (2.2.3) were used in $\hat{\phi}_c(\omega)$, in lieu of (2.2.4). As in that case $E\{\hat{r}(k)\} = r(k)$, the corresponding

$W(\omega)$ function that would appear in (2.4.8) is the DTFT of the *rectangular window*

$$w_R(k) = \begin{cases} 1, & k = 0, \pm 1, \ldots, \pm(N-1) \\ 0, & \text{otherwise} \end{cases} \tag{2.4.16}$$

A straightforward calculation gives

$$\begin{aligned} W_R(\omega) &= \sum_{k=-(N-1)}^{(N-1)} e^{-i\omega k} = 2\,\mathrm{Re}\left[\frac{e^{iN\omega}-1}{e^{i\omega}-1}\right] - 1 \\ &= \frac{2\cos\left[\frac{(N-1)\omega}{2}\right]\sin\left[\frac{N\omega}{2}\right]}{\sin\left[\frac{\omega}{2}\right]} - 1 = \frac{\sin\left[\left(N-\frac{1}{2}\right)\omega\right]}{\sin\left[\frac{\omega}{2}\right]} \end{aligned} \tag{2.4.17}$$

which is displayed in Figure 2.2 (for $N = 25$; to facilitate comparison with $W_B(\omega)$). $W_R(\omega)$ is sometimes called the *Dirichlet kernel.*

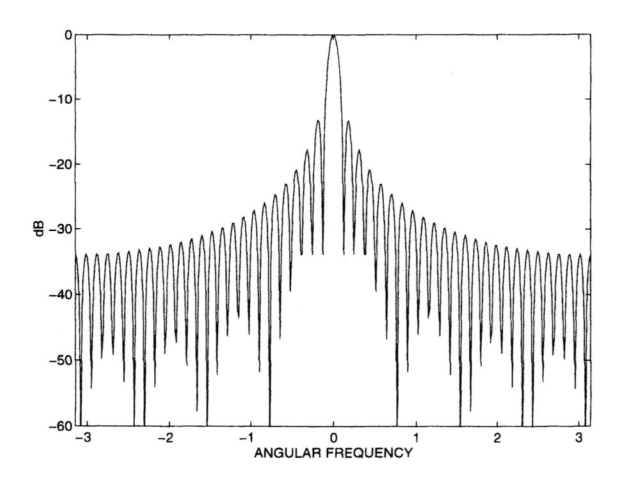

Figure 2.2. $W_R(\omega)/W_R(0)$, for $N = 25$.

As can be seen, there are no "essential" differences between $W_R(\omega)$ and $W_B(\omega)$. For conciseness, in the following we focus on the use of the triangular window.

Since we would like $E\left\{\hat{\phi}_p(\omega)\right\}$ to be as close to $\phi(\omega)$ as possible, it follows from (2.4.8) that $W_B(\omega)$ should be a close approximation to a Dirac impulse. The *half-power (3 dB) width* of the *main lobe* of $W_B(\omega)$ can be shown to be approximately $2\pi/N$ radians (see Exercise 2.13), so in frequency units (with $f = \omega/2\pi$)

$$\boxed{\text{main lobe width in frequency } f \simeq 1/N} \tag{2.4.18}$$

(Also, see the calculation of the time–bandwidth product for windows in the next section, which supports (2.4.18).) It follows from (2.4.18) that $W_B(\omega)$ is a poor approximation of a Dirac impulse for small values of N. In addition, unlike the Dirac delta function, $W_B(\omega)$ has a large number of ***sidelobes***. It follows that the bias of the periodogram spectral estimate can basically be divided into two components. These two components correspond respectively to the nonzero main lobe width and the nonzero sidelobe height of the window function $W_B(\omega)$, as we explain below.

The principal effect of the *main lobe* of $W_B(\omega)$ is to smear or smooth the estimated spectrum. Assume, for instance, that $\phi(\omega)$ has two peaks separated in frequency f by less than $1/N$. Then these two peaks appear as a single broader peak in $\mathrm{E}\{\hat{\phi}_p(\omega)\}$ since (see (2.4.8)) the "response" of the "system" corresponding to $W_B(\omega)$ to the first peak does not get the time to die out before the "response" to the second peak starts. This kind of effect of the main lobe on the estimated spectrum is called *smearing*. Owing to smearing, the periodogram–based methods cannot resolve details in the studied spectrum that are separated by less than $1/N$ in cycles per sampling interval. For this reason, $1/N$ is called the ***spectral resolution limit*** of the periodogram method.

Remark: The previous comments on resolution give us the occasion to stress that, in spite of the fact that we have seen the PSD as a function of the angular frequency (ω), we generally refer to ***the resolution in frequency*** (f) in units of cycles per sampling interval. Of course, the "resolution in angular frequency" is determined from the "resolution in frequency" by the simple relation $\omega = 2\pi f$. ■

The principal effect of the ***sidelobes*** on the estimated spectrum consists of transferring power from the frequency bands that concentrate most of the power in the signal to bands that contain less or no power. This effect is called ***leakage***. For instance, a dominant peak in $\phi(\omega)$ may through convolution with the sidelobes of $W_B(\omega)$ lead to an estimated spectrum that contains power in frequency bands where $\phi(\omega)$ is zero. Note that the smearing effect associated with the main lobe can also be interpreted as a form of leakage from a local peak of $\phi(\omega)$ to neighboring frequency bands.

It follows from the previous discussion that smearing and leakage are particularly critical for spectra with large amplitude ranges, such as peaky spectra. For smooth spectra, these effects are less important. In particular, we see from (2.4.7) that for *white noise* (which has a maximally smooth spectrum) the periodogram is an *unbiased* spectral estimator: $E\{\hat{\phi}_p(\omega)\} = \phi(\omega)$ (see also Exercise 2.7).

The bias of the periodogram estimator, even though it might be severe for spectra with large dynamic ranges when the sample length is small, does not constitute the main limitation of this spectral estimator. In fact, if the bias were the only problem, then by increasing N (assuming this is possible) the bias in $\hat{\phi}_p(\omega)$ would be eliminated. In order to see this, note from (2.4.5), for example, that

$$\lim_{N\to\infty} E\left\{\hat{\phi}_p(\omega)\right\} = \phi(\omega)$$

Hence, the periodogram is an ***asymptotically unbiased spectral estimator***. The main problem of the periodogram method lies in its large variance, as explained next.

2.4.2 Variance Analysis of the Periodogram

The finite–sample variance of $\hat{\phi}_p(\omega)$ can be easily established only in some specific cases, such as in the case of Gaussian white noise. The *asymptotic* variance of $\hat{\phi}_p(\omega)$, however, can be derived for more general signals. In the following, we present an *asymptotic (for $N \gg 1$) analysis* of the variance of $\hat{\phi}_p(\omega)$ since it turns out to be sufficient for showing the poor statistical accuracy of the periodogram (for a finite–sample analysis, see Exercise 2.11).

Some preliminary discussion is required. A sequence $\{e(t)\}$ is called *complex (or circular) white noise* if it satisfies

$$\boxed{\begin{aligned} E\{e(t)e^*(s)\} &= \sigma^2\delta_{t,s} \\ E\{e(t)e(s)\} &= 0, \qquad \text{for all } t \text{ and } s \end{aligned}} \tag{2.4.19}$$

Note that $\sigma^2 = E\left\{|e(t)|^2\right\}$ is the variance (or power) of $e(t)$. Equation (2.4.19) can be rewritten as

$$\left\{\begin{aligned} E\{\mathrm{Re}[e(t)]\,\mathrm{Re}[e(s)]\} &= \tfrac{\sigma^2}{2}\delta_{t,s} \\ E\{\mathrm{Im}[e(t)]\,\mathrm{Im}[e(s)]\} &= \tfrac{\sigma^2}{2}\delta_{t,s} \\ E\{\mathrm{Re}[e(t)]\,\mathrm{Im}[e(s)]\} &= 0 \end{aligned}\right. \tag{2.4.20}$$

Hence, the real and imaginary parts of a complex/circular white noise are real–valued white noise sequences of identical power equal to $\sigma^2/2$, and uncorrelated with one another. See Appendix B for more details on circular random sequences, such as $\{e(t)\}$ above.

In what follows, we shall also make use of the symbol $O(1/N^\alpha)$, for some $\alpha > 0$, to denote a random variable which is such that the square root of its second–order moment goes to zero at least as fast as $1/N^\alpha$, as N tends to infinity.

First, we establish the asymptotic variance/covariance of $\hat{\phi}_p(\omega)$ in the case of *Gaussian complex/circular white noise*. The following result holds.

$$\boxed{\lim_{N\to\infty} E\left\{[\hat{\phi}_p(\omega_1) - \phi(\omega_1)][\hat{\phi}_p(\omega_2) - \phi(\omega_2)]\right\} = \begin{cases} \phi^2(\omega_1), & \omega_1 = \omega_2 \\ 0, & \omega_1 \neq \omega_2 \end{cases}} \tag{2.4.21}$$

Note that, for white noise, $\phi(\omega) = \sigma^2$ (for all ω). Since $\lim_{N\to\infty} \mathrm{E}\{\hat{\phi}_p(\omega)\} = \phi(\omega)$ (*cf.* the analysis in the previous subsection), in order to prove (2.4.21) it suffices to show that

$$\lim_{N\to\infty} E\left\{\hat{\phi}_p(\omega_1)\hat{\phi}_p(\omega_2)\right\} = \phi(\omega_1)\phi(\omega_2) + \phi^2(\omega_1)\delta_{\omega_1,\omega_2} \tag{2.4.22}$$

From (2.2.1), we obtain

$$E\left\{\hat{\phi}_p(\omega_1)\hat{\phi}_p(\omega_2)\right\} = \frac{1}{N^2}\sum_{t=1}^{N}\sum_{s=1}^{N}\sum_{p=1}^{N}\sum_{m=1}^{N} E\{e(t)e^*(s)e(p)e^*(m)\} \cdot e^{-i\omega_1(t-s)}e^{-i\omega_2(p-m)} \tag{2.4.23}$$

For general random processes, the evaluation of the expectation in (2.4.23) is relatively complicated. However, the following general result for Gaussian random variables can be used: If a, b, c, and d are jointly Gaussian (complex or real) random variables, then

$$\boxed{\begin{aligned} E\{abcd\} = E\{ab\}\,E\{cd\} + E\{ac\}\,E\{bd\} + E\{ad\}\,E\{bc\} \\ + E\{a\}\,E\{b\}\,E\{c\}\,E\{d\} \end{aligned}} \tag{2.4.24}$$

For a proof of (2.4.24), see, *e.g.*, [BROCKWELL AND DAVIS 1991; BRILLINGER 1981; JANSSEN AND STOICA 1988]. Thus, if the white noise $e(t)$ is Gaussian as assumed, the fourth–order moment in (2.4.23) is found to be:

$$\begin{aligned} E\{e(t)e^*(s)e(p)e^*(m)\} &= [E\{e(t)e^*(s)\}]\,[E\{e(p)e^*(m)\}] \\ &\quad + [E\{e(t)e(p)\}]\,[E\{e(s)e(m)\}]^* \\ &\quad + [E\{e(t)e^*(m)\}]\,[E\{e^*(s)e(p)\}] \\ &= \sigma^4(\delta_{t,s}\delta_{p,m} + \delta_{t,m}\delta_{s,p}) \end{aligned} \tag{2.4.25}$$

Inserting (2.4.25) in (2.4.23) gives

$$\begin{aligned} E\left\{\hat{\phi}_p(\omega_1)\hat{\phi}_p(\omega_2)\right\} &= \sigma^4 + \frac{\sigma^4}{N^2}\sum_{t=1}^{N}\sum_{s=1}^{N} e^{-i(\omega_1-\omega_2)(t-s)} \\ &= \sigma^4 + \frac{\sigma^4}{N^2}\left|\sum_{t=1}^{N} e^{i(\omega_1-\omega_2)t}\right|^2 \\ &= \sigma^4 + \frac{\sigma^4}{N^2}\left\{\frac{\sin[(\omega_1-\omega_2)N/2]}{\sin[(\omega_1-\omega_2)/2]}\right\}^2 \end{aligned} \tag{2.4.26}$$

The limit of the second term in (2.4.26) is σ^4 when $\omega_1 = \omega_2$ and zero otherwise, and (2.4.22) follows at once.

Remark: Note that in the previous case, it was indeed possible to derive the *finite-sample* variance of $\hat{\phi}_p(\omega)$. For colored noise the above derivation becomes more difficult, and a different approach (presented below) is needed. See Exercise 2.11 for yet another approach that applies to general Gaussian signals. ■

Next, we consider the case of a much more general signal obtained by linearly filtering the Gaussian white noise sequence $\{e(t)\}$ considered above:

$$y(t) = \sum_{k=1}^{\infty} h_k e(t-k) \tag{2.4.27}$$

whose PSD is given by

$$\phi_y(\omega) = |H(\omega)|^2\phi_e(\omega) \tag{2.4.28}$$

(*cf.* (1.4.9)). Here $H(\omega) = \sum_{k=1}^{\infty} h_k e^{-i\omega k}$. The following intermediate result, concerned with signals of the above type, appears to have an independent interest. (Below, we omit the index "p" of $\hat{\phi}_p(\omega)$ in order to simplify the notation.)

For $N \gg 1$,

$$\boxed{\hat{\phi}_y(\omega) = |H(\omega)|^2 \hat{\phi}_e(\omega) + O(1/\sqrt{N})} \tag{2.4.29}$$

Hence, the periodograms approximately satisfy an equation of the form of (2.4.28) that is satisfied by the true PSDs.

In order to prove (2.4.29), first observe that

$$\begin{aligned}
\frac{1}{\sqrt{N}} \sum_{t=1}^{N} y(t) e^{-i\omega t} &= \frac{1}{\sqrt{N}} \sum_{t=1}^{N} \sum_{k=1}^{\infty} h_k e(t-k) e^{-i\omega(t-k)} e^{-i\omega k} \\
&= \frac{1}{\sqrt{N}} \sum_{k=1}^{\infty} h_k e^{-i\omega k} \sum_{p=1-k}^{N-k} e(p) e^{-i\omega p} \\
&= \frac{1}{\sqrt{N}} \sum_{k=1}^{\infty} h_k e^{-i\omega k} \\
&\quad \cdot \left[\sum_{p=1}^{N} e(p) e^{-i\omega p} + \sum_{p=1-k}^{0} e(p) e^{-i\omega p} - \sum_{p=N-k+1}^{N} e(p) e^{-i\omega p} \right] \\
&\triangleq H(\omega) \left[\frac{1}{\sqrt{N}} \sum_{p=1}^{N} e(p) e^{-i\omega p} \right] + \rho(\omega)
\end{aligned} \tag{2.4.30}$$

where

$$\begin{aligned}
\rho(\omega) &= \frac{1}{\sqrt{N}} \sum_{k=1}^{\infty} h_k e^{-i\omega k} \left[\sum_{p=1-k}^{0} e(p) e^{-i\omega p} - \sum_{p=N-k+1}^{N} e(p) e^{-i\omega p} \right] \\
&\triangleq \frac{1}{\sqrt{N}} \sum_{k=1}^{\infty} h_k e^{-i\omega k} \varepsilon_k(\omega)
\end{aligned} \tag{2.4.31}$$

Next, note that

$$\begin{aligned}
E\{\varepsilon_k(\omega)\} &= 0, \\
E\{\varepsilon_k(\omega)\varepsilon_j(\omega)\} &= 0 \text{ for all } k \text{ and } j, \text{ and} \\
E\left\{\varepsilon_k(\omega)\varepsilon_j^*(\omega)\right\} &= 2\sigma^2 \min(k, j)
\end{aligned}$$

which imply

$$E\{\rho(\omega)\} = 0, \qquad E\left\{\rho^2(\omega)\right\} = 0$$

and

$$
\begin{aligned}
E\left\{|\rho(\omega)|^2\right\} &= \frac{1}{N}\left|\sum_{k=1}^{\infty}\sum_{j=1}^{\infty} h_k e^{-i\omega k} h_j^* e^{i\omega j} E\left\{\varepsilon_k(\omega)\varepsilon_j^*(\omega)\right\}\right| \\
&= \frac{2\sigma^2}{N}\left|\sum_{k=1}^{\infty} h_k e^{-i\omega k}\left\{\sum_{j=1}^{k} h_j^* e^{i\omega j} j + \sum_{j=k+1}^{\infty} h_j^* e^{i\omega j} k\right\}\right| \\
&\leq \frac{2\sigma^2}{N}\sum_{k=1}^{\infty}|h_k|\left\{\sum_{j=1}^{\infty}|h_j| j + \sum_{j=1}^{\infty}|h_j| k\right\} \\
&= \frac{4\sigma^2}{N}\left(\sum_{k=1}^{\infty}|h_k|\right)\left(\sum_{j=1}^{\infty}|h_j| j\right)
\end{aligned}
$$

Provided $\sum_{j=1}^{\infty} k|h_k|$ is finite (which, for example, is true if $\{h_k\}$ is exponentially stable; see [SÖDERSTRÖM AND STOICA 1989]), we have

$$E\left\{|\rho(\omega)|^2\right\} \leq \frac{\text{constant}}{N} \tag{2.4.32}$$

Now, from (2.4.30) we obtain

$$\hat{\phi}_y(\omega) = |H(\omega)|^2 \hat{\phi}_e(\omega) + \gamma(\omega) \tag{2.4.33}$$

where

$$\gamma(\omega) = H^*(\omega)E^*(\omega)\rho(\omega) + H(\omega)E(\omega)\rho^*(\omega) + \rho(\omega)\rho^*(\omega)$$

and where

$$E(\omega) = \frac{1}{\sqrt{N}}\sum_{t=1}^{N} e(t)e^{-i\omega t}$$

Since $E(\omega)$ and $\rho(\omega)$ are linear combinations of Gaussian random variables, they are also Gaussian distributed. This means that the fourth–order moment formula (2.4.24) can be used to obtain the second–order moment of $\gamma(\omega)$. By doing so, and also by using (2.4.32) and the fact that, for example,

$$
\begin{aligned}
\left|E\left\{\rho(\omega)E^*(\omega)\right\}\right| &\leq \left[E\left\{|\rho(\omega)|^2\right\}\right]^{1/2}\left[E\left\{|E(\omega)|^2\right\}\right]^{1/2} \\
&= \frac{\text{constant}}{\sqrt{N}} \cdot \left[E\left\{|\hat{\phi}_e(\omega)|^2\right\}\right]^{1/2} = \frac{\text{constant}}{\sqrt{N}}
\end{aligned}
$$

we can verify that $\gamma(\omega) = O(1/\sqrt{N})$, and hence the proof of (2.4.29) is concluded.

The main result of this section is derived by combining (2.4.21) and (2.4.29).

The asymptotic variance/covariance result (2.4.21) is also valid for a general linear signal as defined in (2.4.27).	(2.4.34)

Remark: In the introduction to Chapter 1, we mentioned that the analysis of a complex–valued signal is not always more general than the analysis of the corresponding real–valued signal; we supported this claim by the example of a complex sine wave. Here, we have another instance where the claim is valid. Similarly to the complex sinusoidal signal case, the complex (or circular) white noise does not specialize, in a direct manner, to real white noise. Indeed, if we would let $e(t)$ in (2.4.19) be real, then the two equations in (2.4.19) would conflict with each other (for $t = s$). The *real white noise random process* is a stationary signal which satisfies

$$E\{e(t)e(s)\} = \sigma^2 \delta_{t,s} \tag{2.4.35}$$

If we try to carry out the proof of (2.4.21) under (2.4.35), then we find that the proof has to be modified. This was expected: both $\phi(\omega)$ and $\hat{\phi}_p(\omega)$ are even functions in the real–valued case; hence (2.4.21) should be modified to include the case of both $\omega_1 = \omega_2$ and $\omega_1 = -\omega_2$. ∎

It follows from (2.4.34) that for a fairly general class of signals, the periodogram values are asymptotically (for $N \gg 1$) uncorrelated random variables whose means and standard deviations are both equal to the corresponding true PSD values. Hence, the periodogram is an *inconsistent spectral estimator* which continues to fluctuate around the true PSD, with a nonzero variance, even if the length of the processed sample increases without bound. Furthermore, the fact that the periodogram values $\hat{\phi}_p(\omega)$ are uncorrelated (for large N values) makes the periodogram exhibit an *erratic behavior* (similar to that of a white noise realization). These facts constitute the main limitations of the periodogram approach to PSD estimation. In the next sections, we present several modified periodogram–based methods which attempt to cure the aforementioned difficulties of the basic periodogram approach. As we shall see, the "improved methods" decrease the variance of the estimated spectrum at the expense of increasing its bias (and, hence, decreasing the average resolution).

2.5 The Blackman–Tukey Method

In this section we develop the Blackman–Tukey method [Blackman and Tukey 1959] and compare it to the periodogram. In later sections we consider several other refined periodogram–based methods that, like the Blackman–Tukey (BT) method, seek to reduce the statistical variability of the estimated spectrum; we will compare these methods to one another and to the Blackman–Tukey method.

2.5.1 The Blackman–Tukey Spectral Estimate

As we have seen, the main problem with the periodogram is the high statistical variability of this spectral estimator, even for very large sample lengths. The poor statistical quality of the periodogram PSD estimator has been intuitively explained as arising from both the poor accuracy of $\hat{r}(k)$ in $\hat{\phi}_c(\omega)$ for extreme lags ($|k| \simeq$

N) and the large number of (even if small) covariance estimation errors that are cumulatively summed up in $\hat{\phi}_c(\omega)$. Both these effects may be reduced by truncating the sum in the definition formula of $\hat{\phi}_c(\omega)$, (2.2.2). Following this idea leads to the Blackman–Tukey estimator, which is given by

$$\boxed{\hat{\phi}_{BT}(\omega) = \sum_{k=-(M-1)}^{M-1} w(k)\hat{r}(k)e^{-i\omega k}} \tag{2.5.1}$$

where $\{w(k)\}$ is an even function (*i.e.*, $w(-k) = w(k)$) which is such that $w(0) = 1$, $w(k) = 0$ for $|k| \geq M$, and $w(k)$ decays smoothly to zero with k, and where $M < N$. Since $w(k)$ in (2.5.1) weights the lags of the sample covariance sequence, it is called a *lag window*.

If $w(k)$ in (2.5.1) is selected as the rectangular window, then we simply obtain a truncated version of $\hat{\phi}_c(\omega)$. However, we may choose $w(k)$ in many other ways, and this flexibility may be employed to improve the accuracy of the Blackman–Tukey spectral estimator or to emphasize some of its characteristics that are of particular interest in a given application. In the following subsections, we address the principal issues which concern the problem of window selection. However, before doing so we rewrite (2.5.1) in an alternative form that will be used in several places of the discussion that follows.

Let $W(\omega)$ denote the DTFT of $w(k)$,

$$W(\omega) = \sum_{k=-\infty}^{\infty} w(k)e^{-i\omega k} = \sum_{k=-(M-1)}^{M-1} w(k)e^{-i\omega k} \tag{2.5.2}$$

Making use of the DTFT property that led to (2.4.8), we can then write

$$\begin{aligned}
\hat{\phi}_{BT}(\omega) &= \sum_{k=-\infty}^{\infty} w(k)\hat{r}(k)e^{-i\omega k} \\
&= \text{DTFT of the product of the sequences} \\
&\quad \{\ldots, 0, 0, w(-(M-1)), \ldots, w(M-1), 0, 0, \ldots\} \text{ and} \\
&\quad \{\ldots, 0, 0, \hat{r}(-(N-1)), \ldots, \hat{r}(N-1), 0, 0, \ldots\} \\
&= \{\text{DTFT}(\hat{r}(k))\} * \{\text{DTFT}(w(k))\}
\end{aligned}$$

As $\text{DTFT}\{\ldots, 0, 0, \hat{r}(-(N-1)), \ldots, \hat{r}(N-1), 0, 0, \ldots\} = \hat{\phi}_p(\omega)$, we obtain

$$\boxed{\hat{\phi}_{BT}(\omega) = \hat{\phi}_p(\omega) * W(\omega) = \frac{1}{2\pi}\int_{-\pi}^{\pi} \hat{\phi}_p(\psi)W(\omega - \psi)d\psi} \tag{2.5.3}$$

This equation is analogous to (2.4.8) and can be interpreted in the same way. Hence, since for most windows in common use $W(\omega)$ has a dominant, relatively narrow peak at $\omega = 0$, it follows from (2.5.3) that

The Blackman–Tukey spectral estimator (2.5.1) corresponds to a "locally" weighted average of the periodogram.	(2.5.4)

Since the function $W(\omega)$ in (2.5.3) acts as a window (or weighting) in the frequency domain, it is sometimes called a ***spectral window***. As we shall see, several refined periodogram–based spectral estimators discussed in what follows can be given an interpretation similar to that afforded by (2.5.3).

The form (2.5.3) under which the Blackman–Tukey spectral estimator has been put is quite appealing from an intuitive standpoint. The main problem with the periodogram lies in its large variations about the true PSD. The weighted average in (2.5.3), in the neighborhood of the current frequency point ω, should smooth the periodogram and hence eliminate its large fluctuations.

On the other hand, this smoothing by the spectral window $W(\omega)$ will also have the undesirable effect of reducing the resolution. We may expect that the smaller the M, the larger the reduction in variance and the lower the resolution. These qualitative arguments may be made exact by a statistical analysis of $\hat{\phi}_{BT}(\omega)$, similar to that in the previous section. In fact, it is clear from (2.5.3) that the mean and variance of $\hat{\phi}_{BT}(\omega)$ can be derived from those of $\hat{\phi}_p(\omega)$. Roughly speaking, the results that can be established by the analysis of $\hat{\phi}_{BT}(\omega)$, based on (2.5.3), show that the resolution of this spectral estimator is on the order of $1/M$, whereas its variance is on the order of M/N. The compromise between resolution and variance, which should be considered when choosing the window's length, is clearly seen from the above considerations. We will look at the tradeoff resolution–variance in more detail in what follows. The next discussion addresses some of the main issues which concern window design.

2.5.2 Nonnegativeness of the Blackman–Tukey Spectral Estimate

Since $\phi(\omega) \geq 0$, it is natural to also require that $\hat{\phi}_{BT}(\omega) \geq 0$. The lag window can be selected to achieve this desirable property of the estimated spectrum. The following result holds true.

If the lag window $\{w(k)\}$ is positive semidefinite (*i.e.*, $W(\omega) \geq 0$), then the windowed covariance sequence $\{w(k)\hat{r}(k)\}$ (with $\hat{r}(k)$ given by (2.2.4)) is positive semidefinite, too; which implies that $\hat{\phi}_{BT}(\omega) \geq 0$ for all ω.	(2.5.5)

In order to prove the above result, first note that $\hat{\phi}_{BT}(\omega) \geq 0$ if and only if the sequence $\{\ldots, 0, 0, w(-(M-1))\hat{r}(-(M-1)), \ldots, w(M-1)\hat{r}(M-1), 0, 0, \ldots\}$ is positive semidefinite or, equivalently, the following Toeplitz matrix is positive semidefinite:

$$\begin{bmatrix} w(0)\hat{r}(0) & \cdots & w(M-1)\hat{r}(M-1) & 0 \\ \vdots & \ddots & \vdots & \ddots \\ w(-(M-1))\hat{r}(-(M-1)) & \cdots & w(0)\hat{r}(0) & \\ \ddots & & \ddots & \\ 0 & & & \end{bmatrix} =$$

$$\begin{bmatrix} w(0) & \cdots & w(M-1) & 0 \\ \vdots & \ddots & \vdots & \ddots \\ w(-(M-1)) & \cdots & w(0) & \\ \ddots & & \ddots & \\ 0 & & & \end{bmatrix} \odot \begin{bmatrix} \hat{r}(0) & \cdots & \hat{r}(N-1) & 0 \\ \vdots & \ddots & \vdots & \ddots \\ \hat{r}(-(N-1)) & \cdots & \hat{r}(0) & \\ \ddots & & \ddots & \\ 0 & & & \end{bmatrix}$$

The symbol $\odot$ denotes the Hadamard matrix product (*i.e.*, element–wise multiplication). By a result in matrix theory, the Hadamard product of two positive semidefinite matrices is also a positive semidefinite matrix (see Result R19 in Appendix A). Thus, the proof of (2.5.5) is concluded.

Another, perhaps simpler, proof of (2.5.5) makes use of (2.5.3) in the following way. Since the sequence $\{w(k)\}$ is real and symmetric about the point $k = 0$, its DTFT $W(\omega)$ is an even, real-valued function. Furthermore, if $\{w(k)\}$ is a positive semidefinite sequence then $W(\omega) \geq 0$ for all ω values (see Exercise 1.8). By (2.5.3), $W(\omega) \geq 0$ immediately implies $\hat{\phi}_{BT}(\omega) \geq 0$, as $\hat{\phi}_p(\omega) \geq 0$ by definition.

It should be noted that some lag windows, such as the rectangular window, do not satisfy the assumption made in (2.5.5) and hence their use may lead to estimated spectra that take negative values. The Bartlett window, on the other hand, is positive semidefinite (as can be seen from (2.4.15)).

2.6 Window Design Considerations

The properties of the Blackman–Tukey estimator (and of other refined periodogram methods discussed in the next section) are directly related to the choice of the lag window. In this section, we discuss several relevant properties of windows that are useful in selecting or designing a window to use in a refined spectral estimation procedure.

2.6.1 Time–Bandwidth Product and Resolution–Variance Trade-offs in Window Design

Most windows are such that they take only nonnegative values in both time and frequency domains (or, if they also take negative values, these are much smaller than the positive values of the window). In addition, they peak at the origin in both domains. For this type of window, it is possible to define an *equivalent time width*, N_e, and an *equivalent bandwidth*, β_e, as follows:

$$N_e = \frac{\sum_{k=-(M-1)}^{M-1} w(k)}{w(0)} \tag{2.6.1}$$

and

$$\beta_e = \frac{\frac{1}{2\pi}\int_{-\pi}^{\pi} W(\omega)d\omega}{W(0)} \tag{2.6.2}$$

From the definitions of direct and inverse DTFTs, we obtain

$$W(0) = \sum_{k=-\infty}^{\infty} w(k) = \sum_{k=-(M-1)}^{M-1} w(k) \tag{2.6.3}$$

and

$$w(0) = \frac{1}{2\pi}\int_{-\pi}^{\pi} W(\omega)d\omega \tag{2.6.4}$$

Using (2.6.3) and (2.6.4) in (2.6.1) and (2.6.2) gives the following result.

The (equivalent) time–bandwidth product equals unity: $N_e\beta_e = 1$	(2.6.5)

As already indicated, the result above applies to window–like signals. Some extended results of the time–bandwidth product type, which apply to more general classes of signals, are presented in the Complement in Section 2.8.4.

It is clearly seen from (2.6.5) that a window cannot be both time–limited and band–limited. The more slowly the window decays to zero in one domain, the more concentrated it is in the other domain. The simple result above, (2.6.5), has several other interesting consequences, as explained below.

The equivalent temporal extent (or aperture), N_e, of $w(k)$ is essentially determined by the window's length. For example, for a rectangular window we have $N_e \simeq 2M$, whereas for a triangular window $N_e \simeq M$. This observation, together with (2.6.5), implies that the equivalent bandwidth β_e is basically determined by the window's length. More precisely, $\beta_e = O(1/M)$. This fact lends support to a claim made previously that for a window which concentrates most of its energy in its main lobe, the width of that lobe should be on the order of $1/M$. Since the main lobe's width sets a limit on the spectral resolution achievable (as explained in Section 2.4), the above observation shows that the spectral resolution limit of a windowed method should be on the order of $1/M$. On the other hand, as explained in the previous section, the statistical variance of such a method is essentially proportional to M/N. Hence, we reached the following conclusion.

The choice of window's length should be based on a tradeoff between spectral resolution and statistical variance	(2.6.6)

As a rule of thumb, we should choose $M \leq N/10$ in order to reduce the standard deviation of the estimated spectrum at least three times, compared with the periodogram.

Once M is determined, we cannot decrease simultaneously the energy in the main lobe (to reduce smearing) and the energy in the sidelobes (to reduce leakage). This follows, for example, from (2.6.4) which shows that the area of $W(\omega)$ is fixed once $w(0)$ is fixed (such as $w(0) = 1$). In other words, if we want to decrease the main lobe's width then we should accept an increase in the sidelobe energy and vice versa. In summary:

The selection of window's shape should be based on a tradeoff between smearing and leakage effects.	(2.6.7)

The above tradeoff is usually dictated by the specific application at hand. A number of windows have been developed to address this tradeoff. In some sense, each of these windows can be seen as a design at a specific point in the resolution/leakage tradeoff curve. We consider several such windows in the next subsection.

2.6.2 Some Common Lag Windows

In this section, we list some of the most common lag windows and outline their relevant properties. Our purpose is not to provide a detailed derivation or an exhaustive listing of such windows, but rather to provide a quick reference of common windows. More detailed information on these and other windows can be found in [HARRIS 1978; KAY 1988; MARPLE 1987; OPPENHEIM AND SCHAFER 1989; PRIESTLEY 1989; PORAT 1997], where many of the closed–form windows have been compiled. Table 2.1 lists some common windows along with some useful properties.

In addition to the fixed window designs in Table 2.1, there are windows that contain a design parameter which may be varied to trade between resolution and sidelobe leakage. Two such common designs are the Chebyshev window and the Kaiser window. The Chebyshev window has the property that the peak level of the sidelobe "ripples" is constant. Thus, unlike most other windows, the sidelobe level does not decrease as ω increases. The Kaiser window is defined by

$$w(k) = \frac{I_0\left(\gamma\sqrt{1-[k/(M-1)]^2}\right)}{I_0(\gamma)}, \quad -(M-1) \leq k \leq M-1 \tag{2.6.8}$$

where $I_0(\cdot)$ is the zeroth–order modified Bessel function of the first kind. The parameter γ trades the main lobe width for the sidelobe leakage level; $\gamma = 0$ corresponds to a rectangular window, and $\gamma > 0$ results in lower sidelobe leakage at the expense of a broader main lobe. The approximate value of γ needed to achieve a peak sidelobe level of B dB below the peak value is

$$\gamma \simeq \begin{cases} 0, & B < 21 \\ .584(B-21)^{0.4} + .0789(B-21), & 21 \leq B \leq 50 \\ .11(B-8.7), & B > 50 \end{cases}$$

Table 2.1. Some Common Windows and their Properties
The windows satisfy $w(k) \equiv 0$ for $|k| \geq M$, and $w(k) = w(-k)$; the defining equations below are valid for $0 \leq k \leq (M-1)$.

Window Name	Defining Equation	Approx. Main Lobe Width (radians)	Sidelobe Level (dB)
Rectangular	$w(k) = 1$	$2\pi/M$	-13
Bartlett	$w(k) = (M-k)/M$	$4\pi/M$	-25
Hanning	$w(k) = .5 + .5\cos(\pi k/M)$	$4\pi/M$	-31
Hamming	$w(k) = .54 + .46\cos(\pi k/(M-1))$	$4\pi/M$	-41
Blackman	$w(k) = .42 + .5\cos(\pi k/(M-1))$ $+ .08\cos(2\pi k/(M-1))$	$6\pi/M$	-57

The Kaiser window is an approximation of the optimal window described in the next subsection. It is often chosen over the fixed window designs because it has a lower sidelobe level when γ is selected to have the same main lobe width as the corresponding fixed window (or narrower main lobe width for a given sidelobe level). The optimal window of the next subsection improves on the Kaiser design slightly.

Figure 2.3 shows plots of several windows with $M = 26$. The Kaiser window is shown for $\gamma = 1$ and $\gamma = 4$, and the Chebyshev window is designed to have a -40 dB sidelobe level. Figure 2.4 shows the corresponding normalized window transfer functions $W(\omega)$. Note the constant sidelobe ripple level of the Chebyshev design.

We remark that except for the Bartlett window, none of the windows we have introduced (including the Chebyshev and Kaiser windows) has nonnegative Fourier transform. On the other hand, it is straightforward to produce such a nonnegative definite window by convolving the window with itself. Recall that the Bartlett window is the convolution of a rectangular window with itself. We will make use of the convolution of windows with themselves in the next two subsections, both for window design and for relating temporal windows to covariance lag windows.

2.6.3 Window Design Example

Assume a situation where it is known that the observed signal consists of a useful weak signal and a strong interference, and that both the useful signal and the interference can be assumed to be narrowband signals which are well separated in frequency. However, there is no *a priori* quantitative information available on the frequency separation between the desired signal and the interference. It is required to design a lag window for use in a Blackman–Tukey spectral estimation method, with the purpose of detecting and locating in frequency the useful signal.

The main problem in the application outlined above lies in the fact that the (strong) interference may completely mask the (weak) desired signal through leakage. In order to get rid of this problem, the window design should compromise

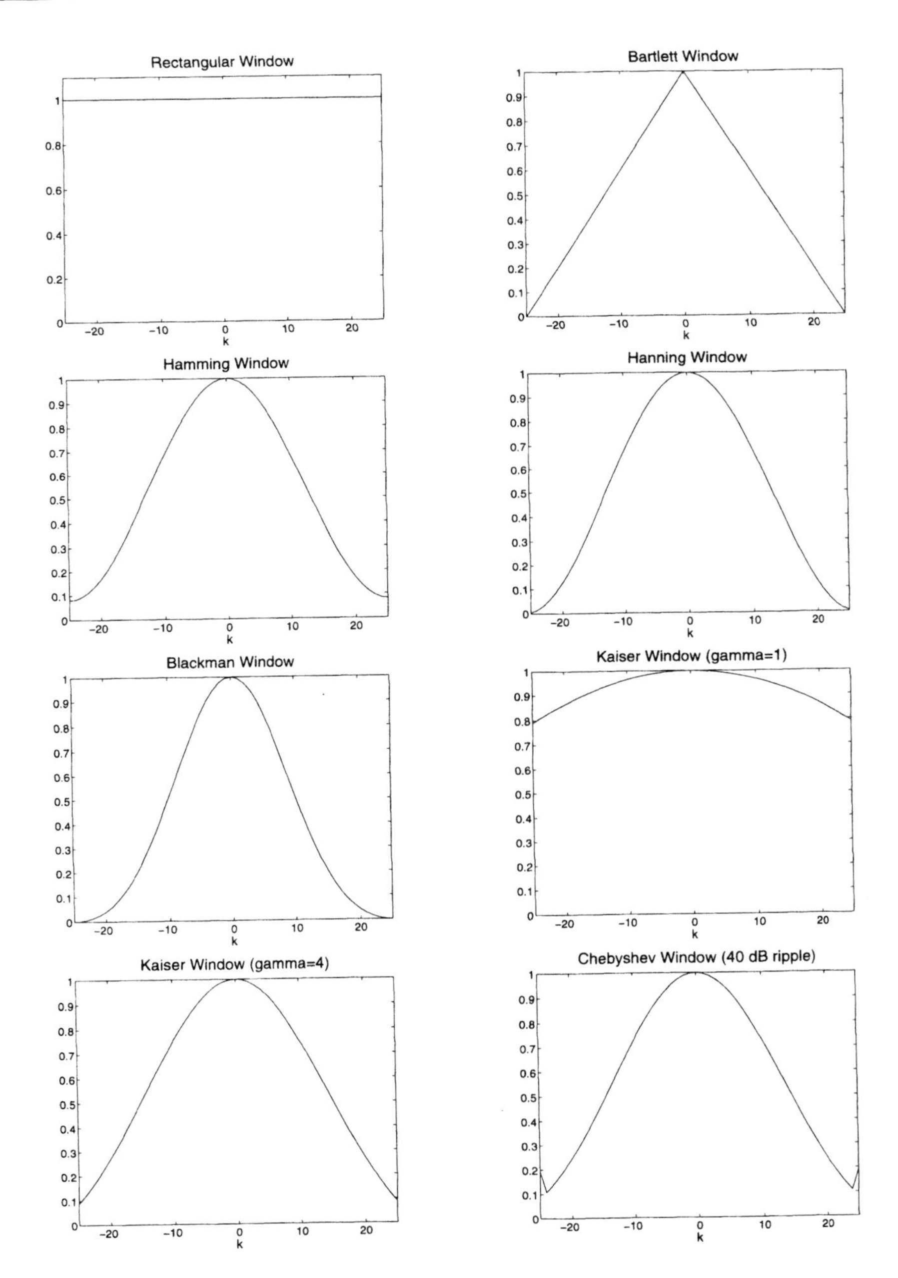

Figure 2.3. Some common window functions (shown for $M = 26$). The Kaiser window uses $\gamma = 1$ and $\gamma = 4$ and the Chebyshev window is designed for a -40 dB sidelobe level.

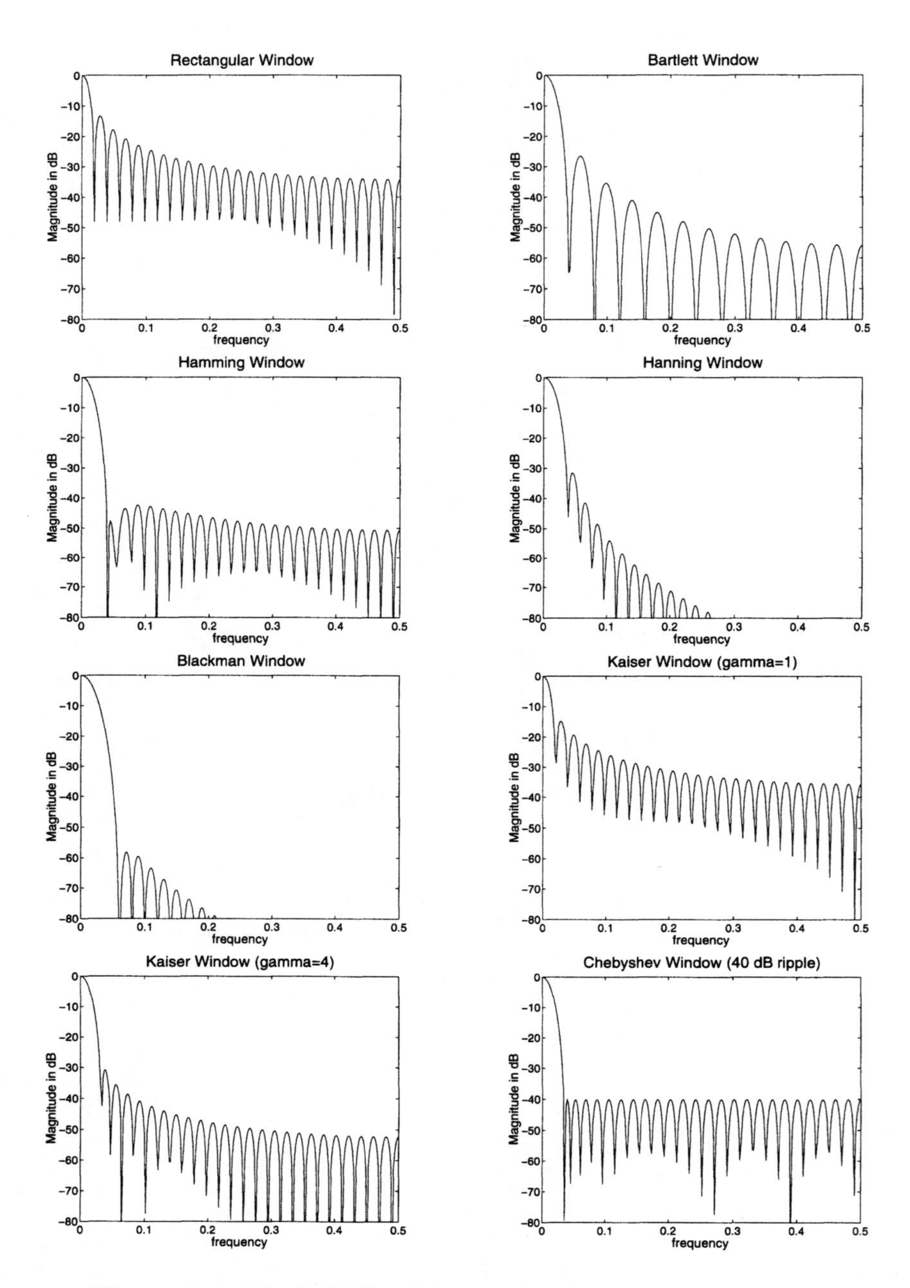

Figure 2.4. The DTFTs of the window functions in Figure 2.3.

smearing for leakage. Note that the smearing effect is not of main concern in this application, as the useful signal and the interference are well separated in frequency. Hence, smearing cannot affect our ability to detect the desired signal; it will only limit, to some degree, our ability to accurately locate in frequency the signal in question.

We consider a window sequence whose DTFT $W(\omega)$ is constructed as the squared magnitude of the DTFT of another sequence $\{v(k)\}$; in this way, we guarantee that the constructed window is positive semidefinite. Mathematically, the above design problem can be formulated as follows. Consider a sequence $\{v(0), \ldots, v(M-1)\}$, and let

$$V(\omega) = \sum_{k=0}^{M-1} v(k) e^{-i\omega k} \tag{2.6.9}$$

The DTFT $V(\omega)$ can be rewritten in the more compact form

$$V(\omega) = v^* a(\omega) \tag{2.6.10}$$

where

$$v = [v(0) \;\ldots\; v(M-1)]^* \tag{2.6.11}$$

and

$$a(\omega) = [1 \; e^{-i\omega} \ldots \; e^{-i(M-1)\omega}]^T \tag{2.6.12}$$

Define the spectral window as

$$W(\omega) = |V(\omega)|^2 \tag{2.6.13}$$

The corresponding lag window can be obtained from (2.6.13) as follows:

$$\begin{aligned}
\sum_{k=-(M-1)}^{M-1} w(k) e^{-i\omega k} &= \sum_{n=0}^{M-1} \sum_{p=0}^{M-1} v(n) v^*(p) e^{-i\omega(n-p)} \\
&= \sum_{n=0}^{M-1} \sum_{k=n}^{n-(M-1)} v(n) v^*(n-k) e^{-i\omega k} \\
&= \sum_{k=-(M-1)}^{M-1} \left[\sum_{n=0}^{M-1} v(n) v^*(n-k) \right] e^{-i\omega k}
\end{aligned} \tag{2.6.14}$$

which gives

$$w(k) = \sum_{n=0}^{M-1} v(n) v^*(n-k) \tag{2.6.15}$$

The last equality in (2.6.14), and hence the equality (2.6.15), are valid under the convention that $v(k) = 0$ for $k < 0$ and $k \geq M$.

As already mentioned, this method of constructing $\{w(k)\}$ from the convolution of the sequence $\{v(k)\}$ with itself has the advantage that the so–obtained lag window is always positive semidefinite or, equivalently, the corresponding spectral window satisfies $W(\omega) \geq 0$ (which is easily seen from (2.6.13)). Besides this, the design of $\{w(k)\}$ can be reduced to the selection of $\{v(k)\}$ which may be more conveniently done, as explained next.

In the present application, the design objective is to reduce the leakage incurred by $\{w(k)\}$ as much as possible. This objective can be formulated as the problem of minimizing the relative energy in the sidelobes of $W(\omega)$ or, equivalently, as the problem of maximizing the relative energy in the main lobe of $W(\omega)$:

$$\max_{v} \left\{ \int_{-\beta\pi}^{\beta\pi} W(\omega)d\omega \Big/ \int_{-\pi}^{\pi} W(\omega)d\omega \right\} \tag{2.6.16}$$

Here, β is a design parameter which quantifies how much smearing (or, basically equivalent, resolution) we can tradeoff for leakage reduction. The larger the β, the more leakage free the optimal window derived from (2.6.16) but also the more diminished the spectral resolution associated with that window.

By writing the criterion in (2.6.16) in the following form

$$\begin{aligned} &\frac{1}{2\pi}\int_{-\beta\pi}^{\beta\pi} |V(\omega)|^2 d\omega \Big/ \frac{1}{2\pi}\int_{-\pi}^{\pi} |V(\omega)|^2 d\omega \\ &= v^* \left[\frac{1}{2\pi}\int_{-\beta\pi}^{\beta\pi} a(\omega)a^*(\omega)d\omega \right] v/v^*v \end{aligned} \tag{2.6.17}$$

(*cf.* (2.6.10) and Parseval's theorem, (1.2.6)), the optimization problem (2.6.16) becomes

$$\max_{v} \frac{v^*\Gamma v}{v^*v} \tag{2.6.18}$$

where

$$\Gamma = \frac{1}{2\pi}\int_{-\beta\pi}^{\beta\pi} a(\omega)a^*(\omega)d\omega \triangleq [\gamma_{m-n}] \tag{2.6.19}$$

and where

$$\gamma_{m-n} = \frac{1}{2\pi}\int_{-\beta\pi}^{\beta\pi} e^{-i(m-n)\omega} d\omega = \frac{\sin[(m-n)\beta\pi]}{(m-n)\pi} \tag{2.6.20}$$

(note that $\gamma_0 = \beta$). By using the function

$$\operatorname{sinc}(x) \triangleq \frac{\sin x}{x}, \qquad (\operatorname{sinc}(0) = 1) \tag{2.6.21}$$

we can write (2.6.20) as

$$\boxed{\gamma_{m-n} = \beta \operatorname{sinc}[(m-n)\beta\pi]} \tag{2.6.22}$$

The solution to the problem (2.6.18) is well known: the maximizing v is given by the dominant eigenvector of Γ, associated with the maximum eigenvalue of this matrix (see Result R13 in Appendix A). To summarize:

> The optimal lag window which minimizes the relative energy in the sidelobe interval $[-\pi, -\beta\pi] \cup [\beta\pi, \pi]$ is given by (2.6.15), where v is the dominant eigenvector of the matrix Γ defined in (2.6.19) and (2.6.22). (2.6.23)

Regarding the choice of the design parameter β, it is clear that β should be larger than $1/M$ in order to allow for a significant reduction of leakage. Otherwise, by selecting for example $\beta \simeq 1/M$, we weigh the resolution issue too much in the design problem, with unfavorable consequences for leakage reduction.

Finally, we remark that a problem quite similar to the above one, although derived from different considerations, will be encountered in Chapter 5 (see also [MULLIS AND SCHARF 1991]).

2.6.4 Temporal Windows and Lag Windows

As we have seen previously, the unwindowed periodogram coincides with the unwindowed correlogram. The Blackman–Tukey estimator is a windowed correlogram obtained using a lag window. Similarly, we can define a windowed periodogram

$$\hat{\phi}_W(\omega) = \frac{1}{N} \left| \sum_{t=1}^{N} v(t) y(t) e^{-i\omega t} \right|^2 \tag{2.6.24}$$

where the weighting sequence $\{v(t)\}$ may be called a *temporal window*. Welch [WELCH 1967] was one of the first researchers who considered windowed periodogram spectral estimators (see Section 2.7.2 for a description of Welch's method), and hence the subscript "W" attached to $\hat{\phi}(\omega)$ in (2.6.24). However, while the reason for windowing the correlogram was clearly motivated, the reason for windowing the periodogram is less obvious. In order to motivate (2.6.24), at least partially, write this equation as

$$\hat{\phi}_W(\omega) = \frac{1}{N} \sum_{t=1}^{N} \sum_{s=1}^{N} v(t) v^*(s) y(t) y^*(s) e^{-i\omega(t-s)} \tag{2.6.25}$$

Next, take expectation of both sides of (2.6.25) to obtain

$$E\left\{\hat{\phi}_W(\omega)\right\} = \frac{1}{N} \sum_{t=1}^{N} \sum_{s=1}^{N} v(t) v^*(s) r(t-s) e^{-i\omega(t-s)} \tag{2.6.26}$$

Inserting

$$r(t-s) = \frac{1}{2\pi} \int_{-\pi}^{\pi} \phi(\omega) e^{i\omega(t-s)} d\omega \tag{2.6.27}$$

in (2.6.26) gives

$$E\left\{\hat{\phi}_W(\omega)\right\} = \frac{1}{N2\pi}\int_{-\pi}^{\pi}\phi(\psi)\left[\sum_{t=1}^{N}\sum_{s=1}^{N}v(t)v^*(s)e^{-i(\omega-\psi)(t-s)}\right]d\psi$$
$$= \frac{1}{N2\pi}\int_{-\pi}^{\pi}\phi(\psi)\left|\sum_{t=1}^{N}v(t)e^{-i(\omega-\psi)t}\right|^2 d\psi \tag{2.6.28}$$

Define

$$W(\omega) = \frac{1}{N}\left|\sum_{t=1}^{N}v(t)e^{-i\omega t}\right|^2 \tag{2.6.29}$$

By using this notation, we can write (2.6.28) as

$$E\left\{\hat{\phi}_W(\omega)\right\} = \frac{1}{2\pi}\int_{-\pi}^{\pi}\phi(\psi)W(\omega-\psi)d\psi \tag{2.6.30}$$

As the equation (2.6.29) is similar to (2.6.13), the sequence whose DTFT is equal to $W(\omega)$ immediately follows from (2.6.15):

$$\boxed{w(k) = \frac{1}{N}\sum_{n=1}^{N}v(n)v^*(n-k)} \tag{2.6.31}$$

Next, by comparing (2.6.30) and (2.5.3), we get the following result.

The windowed periodogram and the windowed correlogram have the same *average* behavior, provided the temporal and lag windows are related as in (2.6.31).	(2.6.32)

Hence $E\{\hat{\phi}_W(\omega)\} = E\{\hat{\phi}_{BT}(\omega)\}$, provided the temporal and lag windows are matched to one another. A similarly simple relationship between $\hat{\phi}_W(\omega)$ and $\hat{\phi}_{BT}(\omega)$, however, does not seem to exist. This makes it somewhat difficult to motivate the windowed periodogram as defined in (2.6.24). The Welch periodogram, though, does not weigh all data samples as in (2.6.24), and is a useful spectral estimator (see the next section).

2.7 Other Refined Periodogram Methods

In Section 2.5 we introduced the Blackman–Tukey estimator as an alternative to the periodogram. In this section we present three other modified periodograms: the Bartlett, Welch, and Daniell methods. Like the Blackman–Tukey method, they seek to reduce the variance of the periodogram by smoothing or averaging the periodogram estimates in some way. We will relate these methods to one another and to the Blackman–Tukey method.

2.7.1 Bartlett Method

The basic idea of the Bartlett method [BARTLETT 1948; BARTLETT 1950] is simple: to reduce the large fluctuations of the periodogram, split up the available sample of N observations into $L = N/M$ subsamples of M observations each, and then average the periodograms obtained from the subsamples for each value of ω. Mathematically, the Bartlett method can be described as follows. Let

$$y_j(t) = y((j-1)M + t), \qquad \begin{array}{rcl} t & = & 1, \ldots, M \\ j & = & 1, \ldots, L \end{array} \tag{2.7.1}$$

denote the observations of the jth subsample, and let

$$\hat{\phi}_j(\omega) = \frac{1}{M} \left| \sum_{t=1}^{M} y_j(t) e^{-i\omega t} \right|^2 \tag{2.7.2}$$

denote the corresponding periodogram. The Bartlett spectral estimate is then given by

$$\hat{\phi}_B(\omega) = \frac{1}{L} \sum_{j=1}^{L} \hat{\phi}_j(\omega) \tag{2.7.3}$$

Since the Bartlett method operates on data segments of length M, the resolution afforded should be on the order of $1/M$. Hence, the spectral resolution of the Bartlett method is reduced by a factor L, compared to the resolution of the original periodogram method. In return for this reduction in resolution, we can expect that the Bartlett method has a reduced variance. It can, in fact, be shown that the Bartlett method reduces the variance of the periodogram by the same factor L (see below). The compromise between resolution and variance when selecting M (or L) is thus evident.

An interesting way to look at the Bartlett method and its properties is by relating it to the Blackman–Tukey method which was discussed before in a rather detailed manner. As we know, $\hat{\phi}_j(\omega)$ of (2.7.2) can be rewritten as

$$\hat{\phi}_j(\omega) = \sum_{k=-(M-1)}^{M-1} \hat{r}_j(k) e^{-i\omega k} \tag{2.7.4}$$

where $\{\hat{r}_j(k)\}$ is the sample covariance sequence corresponding to the jth subsample. Inserting (2.7.4) in (2.7.3) gives

$$\hat{\phi}_B(\omega) = \sum_{k=-(M-1)}^{M-1} \left[\frac{1}{L} \sum_{j=1}^{L} \hat{r}_j(k) \right] e^{-i\omega k} \tag{2.7.5}$$

We see that $\hat{\phi}_B(\omega)$ is similar in form to the Blackman–Tukey estimator that uses a rectangular window. The average, over j, of the subsample covariance $\hat{r}_j(k)$ is an estimate of the ACS $r(k)$. However, the ACS estimate in (2.7.5) does not make efficient use of available data lag products $y(t)y^*(t-k)$, especially for $|k|$ near $M-1$ (see Exercise 2.12). In fact, for $k = M-1$, only about $1/M$th of the available lag products are used to form the ACS estimate in (2.7.5). We expect that the variance of these lags is higher than for the corresponding $\hat{r}(k)$ lags used in the Blackman–Tukey estimate, and similarly, the variance of $\hat{\phi}_B(\omega)$ is higher than that of $\hat{\phi}_{BT}(\omega)$. In addition, the Bartlett method uses a fixed rectangular lag window, and thus has less flexibility in resolution–leakage tradeoff than does the Blackman–Tukey method. For these reasons, we conclude that

> The Bartlett estimate, as defined in (2.7.1)–(2.7.3), is similar in form to, but typically has a slightly higher variance than, the Blackman–Tukey estimate with a rectangular lag window of length M. (2.7.6)

The reduction in resolution and the decrease of variance (both by a factor $L = N/M$) for the Bartlett estimate, as compared to the basic periodogram method, follows from (2.7.6) and the properties of the Blackman–Tukey spectral estimator given previously.

The main lobe of the rectangular window is narrower than that associated with most other lag windows (this follows from the observation that the rectangular window clearly has the largest equivalent time width , and the fact that the time–bandwidth product is constant, see (2.6.5)). Thus, it follows from (2.7.6) that in the class of Blackman–Tukey estimates, the Bartlett estimator can be expected to have the least smearing (and hence the best resolution) but the most significant leakage.

2.7.2 Welch Method

The Welch method [WELCH 1967] is obtained by refining the Bartlett method in two respects. First, the data segments in the Welch method are allowed to overlap. Second, each data segment is windowed prior to computing the periodogram. To describe the Welch method in a mathematical form, let

$$y_j(t) = y((j-1)K + t), \qquad \begin{array}{rcl} t &=& 1,\ldots,M \\ j &=& 1,\ldots,S \end{array} \tag{2.7.7}$$

denote the jth data segment. In (2.7.7), $(j-1)K$ is the starting point for the jth sequence of observations. If $K = M$, then the sequences do not overlap (but are contiguous) and we get the sample splitting used by the Bartlett method (which leads to $S = L = N/M$ data subsamples). However, the value recommended for K in the Welch method is $K = M/2$, in which case $S \simeq 2M/N$ data segments (with 50% overlap between successive segments) are obtained.

The windowed periodogram corresponding to $y_j(t)$ is computed as

$$\hat{\phi}_j(\omega) = \frac{1}{MP}\left|\sum_{t=1}^{M} v(t)y_j(t)e^{-i\omega t}\right|^2 \tag{2.7.8}$$

where P denotes the "power" of the temporal window $\{v(t)\}$:

$$P = \frac{1}{M}\sum_{t=1}^{M}|v(t)|^2 \tag{2.7.9}$$

The Welch estimate of PSD is determined by averaging the windowed periodograms in (2.7.8):

$$\hat{\phi}_W(\omega) = \frac{1}{S}\sum_{j=1}^{S}\hat{\phi}_j(\omega) \tag{2.7.10}$$

The reasons for the above modifications to the Bartlett method, which led to the Welch method, are simple to explain. By allowing overlap between the data segments and hence by getting more periodograms to be averaged in (2.7.10), we hope to decrease the variance of the estimated PSD. By introducing the window in the periodogram computation it may be hoped to get more control over the bias/resolution properties of the estimated PSD (see Section 2.6.4). Additionally, the temporal window may be used to give less weight to the data samples at the ends of each subsample, hence making the consecutive subsample sequences less correlated to one another, even though they are overlapping. The principal effect of this "decorrelation" should be a more effective reduction of variance via the averaging in (2.7.10).

The analysis that led to the results (2.6.30)–(2.6.32) can be modified to show that the use of windowed periodograms in the Welch method, as contrasted to the unwindowed periodograms in the Bartlett method, indeed offers more flexibility in controlling the bias properties of the estimated spectrum. The variance of the Welch spectral estimator is more difficult to analyze (except in some special cases). However, there is empirical evidence that the Welch method can offer lower variance than the Bartlett method but the difference in the variances corresponding to the two methods is not dramatic.

We can relate the Welch estimator to the Blackman–Tukey spectral estimator by a straightforward calculation as we did for the Bartlett method. By inserting (2.7.8) in (2.7.10), we obtain

$$\hat{\phi}_W(\omega) = \frac{1}{S}\sum_{j=1}^{S}\frac{1}{MP}\sum_{t=1}^{M}\sum_{k=1}^{M}v(t)v^*(k)y_j(t)y_j^*(k)e^{-i\omega(t-k)} \tag{2.7.11}$$

For large values of N and for $K = M/2$ or smaller, S results sufficiently large for the average $(1/S)\sum_{j=1}^{S}y_j(t)y_j^*(k)$ to be close to the covariance $r(t-k)$. We do not replace the previous sum by the true covariance lag. However, we assume that this sum does not depend on both t and k, but only on their difference $(t-k)$, at least approximately; say

$$\tilde{r}(t,k) = \frac{1}{S}\sum_{j=1}^{S}y_j(t)y_j^*(k) \simeq \tilde{r}(t-k) \tag{2.7.12}$$

Using (2.7.12) in (2.7.11) gives

$$\begin{aligned}\hat{\phi}_W(\omega) &\simeq \frac{1}{MP}\sum_{t=1}^{M}\sum_{k=1}^{M} v(t)v^*(k)\tilde{r}(t-k)e^{-i\omega(t-k)} \\ &= \frac{1}{MP}\sum_{t=1}^{M}\sum_{\tau=t-1}^{t-M} v(t)v^*(t-\tau)\tilde{r}(\tau)e^{-i\omega\tau} \\ &= \sum_{\tau=-(M-1)}^{M-1}\left[\frac{1}{MP}\sum_{t=1}^{M} v(t)v^*(t-\tau)\right]\tilde{r}(\tau)e^{-i\omega\tau}\end{aligned} \tag{2.7.13}$$

By introducing

$$w(\tau) = \frac{1}{MP}\sum_{t=1}^{M} v(t)v^*(t-\tau) \tag{2.7.14}$$

(under the convention that $v(k) = 0$ for $k < 1$ and $k > M$), we can write (2.7.13) as

$$\hat{\phi}_W(\omega) \simeq \sum_{\tau=-(M-1)}^{M-1} w(\tau)\tilde{r}(\tau)e^{-i\omega\tau} \tag{2.7.15}$$

which is to be compared to the form of the Blackman–Tukey estimator. To summarize, the Welch estimator has been shown to approximate a Blackman–Tukey–*type* estimator for the estimated covariance sequence (2.7.12) (which may be expected to have finite–sample properties different from those of $\hat{r}(k)$).

The Welch estimator can be efficiently computed via the FFT, and is one of the most frequently used PSD estimation methods. Its previous interpretation is pleasing, even if approximate, since the Blackman–Tukey form of spectral estimator is theoretically the most favored one. This interpretation also shows that we may think of replacing the usual covariance estimates $\{\hat{r}(k)\}$ in the Blackman–Tukey estimator by other sample covariances, with the purpose of either reducing the computational burden or improving the statistical accuracy.

2.7.3 Daniell Method

As shown in (2.4.21), the periodogram values $\hat{\phi}(\omega_k)$ corresponding to different frequency values ω_k are (asymptotically) uncorrelated random variables. One may then think of reducing the large variance of the basic periodogram estimator by averaging the periodogram over small intervals centered on the current frequency ω. This is the idea behind the Daniell method [DANIELL 1946]. The practical form of the Daniell estimate, which can be implemented by means of the FFT, is the following:

$$\hat{\phi}_D(\omega_k) = \frac{1}{2J+1}\sum_{j=k-J}^{k+J} \hat{\phi}_p(\omega_j) \tag{2.7.16}$$

where

$$\omega_k = \frac{2\pi}{\tilde{N}}k, \qquad k = 0, \ldots, \tilde{N} - 1 \tag{2.7.17}$$

and where $\tilde{N}$ is (much) larger than N to ensure a fine sampling of $\hat{\phi}_p(\omega)$. The periodogram samples needed in (2.7.16) can be obtained, for example, by using a radix–2 FFT algorithm applied to the zero–padded data sequence, as described in Section 2.3. The parameter J in the Daniell method should be chosen sufficiently small to guarantee that $\phi(\omega)$ is nearly constant on the interval(s):

$$\left[\omega - \frac{2\pi}{\tilde{N}}J,\ \omega + \frac{2\pi}{\tilde{N}}J\right] \tag{2.7.18}$$

Since $\tilde{N}$ can in principle be chosen as large as we want, we can choose J fairly large without violating the above requirement that $\phi(\omega)$ is nearly constant over the interval in (2.7.18). For the sake of illustration, let us assume that we keep the ratio $J/\tilde{N}$ constant, but increase both J and $\tilde{N}$ significantly. As $J/\tilde{N}$ is constant, the resolution/bias properties of the Daniell estimator should be basically unaffected. On the other hand, the fact that the number of periodogram values averaged in (2.7.16) increases with increased J might suggest that the variance decreases. However, we know that this should not be possible, as the variance can be decreased only at the expense of increasing the bias (and vice versa). Indeed, in the case under discussion the periodogram values averaged in (2.7.16) become more and more correlated as $\tilde{N}$ increases and hence the variance of $\hat{\phi}_D(\omega)$ does not necessarily decrease with J if $\tilde{N}$ is larger than N (see, *e.g.*, Exercise 2.11). We will return to the bias and variance properties of the Daniell method a bit later.

By introducing $\beta = 2J/\tilde{N}$, one can write (2.7.18) in a form that is more convenient for the discussion that follows, namely

$$[\omega - \pi\beta, \omega + \pi\beta] \tag{2.7.19}$$

Equation (2.7.16) is a discrete approximation of the theoretical version of the Daniell estimator, which is given by

$$\hat{\phi}_D(\omega) = \frac{1}{2\pi\beta}\int_{\omega-\beta\pi}^{\omega+\beta\pi} \hat{\phi}_p(\psi)d\psi \tag{2.7.20}$$

The larger the $\tilde{N}$, the smaller the difference between the approximation (2.7.16) and the continuous version, (2.7.20), of the Daniell spectral estimator.

It is intuitively clear from (2.7.20) that as β increases, the resolution of the Daniell estimator decreases (or, essentially equivalent, the bias increases) and the variance gets lower. In fact, if we introduce

$$M = 1/\beta \tag{2.7.21}$$

(in an approximate sense, as $1/\beta$ is not necessarily an integer) then we may expect that the resolution and the variance of the Daniell estimator are both decreased by a factor M, compared to the basic periodogram method. In order to support this claim, we relate the Daniell estimator to the Blackman–Tukey estimation technique. By simply comparing (2.7.20) and (2.5.3), we obtain the following result.

> The Daniell estimator is a particular case of the Blackman–Tukey class of spectral estimators, corresponding to a rectangular *spectral* window:
> $$W(\omega) = \begin{cases} 1/\beta, & \omega \in [-\beta\pi, \beta\pi] \\ 0, & \text{otherwise} \end{cases} \tag{2.7.22}$$

The above observation, along with the time–bandwidth product result and the properties of the Blackman–Tukey spectral estimator, lends support to the previously made claim on the Daniell estimator. Note that the Daniell estimate of PSD is a nonnegative function by its very definition, (2.7.20), which is not necessarily the case for several members of the Blackman–Tukey class of PSD estimators.

The lag window corresponding to the $W(\omega)$ in (2.7.22) is readily evaluated as follows:

$$\begin{aligned} w(k) &= \frac{1}{2\pi}\int_{-\pi}^{\pi} W(\omega)e^{i\omega k}d\omega = \frac{1}{2\pi\beta}\int_{-\pi\beta}^{\pi\beta} e^{i\omega k}d\omega \\ &= \frac{\sin(k\pi\beta)}{k\pi\beta} = \operatorname{sinc}(k\pi\beta) \end{aligned} \tag{2.7.23}$$

Note that $w(k)$ does not vanish as k increases, which leads to a subtle (but not essential) difference between the lag windowed forms of the Daniell and Blackman–Tukey estimators. Since the inverse DTFT of $\hat{\phi}_p(\omega)$ is given by the sequence $\{\ldots, 0, 0, \hat{r}(-(N-1)), \ldots, \hat{r}(N-1), 0, 0, \ldots\}$, it follows immediately from (2.7.20) that $\hat{\phi}_D(\omega)$ can also be written as

$$\hat{\phi}_D(\omega) = \sum_{k=-(N-1)}^{N-1} w(k)\hat{r}(k)e^{-i\omega k} \tag{2.7.24}$$

It is seen from (2.7.24) that, like the Blackman–Tukey estimator, $\hat{\phi}_D(\omega)$ is a windowed version of the correlogram but, unlike the Blackman–Tukey estimator, the sum in (2.7.24) is not truncated to a value $M < N$. Hence, contrary to what might have been expected intuitively, the parameter M defined in (2.7.21) cannot be exactly interpreted as a "truncation point" for the lag windowed version of $\hat{\phi}_D(\omega)$. However, since the equivalent bandwidth of $W(\omega)$ is clearly equal to β,

$$\beta_e = \beta$$

it follows that the equivalent time width of $w(k)$ is

$$N_e = 1/\beta_e = M$$

which shows that M plays essentially the same role here as the "truncation point" in the Blackman–Tukey estimator (and, indeed, it can be verified that $w(k)$ in (2.7.23) takes small values for $|k| > M$).

In closing this section and this chapter, we point out that the periodogram–based methods for spectrum estimation are all variations on a theme. These methods attempt to reduce the variance of the basic periodogram estimator, at the expense of some reduction in resolution, by various means such as: averaging periodograms derived from data subsamples (Bartlett and Welch methods); averaging periodogram values locally around the frequency of interest (Daniell method); and smoothing the periodogram (Blackman–Tukey method). The unifying theme of these methods is seen in that they are essentially special forms of the Blackman–Tukey approach. In Chapter 5 we will push the unifying theme one step further by showing that the periodogram–based methods can also be obtained as special cases of the filter bank approach to spectrum estimation described there (see also [MULLIS AND SCHARF 1991]).

Finally, it is interesting to note that, while the modifications of the periodogram described in this chapter are indeed required when estimating a continuous PSD, the *unmodified periodogram* can be shown to be a satisfactory estimator (actually, the best one in large samples) for discrete (or line) spectra corresponding to sinusoidal signals. This is shown in Chapter 4.

2.8 Complements

2.8.1 Sample Covariance Computation via FFT

Computation of the sample covariances is a ubiquitous problem in spectral estimation and signal processing applications. Make use of the DTFT–like formula (2.2.2), relating the periodogram and the sample covariance sequence, to devise an FFT–based algorithm for computation of the $\{\hat{r}(k)\}_{k=0}^{N-1}$. Compare the computational requirements of such an algorithm with those corresponding to the evaluation of $\{\hat{r}(k)\}$ via the temporal averaging formula (2.2.4), and show that the former may be computationally more efficient than the latter if N is larger than a certain value.

Solution: From (2.2.2) and (2.2.6) we have that (we omit the subscript p of $\hat{\phi}_p(\omega)$ for notational simplicity):

$$\hat{\phi}(\omega) = \sum_{k=-N+1}^{N-1} \hat{r}(k)e^{-i\omega k} = \sum_{p=1}^{2N-1} \hat{r}(p-N)e^{-i\omega(p-N)}$$

or, equivalently,

$$e^{-i\omega N}\,\hat{\phi}(\omega) = \sum_{p=1}^{2N-1} \rho(p)e^{-i\omega p} \tag{2.8.1}$$

where $\rho(p) \triangleq \hat{r}(p-N)$. Equation (2.8.1) has the standard form of a DFT. It is evident from (2.8.1) that in order to determine the sample covariance sequence we need at least $(2N-1)$ values of the periodogram. This was expected: The sequence $\{\hat{r}(k)\}_{k=0}^{N-1}$ contains $(2N-1)$ real–valued unknowns for the determination of which at least $(2N-1)$ periodogram values should be necessary (as $\hat{\phi}(\omega)$ is real valued).

Let

$$\omega_k = \frac{2\pi}{2N-1}(k-1), \qquad k = 1, \ldots, 2N-1$$

Also, let the sequence $\{y(t)\}_{t=1}^{2N-1}$ be obtained by padding the raw data sequence with $(N-1)$ zeroes. Compute

$$Y_k = \sum_{t=1}^{2N-1} y(t)e^{-i\omega_k t} \qquad (k = 1, 2, \ldots, 2N-1) \tag{2.8.2}$$

by means of a $(2N-1)$–point FFT algorithm. Next, evaluate

$$\tilde{\phi}_k = e^{-i\omega_k N}\, |Y_k|^2/N \qquad (k = 1, \ldots, 2N-1) \tag{2.8.3}$$

Finally, determine the sample covariances via the "inversion" of (2.8.1):

$$\begin{aligned} \rho(p) &= \sum_{k=1}^{2N-1} \tilde{\phi}_k e^{i\omega_k p}/(2N-1) \\ &= \sum_{k=1}^{2N-1} \tilde{\phi}_k e^{i\omega_p k}/(2N-1) \end{aligned} \tag{2.8.4}$$

The previous computation may once again be done by using a $(2N-1)$–point FFT algorithm. The bulk of the procedure outlined above consists of the FFT–based computation of (2.8.2) and (2.8.4). That computation requires about $2N \log_2(2N)$ flops (assuming that the radix–2 FFT algorithm is used; the required number of operations is larger than the one previously given whenever N is not a power of two.) The direct evaluation of the sample covariance sequence via (2.2.4) requires

$$N + (N-1) + \cdots + 1 \simeq N^2/2 \qquad \text{flops}$$

Hence, the FFT–based computation would be more efficient whenever

$$N > 4\log_2(2N)$$

This inequality is satisfied for $N \geq 32$. (Actually, N needs to be greater than 32 because we neglected the operations needed to implement equation (2.8.3).)

The previous discussion assumes that N is a power of two. If this is not the case then the relative computational efficiency of the two procedures may be different. Note, also, that there are several other issues that may affect this comparison. For instance, if only the lags $\{\hat{r}(k)\}_{k=0}^{M-1}$ (with $M \ll N$) are required, then the number of computations required by (2.2.4) is drastically reduced. On the other hand, the

FFT–based procedure can also be implemented in a more efficient way in such a case, so that it remains computationally more efficient than a direct calculation, for instance, for $N \geq 100$ [OPPENHEIM AND SCHAFER 1989]. We conclude that the various implementation details may change the value of N beyond which the FFT–based procedure is more efficient than the direct approach, and hence may influence the decision as to which of the two procedures should be used in a given application.

2.8.2 FFT–Based Computation of Windowed Blackman–Tukey Periodograms

The windowed Blackman–Tukey periodogram (2.5.1), unlike its unwindowed version, is not amenable to a direct computation via a single FFT. Show that three FFTs are sufficient to evaluate (2.5.1): two FFTs for the computation of the sample covariance sequence entering the equation (2.5.1) (as described in the Complement in Section 2.8.1); and one FFT for the evaluation of (2.5.1). Alternatively, show that the computational formula for $\{\hat{r}(k)\}$ derived in the Complement in Section 2.8.1 can be used to obtain an FFT–based algorithm for evaluation of (2.5.1) directly in terms of $\hat{\phi}_p(\omega)$. Relate the latter way of computing (2.5.1) to the evaluation of $\hat{\phi}_{BT}(\omega)$ from the integral equation (2.5.3). Finally, compare the two ways outlined above for evaluating the windowed Blackman–Tukey periodogram.

Solution: The windowed Blackman–Tukey periodogram can be written as

$$\begin{aligned}\hat{\phi}_{BT}(\omega) &= \sum_{k=-(N-1)}^{N-1} w(k)\hat{r}(k)e^{-i\omega k} \\ &= \sum_{k=0}^{N-1} w(k)\hat{r}(k)e^{-i\omega k} + \sum_{k=0}^{N-1} w(k)\hat{r}^*(k)e^{i\omega k} - w(0)\hat{r}(0) \\ &= 2\,\mathrm{Re}\left\{\sum_{k=0}^{N-1} w(k)\hat{r}(k)e^{-i\omega k}\right\} - w(0)\hat{r}(0) \end{aligned} \tag{2.8.5}$$

where we made use of the facts that the window sequence is even and $\hat{r}(-k) = \hat{r}^*(k)$. It is now evident that an N–point FFT can be used to evaluate $\hat{\phi}_{BT}(\omega)$ at $\omega = 2\pi k/N$ ($k = 0, \ldots, N-1$). This requires about $\frac{1}{2}N\log_2(N)$ flops that should be added to the $2N\log_2(2N)$ flops required to compute $\{\hat{r}(k)\}$ (as in Section 2.8.1), hence giving a total of about $N[\frac{1}{2}\log_2(N) + 2\log_2(2N)]$ flops for this way of evaluating $\hat{\phi}_{BT}(\omega)$.

Next, we make use of the expression (2.8.4) for $\{\hat{r}(k)\}$ that was derived in the Complement in Section 2.8.1,

$$\hat{r}(p-N) = \frac{1}{2N-1}\sum_{k=1}^{2N-1} \hat{\phi}(\bar{\omega}_k)e^{i\bar{\omega}_k(p-N)} \qquad (p = 1, \ldots, 2N-1) \tag{2.8.6}$$

where $\bar{\omega}_k = 2\pi(k-1)/(2N-1)$, $(k = 1, \ldots, 2N-1)$, and where $\hat{\phi}(\omega)$ is the unwindowed periodogram. Inserting (2.8.6) into (2.5.1), we obtain

$$\hat{\phi}_{BT}(\omega) = \frac{1}{2N-1} \sum_{s=-(N-1)}^{N-1} w(s)e^{-i\omega s} \sum_{k=1}^{2N-1} \hat{\phi}(\bar{\omega}_k)e^{i\bar{\omega}_k s} =$$

$$= \frac{1}{2N-1} \sum_{k=1}^{2N-1} \hat{\phi}(\bar{\omega}_k) \left[\sum_{s=-(N-1)}^{N-1} w(s)e^{-i(\omega-\bar{\omega}_k)s} \right] \tag{2.8.7}$$

which gives

$$\boxed{\hat{\phi}_{BT}(\omega) = \frac{1}{2N-1} \sum_{k=1}^{2N-1} \hat{\phi}(\bar{\omega}_k)\, W(\omega - \bar{\omega}_k)} \tag{2.8.8}$$

where $W(\omega)$ is the spectral window.

It might be thought that the last step in the above derivation requires that $\{w(k)\}$ is a "truncated–type" window (*i.e.*, $w(k) = 0$ for $|k| \geq N$). However, no such requirement on $\{w(k)\}$ is needed, as explained next. By inserting the usual expression for $\hat{\phi}(\omega)$ into (2.8.6) we obtain:

$$\hat{r}(p-N) = \frac{1}{2N-1} \sum_{k=1}^{2N-1} \left[\sum_{s=-(N-1)}^{N-1} \hat{r}(s)e^{-i\bar{\omega}_k s} \right] e^{i\bar{\omega}_k(p-N)}$$

$$= \frac{1}{2N-1} \sum_{s=-(N-1)}^{N-1} \hat{r}(s) \left[\sum_{k=1}^{2N-1} e^{i\bar{\omega}_k(p-N-s)} \right]$$

$$\triangleq \frac{1}{2N-1} \sum_{s=-(N-1)}^{N-1} \hat{r}(s)\Delta(s,p)$$

where

$$\Delta(s,p) = \sum_{k=1}^{2N-1} e^{i\bar{\omega}_{p-N-s}k} = e^{i\bar{\omega}_{p-N-s}} \frac{e^{i(2N-1)\bar{\omega}_{p-N-s}} - 1}{e^{i\bar{\omega}_{p-N-s}} - 1}$$

As $(2N-1)\bar{\omega}_{p-N-s} = 2\pi(p-N-s)$, it follows that

$$\Delta(s,p) = (2N-1)\delta_{p-N,s}$$

from which we immediately get

$$\frac{1}{2N-1} \sum_{s=-(N-1)}^{N-1} \hat{r}(s)\Delta(s,p) = \begin{cases} \hat{r}(p-N) & p = 1, \ldots, 2N-1 \\ 0, & \text{otherwise} \end{cases} \tag{2.8.9}$$

First, the above calculation provides a cross–checking of the derivation of equation (2.8.6) in the Complement in Section 2.8.1. Second, the result (2.8.9) implies that the values of $\hat{r}(p-N)$ calculated with the formula (2.8.6) are equal to zero for $p < 1$ or $p > 2N - 1$. It follows that the limits for the summation over s in (2.8.7) can be extended to $\pm\infty$, hence showing that (2.8.8) is valid for an arbitrary window. In the general case there seems to be no way for evaluating (2.8.8) by means of an FFT algorithm. Hence, it appears that for a general window it is more efficient to base the computation of $\hat{\phi}_{BT}(\omega)$ on (2.8.5) rather than on (2.8.8). For certain windows, however, (2.8.8) may be computationally more efficient than (2.8.5). For instance, in the case of the Daniell method, which corresponds to a rectangular spectral window, (2.8.8) takes a very convenient computational form and should be preferred to (2.8.5). It should be noted that (2.8.8) can be viewed as an *exact formula* for evaluation of the integral in equation (2.5.3). In particular, (2.8.8) provides an *exact* implementation formula for the Daniell periodogram (2.7.20) (whereas (2.7.16) is only an approximation of the integral (2.7.20) that is valid for sufficiently large values of N).

2.8.3 Estimation of Cross–Spectra and Coherency Spectra

As can be seen from the Complement in Section 1.6.1, the estimation of the cross–spectrum $\phi_{yu}(\omega)$ of two stationary signals, $y(t)$ and $u(t)$, is a useful operation when studying possible linear (dynamic) relations between $y(t)$ and $u(t)$. Let $z(t)$ denote the bivariate signal

$$z(t) = [y(t)\ u(t)]^T$$

and let

$$\hat{\phi}(\omega) = \frac{1}{N} Z(\omega)Z^*(\omega) \tag{2.8.10}$$

denote the unwindowed periodogram estimate of the spectral density matrix of $z(t)$. In equation (2.8.10),

$$Z(\omega) = \sum_{t=1}^{N} z(t)e^{-i\omega t}$$

is the DTFT of $\{z(t)\}_{t=1}^{N}$. Partition $\hat{\phi}(\omega)$ as

$$\hat{\phi}(\omega) = \begin{bmatrix} \hat{\phi}_{yy}(\omega) & \hat{\phi}_{yu}(\omega) \\ \hat{\phi}_{yu}^*(\omega) & \hat{\phi}_{uu}(\omega) \end{bmatrix} \tag{2.8.11}$$

As indicated by the notation previously used, estimates of $\phi_{yy}(\omega)$, $\phi_{uu}(\omega)$ and of the cross–spectrum $\phi_{yu}(\omega)$ may be obtained from the corresponding elements of $\hat{\phi}(\omega)$.

Show that the estimate of the coherency spectrum obtained from (2.8.11) is always such that

$$|\hat{C}_{yu}(\omega)| = 1 \qquad \text{for all } \omega \tag{2.8.12}$$

and hence it is useless. Conclude also from this fact that the unwindowed periodogram is a poor estimate of the PSD.

Next consider a windowed Blackman–Tukey periodogram estimate of the cross–spectrum:

$$\hat{\phi}_{yu}(\omega) = \sum_{k=-M}^{M} w(k)\hat{r}_{yu}(k)e^{-i\omega k} \tag{2.8.13}$$

where $w(k)$ is the lag window, and $\hat{r}_{yu}(k)$ is some usual estimate of $r_{yu}(k)$. Unlike $r_{yy}(k)$ or $r_{uu}(k)$, $r_{yu}(k)$ does not necessarily peak at $k = 0$ and, moreover, is not an even function in general. The choice of the lag window for estimating cross–spectra may hence be governed by different rules from those commonly used in the autospectrum estimation. Address these possible differences and obtain some guidelines for selection of $w(k)$ in (2.8.13). Finally, consider the following definition for $\hat{r}_{yu}(k)$:

$$\hat{r}_{yu}(k) = \begin{cases} \dfrac{1}{N}\displaystyle\sum_{t=k+1}^{N} y(t)u^*(t-k), & k = 0, 1, 2, \ldots \\ \dfrac{1}{N}\displaystyle\sum_{t=1}^{N+k} y(t)u^*(t-k), & k = 0, -1, -2, \ldots \end{cases}$$

in (2.8.13). Relate the so–obtained Blackman–Tukey–type cross–spectral estimator to a spectrally windowed periodogram estimator of the cross–spectrum (as we did in Section 2.5 for the autospectrum estimators).

Solution: Since the rank of the 2×2 matrix in (2.8.11) is equal to one (see Result R22 in Appendix A), we must have

$$\hat{\phi}_{uu}(\omega)\hat{\phi}_{yy}(\omega) = |\hat{\phi}_{yu}(\omega)|^2$$

which readily leads to the conclusion that the coherency spectrum estimate obtained from the elements of $\hat{\phi}(\omega)$ is bound to satisfy (2.8.12), and hence is meaningless.

Next, consider (2.8.13). The main task of a lag window is to retain the "essential part" of the covariance sequence in the defining equation for the spectral density. In this way the bias is kept small and the variance is also reduced as the noisy tails of the sample covariance sequence are weighted out. For simplicity of discussion, assume that most of the area under the plot of $\hat{r}_{yu}(k)$ is concentrated about $k = k_0$, with $|k_0| \ll N$. As $\hat{r}_{yu}(k)$ is a reasonably accurate estimate of $r_{yu}(k)$, provided $|k| \ll N$, we can assume that $\{\hat{r}_{yu}(k)\}$ and $\{r_{yu}(k)\}$ have similar shapes. In such a case, one can redefine (2.8.13) as

$$\hat{\phi}_{yu}(\omega) = \sum_{k=-M}^{M} w(k-k_0)\hat{r}_{yu}(k)e^{-i\omega k}$$

where the lag window $w(s)$ is of the type recommended for autospectrum estimation. The choice of an appropriate value for k_0 in the above cross–spectral estimator is essential, for if k_0 is poorly selected the following situations can occur:

- If M is chosen small to reduce the variance, the bias may be significant as "essential" lags of the cross–covariance sequence may be left out.
- If M is chosen large to reduce the bias, the variance may significantly be inflated as poorly estimated high–order "nonessential" lags are included into the spectral estimation formula.

Finally, let us look at the cross–spectrum estimators derived from (2.8.11) and (2.8.13), respectively, with a view of establishing a relation between them. Partition $Z(\omega)$ as

$$Z(\omega) = \begin{bmatrix} Y(\omega) \\ U(\omega) \end{bmatrix}$$

and observe that

$$\begin{aligned}
&\frac{1}{2\pi N}\int_{-\pi}^{\pi} Y(\omega)U^*(\omega)e^{i\omega k}\, d\omega \\
&= \frac{1}{2\pi N}\int_{-\pi}^{\pi} \sum_{t=1}^{N}\sum_{s=1}^{N} y(t)u^*(s)e^{-i\omega(t-s)}\, e^{i\omega k}\, d\omega \\
&= \frac{1}{N}\sum_{t=1}^{N}\sum_{s=1}^{N} y(t)u^*(s)\delta_{k,t-s} \\
&= \frac{1}{N}\sum_{t\in[1,N]\cap[1+k,N+k]} y(t)u^*(t-k) = \hat{r}_{yu}(k)
\end{aligned} \tag{2.8.14}$$

Let

$$\hat{\phi}^p_{yu}(\omega) = \frac{1}{N}\, Y(\omega)U^*(\omega)$$

denote the unwindowed cross–spectral periodogram–like estimator, given by the off–diagonal element of $\hat{\phi}(\omega)$ in (2.8.11). With this notation, (2.8.14) can be written more compactly as

$$\hat{r}_{yu}(k) = \frac{1}{2\pi}\int_{-\pi}^{\pi} \hat{\phi}^p_{yu}(\mu)e^{i\mu k}\, d\mu$$

By using the above equation in (2.8.13), we obtain:

$$\begin{aligned}
\hat{\phi}_{yu}(\omega) &= \frac{1}{2\pi}\int_{-\pi}^{\pi} \hat{\phi}^p_{yu}(\mu)\sum_{k=-M}^{M} w(k)e^{-i(\omega-\mu)k}\, d\mu \\
&= \frac{1}{2\pi}\int_{-\pi}^{\pi} W(\omega-\mu)\hat{\phi}^p_{yu}(\mu)\, d\mu
\end{aligned} \tag{2.8.15}$$

where $W(\omega) = \sum_{k=-\infty}^{\infty} w(k)e^{-i\omega k}$ is the spectral window. The previous equation should be compared with the similar equation, (2.5.3), that holds in the case of autospectra.

For implementation purposes, one can use the following discrete approximation of (2.8.15):

$$\hat{\phi}_{yu}(\omega) = \frac{1}{N} \sum_{k=-N}^{N} W(\omega - \omega_k)\hat{\phi}_{yu}^{p}(\omega_k)$$

where $\omega_k = \frac{2\pi}{N}k$ are the Fourier frequencies. The periodogram (cross–spectral) estimate that appears in the above equation can be efficiently computed by means of an FFT algorithm.

2.8.4 More Time–Bandwidth Product Results

The time (or duration)–bandwidth product result (2.6.16) relies on the assumptions that both $w(t)$ and $W(\omega)$ have a dominant peak at the origin, that they both are real–valued, and that they take on nonnegative values only. While most window–like signals (nearly) satisfy these assumptions, many other signals do not satisfy them. In this complement the problem is to obtain time–bandwidth product results that apply to a much broader class of signals.

Solution: We begin by showing how the result (2.6.16) can be extended to a more general class of signals. Let $x(t)$ denote a general discrete–time sequence and let $X(\omega)$ denote its DTFT. Both $x(t)$ and $X(\omega)$ are allowed to take negative or complex values, and neither is required to peak at the origin. Let t_0 and ω_0 denote the maximum points of $|x(t)|$ and $|X(\omega)|$, respectively. The time width (or duration) and bandwidth definitions in (2.6.1) and (2.6.2) are modified as follows:

$$\bar{N}_e = \frac{\sum_{t=-\infty}^{\infty} |x(t)|}{|x(t_0)|}$$

and

$$\bar{\beta}_e = \frac{\frac{1}{2\pi}\int_{-\pi}^{\pi} |X(\omega)| d\omega}{|X(\omega_0)|}$$

Because $x(t)$ and $X(\omega)$ form a Fourier transform pair, we obtain

$$|X(\omega_0)| = \left| \sum_{t=-\infty}^{\infty} x(t)e^{-i\omega_0 t} \right| \leq \sum_{t=-\infty}^{\infty} |x(t)|$$

and

$$|x(t_0)| = \left| \frac{1}{2\pi} \int_{-\pi}^{\pi} X(\omega)e^{i\omega t_0} d\omega \right| \leq \frac{1}{2\pi} \int_{-\pi}^{\pi} |X(\omega)| d\omega$$

which implies that

$$\boxed{\bar{N}_e \bar{\beta}_e \geq 1} \tag{2.8.16}$$

The above result, similar to (2.6.5), can be used to conclude that:

$$\boxed{\text{A sequence } \{x(t)\} \text{ cannot be narrow in both time and frequency.}} \tag{2.8.17}$$

More precisely, if $x(t)$ is narrow in one domain it must be wide in the other domain. However, the inequality result (2.8.16), unlike (2.6.5), does not necessarily imply that $\bar{\beta}_e$ decreases whenever $\bar{N}_e$ increases (or vice versa). Furthermore, the result (2.8.16) — again unlike (2.6.5) — does not exclude the possibility that the signal is broad in both domains. In fact, in the general class of signals to which (2.8.16) applies there are signals which are broad in both the time and frequency domains (for such signals $\bar{N}_e \bar{\beta}_e \gg 1$); see, *e.g.*, [PAPOULIS 1977]. Evidently, the significant consequence of (2.8.16) is (2.8.17), which is precisely what makes the duration–bandwidth result an important one.

The duration–bandwidth product type of result (such as (2.6.5) or (2.8.16), and (2.8.21) below) has been sometimes referred to by using the generic name of *uncertainty principle*, in an attempt to relate it to the Heisenberg Uncertainty Principle in quantum mechanics. (Briefly stated, the Heisenberg Uncertainty Principle asserts that the position and velocity of a particle cannot be simultaneously specified to arbitrary precision.) To support the relationship, one can argue as follows: Suppose that we are given a sequence with (equivalent) duration equal to N_e and that we are asked to use a linear filtering device to determine the sequence's spectral content in a certain narrow band. Because the filter impulse response cannot be longer than N_e (in fact, it should be (much) shorter!), it follows from the time–bandwidth product result that the filter's bandwidth can be on the order of $1/N_e$ but not smaller. Hence, the sequence's spectral content in fine bands on an order smaller than $1/N_e$ cannot be exactly determined and therefore is "uncertain". This is in effect the type of limitation that applies to the nonparametric spectral methods discussed in this chapter. However, this way of arguing is related to a *specific approach to spectral estimation and not to a fundamental limitation associated with the signal itself.* (As we will see in later chapters of this text, there are parametric methods of spectral analysis that can provide the "high resolution" necessary to determine the spectral content in bands that are on an order less than $1/N_e$).

Next, we present another, slightly more general form of time–bandwidth product result. The definitions of duration and bandwidth used to obtain (2.8.16) make full sense whenever $|x(t)|$ and $|X(\omega)|$ are single pulse–like waveforms, though these definitions may give reasonable results in many other instances as well. There are several other possible definitions of the broadness of a waveform in either the time or frequency domain. The definition used below and the corresponding time–bandwidth product result appear to be among the most general.

Let

$$\tilde{x}(t) = \frac{x(t)}{\sqrt{\sum_{t=-\infty}^{\infty} |x(t)|^2}} \tag{2.8.18}$$

and

$$\tilde{X}(\omega) = \frac{X(\omega)}{\sqrt{\frac{1}{2\pi}\int_{-\pi}^{\pi} |X(\omega)|^2 d\omega}} \tag{2.8.19}$$

By Parseval's theorem (see (1.2.6)) the denominators in (2.8.18) and (2.8.19) are equal to each other. Therefore, $\tilde{X}(\omega)$ is the DTFT of $\tilde{x}(t)$ as is already indicated by notation. Observe that

$$\sum_{t=-\infty}^{\infty} |\tilde{x}(t)|^2 = \frac{1}{2\pi}\int_{-\pi}^{\pi} |\tilde{X}(\omega)|^2 d\omega = 1$$

Hence, both $\{|\tilde{x}(t)|^2\}$ and $\{|\tilde{X}(\omega)|^2/2\pi\}$ can be interpreted as probability density functions in the sense that they are nonnegative and that they sum or integrate to one. The means and variances associated with these two "probability" densities are given by the following equations.

Time Domain:

$$\mu = \sum_{t=-\infty}^{\infty} t|\tilde{x}(t)|^2$$

$$\sigma^2 = \sum_{t=-\infty}^{\infty} (t-\mu)^2|\tilde{x}(t)|^2$$

Frequency Domain:

$$\nu = \frac{1}{(2\pi)^2}\int_{-\pi}^{\pi} \omega|\tilde{X}(\omega)|^2 d\omega$$

$$\rho^2 = \frac{1}{(2\pi)^3}\int_{-\pi}^{\pi} (\omega - 2\pi\nu)^2|\tilde{X}(\omega)|^2 d\omega$$

The values of the "standard deviations" σ and ρ show whether the normalized functions $\{|\tilde{x}(t)|\}$ and $\{|\tilde{X}(\omega)|\}$, respectively, are narrow or broad. Hence, *we can use σ and ρ as definitions for the duration and bandwidth, respectively,* of the original functions $\{x(t)\}$ and $\{X(\omega)\}$.

In what follows, we assume that:

$$\mu = 0, \qquad \nu = 0 \tag{2.8.20}$$

For continuous–time signals, the zero–mean assumptions can always be made to hold by appropriately translating the origin on the time and frequency axes (see, *e.g.*, [COHEN 1995]). However, doing the same in the case of the discrete–time sequences considered here does not appear to be possible. Indeed, μ may not be integer–valued, and the support of $X(\omega)$ is finite and hence is affected by translation. Consequently, in the present case the zero–mean assumption introduces some restriction; nevertheless we impose it to simplify the analysis.

According to the discussion above and assumption (2.8.20), we define the (equivalent) time width and bandwidth of $x(t)$ as follows:

$$\tilde{N}_e = \left[\sum_{t=-\infty}^{\infty} t^2 |\tilde{x}(t)|^2 \right]^{1/2}$$

$$\tilde{\beta}_e = \frac{1}{2\pi} \left[\frac{1}{2\pi} \int_{-\pi}^{\pi} \omega^2 |\tilde{X}(\omega)|^2 d\omega \right]^{1/2}$$

In the remainder of this complement, we prove the following time–bandwidth product result:

$$\boxed{\tilde{N}_e \tilde{\beta}_e \geq \frac{1}{4\pi}} \tag{2.8.21}$$

which holds true under (2.8.20) and the weak additional assumption that

$$|\tilde{X}(\pi)| = 0 \tag{2.8.22}$$

To prove (2.8.21), first we note that

$$\tilde{X}'(\omega) \triangleq \frac{d\tilde{X}(\omega)}{d\omega} = -i \sum_{t=-\infty}^{\infty} t\tilde{x}(t) e^{-i\omega t}$$

Hence, $i\tilde{X}'(\omega)$ is the DTFT of $\{t\tilde{x}(t)\}$, which implies (by Parseval's theorem) that

$$\sum_{t=-\infty}^{\infty} t^2 |\tilde{x}(t)|^2 = \frac{1}{2\pi} \int_{-\pi}^{\pi} |\tilde{X}'(\omega)|^2 d\omega \tag{2.8.23}$$

Consequently, by the Cauchy–Schwartz inequality for functions (see Result R23 in Appendix A),

$$\begin{aligned} \tilde{N}_e \tilde{\beta}_e &= \left[\frac{1}{2\pi} \int_{-\pi}^{\pi} |\tilde{X}'(\omega)|^2 d\omega \right]^{1/2} \left[\frac{1}{(2\pi)^3} \int_{-\pi}^{\pi} \omega^2 |\tilde{X}(\omega)|^2 d\omega \right]^{1/2} \\ &\geq \frac{1}{(2\pi)^2} \left| \int_{-\pi}^{\pi} \omega \tilde{X}^*(\omega) \tilde{X}'(\omega) d\omega \right| \\ &= \frac{1}{2(2\pi)^2} \left\{ \left| \int_{-\pi}^{\pi} \omega \tilde{X}^*(\omega) \tilde{X}'(\omega) d\omega \right| \right. \\ &\qquad \left. + \left| \int_{-\pi}^{\pi} \omega \tilde{X}(\omega) \tilde{X}^{*\prime}(\omega) d\omega \right| \right\} \end{aligned} \tag{2.8.24}$$

(the first equality above follows from (2.8.23) and the last one from a simple calculation). Hence

$$\begin{aligned}\tilde{N}_e\tilde{\beta}_e &\geq \frac{1}{2(2\pi)^2}\left|\int_{-\pi}^{\pi} \omega \left[\tilde{X}^*(\omega)\tilde{X}'(\omega) + \tilde{X}(\omega)\tilde{X}^{*\prime}(\omega)\right] d\omega\right| \\ &= \frac{1}{2(2\pi)^2}\left|\int_{-\pi}^{\pi} \omega \left[|\tilde{X}(\omega)|^2\right]' d\omega\right|\end{aligned}$$

which, after integrating by parts and using (2.8.22), yields

$$\tilde{N}_e\tilde{\beta}_e \geq \frac{1}{2(2\pi)^2}\left|\omega|\tilde{X}(\omega)|^2\Big|_{-\pi}^{\pi} - \int_{-\pi}^{\pi} |\tilde{X}(\omega)|^2 d\omega\right| = \frac{1}{2(2\pi)}$$

and the proof is concluded.

Remark: There is an alternative way to complete the proof above, starting from the inequality in (2.8.24). In fact, as we will see, this alternative proof yields a tighter inequality than (2.8.21). Let $\varphi(\omega)$ denote the phase of $\tilde{X}(\omega)$:

$$\tilde{X}(\omega) = |\tilde{X}(\omega)|e^{i\varphi(\omega)}$$

Then,

$$\begin{aligned}\omega\tilde{X}^*(\omega)\tilde{X}'(\omega) &= \omega|\tilde{X}(\omega)|\left[|\tilde{X}(\omega)|\right]' + i\omega\varphi'(\omega)|\tilde{X}(\omega)|^2 \\ &= \frac{1}{2}\left[\omega|\tilde{X}(\omega)|^2\right]' - \frac{1}{2}|\tilde{X}(\omega)|^2 + i\omega\varphi'(\omega)|\tilde{X}(\omega)|^2 \qquad (2.8.25)\end{aligned}$$

Inserting (2.8.25) into (2.8.24) yields

$$\tilde{N}_e\tilde{\beta}_e \geq \frac{1}{(2\pi)^2}\left|\frac{\omega}{2}|\tilde{X}(\omega)|^2\Big|_{-\pi}^{\pi} - \pi + i2\pi\gamma\right| \qquad (2.8.26)$$

where

$$\gamma = \frac{1}{2\pi}\int_{-\pi}^{\pi} \omega\varphi'(\omega)|\tilde{X}(\omega)|^2 d\omega$$

can be interpreted as the "covariance" of ω and $\varphi'(\omega)$ under the "probability density function" given by $|\tilde{X}(\omega)|^2/(2\pi)$. From (2.8.26) we obtain at once

$$\boxed{\tilde{N}_e\tilde{\beta}_e \geq \frac{1}{4\pi}\sqrt{1+4\gamma^2}} \qquad (2.8.27)$$

which is a slightly stronger result than (2.8.21). ■

The results (2.8.21) and (2.8.27) are similar to (2.8.16), and hence the type of comments previously made about (2.8.16) applies to (2.8.21) and (2.8.27) as well.

2.9 Exercises

Exercise 2.1: Covariance Estimation for Signals with Unknown Means

The sample covariance estimators (2.2.3) and (2.2.4) are based on the assumption that the signal mean is equal to zero. A simple calculation shows that, under the zero–mean assumption,

$$E\{\tilde{r}(k)\} = r(k) \tag{2.9.1}$$

and

$$E\{\hat{r}(k)\} = \frac{N - |k|}{N} r(k) \tag{2.9.2}$$

where $\{\tilde{r}(k)\}$ denotes the sample covariance estimate in (2.2.3). Equations (2.9.1) and (2.9.2) show that $\tilde{r}(k)$ is an unbiased estimate of $r(k)$, whereas $\hat{r}(k)$ is a biased one (note, however, that the bias in $\hat{r}(k)$ is small for $N \gg |k|$). For this reason, $\{\tilde{r}(k)\}$ and $\{\hat{r}(k)\}$ are often called the unbiased and, respectively, biased sample covariances.

Whenever the signal mean is unknown, a most natural modification of the covariance estimators (2.2.3) and (2.2.4) is as follows:

$$\boxed{\tilde{r}(k) = \frac{1}{N-k} \sum_{t=k+1}^{N} [y(t) - \bar{y}]\,[y(t-k) - \bar{y}]^*} \tag{2.9.3}$$

and

$$\boxed{\hat{r}(k) = \frac{1}{N} \sum_{t=k+1}^{N} [y(t) - \bar{y}]\,[y(t-k) - \bar{y}]^*} \tag{2.9.4}$$

where $\bar{y}$ is the sample mean

$$\bar{y} = \frac{1}{N} \sum_{t=1}^{N} y(t) \tag{2.9.5}$$

Show that in the unknown mean case, the usual names of unbiased and biased sample covariances associated with (2.9.3) and (2.9.4), respectively, may no longer be appropriate. Indeed, in such a case *both estimators may be biased*; furthermore, $\hat{r}(k)$ *may be less biased than* $\tilde{r}(k)$. To simplify the calculations, assume that $y(t)$ is white noise.

Exercise 2.2: Covariance Estimation for Signals with Unknown Means (cont'd)

Show that the sample covariance sequence $\{\hat{r}(k)\}$ in equation (2.9.4) of Exercise 2.1 satisfies the following equality:

$$\boxed{\sum_{k=-(N-1)}^{N-1} \hat{r}(k) = 0} \tag{2.9.6}$$

The above equality may seem somewhat surprising. (Why should the $\{\tilde{r}(k)\}$ satisfy such a constraint, which the true covariances do not necessarily satisfy? Note, for instance, that the latter covariance sequence may well comprise only positive elements.) However, the equality in (2.9.6) has a natural explanation when viewed in the context of periodogram–based spectral estimation. Derive and explain formula (2.9.6) in the aforementioned context.

Exercise 2.3: Unbiased ACS Estimates may lead to Negative Spectral Estimates

We stated in Section 2.2.2 that if unbiased ACS estimates, given by equation (2.2.3), are used in the correlogram spectral estimate (2.2.2), then negative spectral estimates may result. Find an example data sequence $\{y(t)\}_{t=1}^{N}$ that gives such a negative spectral estimate.

Exercise 2.4: Variance of Estimated ACS

Let $\{y(t)\}_{t=1}^{N}$ be real Gaussian (for simplicity), with zero mean, ACS equal to $\{r(k)\}$, and ACS estimate (either biased or unbiased) equal to $\{\hat{r}(k)\}$ (given by equation (2.2.3) or (2.2.4); we treat both cases simultaneously). Assume, without loss of generality, that $k \geq 0$.

(a) Make use of equation (2.4.24) to show that

$$\text{var}\{\hat{r}(k)\} = \alpha^2(k) \sum_{m=-(N-k-1)}^{N-k-1} (N-k-|m|)\left[r^2(m) + r(m+k)r(m-k)\right]$$

where

$$\alpha(k) = \begin{cases} \dfrac{1}{N-k} & \text{for unbiased ACS estimates} \\ \dfrac{1}{N} & \text{for biased ACS estimates} \end{cases}$$

Hence, for large N, the standard deviation of the ACS estimate is $O(1/\sqrt{N})$ under weak conditions on the true ACS $\{r(k)\}$.

(b) For the special case that $y(t)$ is white Gaussian noise, show that $\text{cov}\{\hat{r}(k), \hat{r}(l)\} = 0$ for $k \neq l$, and find a simple expression for $\text{var}\{\hat{r}(k)\}$.

Exercise 2.5: Another Proof of the Equality $\hat{\phi}_p(\omega) = \hat{\phi}_c(\omega)$

The proof of the result (2.2.6) in the text introduces an auxiliary random sequence and treats the original data sequence as deterministic (nonrandom). That proof relies on several results previously derived. A more direct proof of (2.2.6) can be found using only (2.2.1), (2.2.2), and (2.2.4). Find such a proof.

Exercise 2.6: Linear Transformation Interpretation of the DFT

Let F be the $N \times N$ matrix whose (k,t)th element is given by W^{kt}, where W is as defined in (2.3.2). Then the DFT, (2.3.3), can be written as a linear transformation of the data vector $y \triangleq [y(1) \dots y(N)]^T$,

$$Y \triangleq [Y(0) \dots Y(N-1)]^T = Fy \tag{2.9.7}$$

Show that F is an orthogonal matrix that satisfies

$$\frac{1}{N} FF^* = I \tag{2.9.8}$$

and, as a result, that the *inverse transform* is

$$y = \frac{1}{N} F^* Y \tag{2.9.9}$$

Deduce from the above that the DFT is nothing but a representation of the data vector y via an orthogonal basis in $\mathcal{C}^n$ (the basis vectors are the columns of F^*). Also, deduce that if the sequence $\{y(t)\}$ is periodic with a period equal to N, then the Fourier coefficient vector, Y, determines the whole sequence $\{y(t)\}_{t=1,2,\dots}$, and that in effect the inverse transform (2.9.9) can be extended to include all samples $y(1), \dots, y(N)$, $y(N+1)$, $y(N+2), \dots$

Exercise 2.7: For White Noise the Periodogram is an Unbiased PSD Estimator

Let $y(t)$ be a zero-mean white noise with variance σ^2 and let

$$Y(\omega_k) = \frac{1}{\sqrt{N}} \sum_{t=0}^{N-1} y(t) e^{-i\omega_k t}\ ; \quad \omega_k = \frac{2\pi}{N} k \qquad (k = 0, \dots, N-1)$$

denote its (normalized) DFT evaluated at the Fourier frequencies.

(a) Derive the covariances

$$E\{Y(\omega_k) Y^*(\omega_r)\}, \qquad k, r = 0, \dots, N-1$$

(b) Use the result of the previous calculation to conclude that the periodogram $\hat{\phi}(\omega_k) = |Y(\omega_k)|^2$ is an unbiased estimator of the PSD of $y(t)$.

(c) Explain whether the unbiasedness property holds for $\omega \neq \omega_k$ as well. Present an intuitive explanation for your finding.

Exercise 2.8: Shrinking the Periodogram

First, we introduce a simple general result on mean squared error (MSE) reduction by shrinking. Let $\hat{x}$ be some estimate of a true (and unknown) parameter x. Assume that $\hat{x}$ is unbiased, *i.e.*, $E(\hat{x}) = x$, and let $\sigma_{\hat{x}}^2$ denote the MSE of $\hat{x}$

$$\sigma_{\hat{x}}^2 = E\left\{(\hat{x} - x)^2\right\}$$

(Since $\hat{x}$ is unbiased, $\sigma_{\hat{x}}^2$ also equals the variance of $\hat{x}$.) For a fixed (nonrandom) ρ, let

$$\tilde{x} = \rho\hat{x}$$

be another estimate of x. The "shrinkage coefficient" ρ can be chosen so as to make the MSE of $\tilde{x}$ (much) smaller than $\sigma_{\hat{x}}^2$. (Note that $\tilde{x}$, for $\rho \neq 1$, is a biased estimate of x; hence $\tilde{x}$ trades off bias for variance.) More precisely, show that the MSE of $\tilde{x}$, $\sigma_{\tilde{x}}^2$, achieves its minimum value (with respect to ρ) of

$$\sigma_{\tilde{x}_o}^2 = \rho_o\, \sigma_{\hat{x}}^2$$

for

$$\rho_o = \frac{x^2}{x^2 + \sigma_{\hat{x}}^2}$$

Next, consider the application of the previous result to the periodogram. As we explained in the chapter, the periodogram–based spectral estimate is asymptotically unbiased and has an asymptotic MSE equal to the squared PSD value:

$$E\left\{\hat{\phi}_p(\omega)\right\} \to \phi(\omega), \qquad E\left\{(\hat{\phi}_p(\omega) - \phi(\omega))^2\right\} \to \phi^2(\omega) \qquad \text{as } N \to \infty$$

Show that the "optimally shrunk" periodogram estimate is

$$\tilde{\phi}(\omega) = \hat{\phi}_p(\omega)/2$$

and that the MSE of $\tilde{\phi}(\omega)$ is half the MSE of $\hat{\phi}_p(\omega)$.

Finally, comment on the general applicability of this extremely simple tool for MSE reduction.

Exercise 2.9: Asymptotic Maximum Likelihood Estimation of $\phi(\omega)$ from $\hat{\phi}_p(\omega)$

It follows from the calculations in Section 2.4 that, asymptotically in N, $\hat{\phi}_p(\omega)$ has mean $\phi(\omega)$ and variance $\phi^2(\omega)$. In this exercise we assume that $\hat{\phi}_p(\omega)$ is (asymptotically) Gaussian distributed (which is *not* necessarily the case; however, the spectral estimator derived here under the Gaussian assumption may also be used when this assumption does not hold). Hence, the asymptotic probability density function of $\hat{\phi}_p(\omega)$ is (we omit the index p as well as the dependence on ω to simplify the notation):

$$p_\phi(\hat{\phi}) = \frac{1}{(2\pi)^{1/2}\phi}\, \exp[-(\hat{\phi} - \phi)^2/2\phi^2]$$

Show that the maximum likelihood estimate (MLE) of ϕ based on $\hat{\phi}$, which by definition is equal to the maximizer of $p_\phi(\hat{\phi})$ (see Appendix B for a short introduction of maximum likelihood estimation) is given by

$$\tilde{\phi} = (\sqrt{5} - 1)\hat{\phi}/2$$

Compare $\tilde{\phi}$ with the "optimally shrunk" estimate of ϕ derived in Exercise 2.8.

Exercise 2.10: Plotting the Spectral Estimates in dB

It has been shown in this chapter that the spectral estimate $\hat{\phi}(\omega)$, obtained via an improved periodogram method, is asymptotically unbiased with a variance of the form $\mu^2\phi^2(\omega)$, where μ is a constant that can be made (much) smaller than one by appropriately choosing the window. This fact implies that the *confidence interval* $\hat{\phi}(\omega) \pm \mu\phi(\omega)$, constructed around the estimated PSD, should include the true (and unknown) PSD with a large probability. Now, obtaining a confidence interval as above has a twofold drawback: first, $\phi(\omega)$ is unknown; secondly, the interval may have significantly different widths for different frequency values.

Show that plotting $\hat{\phi}(\omega)$ in decibels eliminates the previous drawbacks. More precisely, show that when $\hat{\phi}(\omega)$ is expressed in dB, its asymptotic variance is $c^2\mu^2$ (with $c = 10\log_{10} e$), and hence that the confidence interval for a log–scale plot has the same width (independent of $\phi(\omega)$) for all ω.

Exercise 2.11: Finite–Sample Variance/Covariance Analysis of the Periodogram

This exercise has two aims. *First*, it shows that in the Gaussian case the variance/covariance analysis of the periodogram can be done in an extremely simple manner (even *without* the assumption that the data comes from a linear process, as in (2.4.26)). *Secondly*, the exercise asks for a finite–sample analysis which, for some purposes, may be more useful than the asymptotic analysis presented in the text. Indeed, the asymptotic analysis result (2.4.21) may be misleading if not interpreted with care. For instance, (2.4.21) says that asymptotically (for $N \to \infty$) $\hat{\phi}(\omega_1)$ and $\hat{\phi}(\omega_2)$ are uncorrelated with one another, no matter how close ω_1 and ω_2 are. This cannot be true in finite samples, and hence the following question naturally arises: For a given N, how close can ω_1 be to ω_2 such that $\hat{\phi}(\omega_1)$ and $\hat{\phi}(\omega_2)$ are (nearly) uncorrelated with each other? The finite–sample analysis of this exercise can provide an answer to such questions, whereas the asymptotic analysis cannot.

Let

$$a(\omega) = [e^{i\omega} \dots e^{iN\omega}]^T$$

$$y = [y(1) \dots y(N)]^T$$

Then the periodogram, (2.2.1), can be written as (we omit the subindex p of $\hat{\phi}_p(\omega)$ in this exercise):

$$\hat{\phi}(\omega) = |a^*(\omega)y|^2/N \tag{2.9.10}$$

Assume that $\{y(t)\}$ is a zero mean, stationary circular Gaussian process. The "circular Gaussianity" assumption (see, *e.g.*, Appendix B) allows us to write the fourth–order moments of $\{y(t)\}$ as (see equation (2.4.24)):

$$\begin{aligned} E\{y(t)y^*(s)y(u)y^*(v)\} &= E\{y(t)y^*(s)\}E\{y(u)y^*(v)\} \\ &\quad + E\{y(t)y^*(v)\}E\{y(u)y^*(s)\} \end{aligned} \tag{2.9.11}$$

Make use of (2.9.10) and (2.9.11) to show that

$$\boxed{\begin{aligned} \text{cov}\{\hat{\phi}(\mu),\, \hat{\phi}(\nu)\} &\triangleq E\left\{\left[\hat{\phi}(\mu) - E\left\{\hat{\phi}(\mu)\right\}\right]\left[\hat{\phi}(\nu) - E\left\{\hat{\phi}(\nu)\right\}\right]\right\} \\ &= |a^*(\mu)Ra(\nu)|^2/N^2 \end{aligned}} \tag{2.9.12}$$

where $R = E\{yy^*\}$. Deduce from (2.9.12) that

$$\text{var}\{\hat{\phi}(\mu)\} = |a^*(\mu)Ra(\mu)|^2/N^2 \tag{2.9.13}$$

Use (2.9.13) to readily rederive the variance part of the asymptotic result (2.4.21). Next, use (2.9.13) to show that *the covariance between $\hat{\phi}(\mu)$ and $\hat{\phi}(\nu)$ is not significant if*

$$|\mu - \nu| > 4\pi/N$$

and also that it may be significant otherwise. **Hint:** To show the inequality above, make use of the Carathéodory parameterization of a covariance matrix in Section 4.9.2.

Exercise 2.12: Data–Weighted ACS Estimate Interpretation of Bartlett and Welch Methods

Consider the Bartlett estimator, and assume $LM = N$.

(a) Show that the Bartlett spectral estimate can be written as:

$$\hat{\phi}_B(\omega) = \sum_{k=-(M-1)}^{M-1} \tilde{r}(k)e^{-i\omega k}$$

where

$$\tilde{r}(k) = \sum_{t=k+1}^{N} \alpha(k,t)y(t)y^*(t-k), \qquad 0 \le k < M$$

for some $\alpha(k,t)$ to be derived. Note that this is nearly of the form of the Blackman–Tukey spectral estimator, with the exception that the "standard" biased ACS estimate that is used in the Blackman–Tukey estimator is replaced by the "generalized" ACS estimate $\tilde{r}(k)$.

(b) Make use of the derived expression for $\alpha(k,t)$ to conclude that the Bartlett estimator is inferior to the Blackman–Tukey estimator (especially for small N) because it fails to use all available lag products in forming ACS estimates.

(c) Find $\alpha(k,t)$ for the Welch method. What overlap values (K in equation (2.7.7)) give lag product usage similar to the Blackman–Tukey method?

Exercise 2.13: Approximate Formula for Bandwidth Calculation

Let $W(\omega)$ denote a general spectral window that has a peak at $\omega = 0$ and is symmetric about that point. In addition, assume that the peak of $W(\omega)$ is narrow (as usually it should be). Under these assumptions, make use of a Taylor series expansion to show that an approximate formula for calculating *the bandwidth B of the peak of $W(\omega)$* is the following:

$$\boxed{B \simeq 2\sqrt{|W(0)/W''(0)|}} \tag{2.9.14}$$

The spectral peak bandwidth B is mathematically defined as follows. Let ω_1 and ω_2 denote the "half–power points," defined through

$$W(\omega_1) = W(\omega_2) = W(0)/2, \qquad \omega_1 < \omega_2$$

(hence the ratio $10\log_{10}(W(0)/W(\omega_j)) \simeq 3\text{dB}$ for $j = 1, 2$; we use $10\log_{10}$ rather than $20\log_{10}$ because the spectral window is applied to a power quantity, $\phi(\omega)$). Then since $W(\omega)$ is symmetric, so $\omega_2 = -\omega_1$,

$$B \triangleq \omega_2 - \omega_1 = 2\omega_2$$

As an application of (2.9.14), show that

$$B \simeq 0.78 \cdot 2\pi/N \text{ (in radians per sampling interval)}$$

or, equivalently, that

$$B \simeq 0.78/N \text{ (in cycles per sampling interval)}$$

for the Bartlett window (2.4.15).

Note that this formula remains approximate even as $N \to \infty$. Even though the half power bandwidth of the window gets smaller as N increases (so that one would expect the Taylor series expansion to be more accurate), the curvature of the window at $\omega = 0$ increases without bound as N increases. For the Bartlett window, verify that $B \simeq 0.9 \cdot 2\pi/N$ for N large, which differs from the prediction in this exercise by about 16%.

Exercise 2.14: A Further Look at the Time–Bandwidth Product

We saw in Section 2.6.1 that the product between the equivalent time and frequency widths of a regular window equals unity. Use the formula (2.9.14) derived in Exercise 2.13 to show that the spectral peak bandwidth B of a window $w(k)$ that is nonzero only for $|k| < N$, satisfies

$$\boxed{B \cdot N \geq 1/\pi, \quad \text{(in cycles per sampling interval)}} \tag{2.9.15}$$

This once again illustrates the "time–bandwidth product" type of result. Note that (2.9.15) involves the *effective* window time length and spectral *peak width*, as opposed to (2.6.5) which was concerned with *equivalent* time and frequency widths.

Exercise 2.15: Bias Considerations in Blackman–Tukey Window Design

The discussion in this chapter treated the bias of a spectral estimator and its resolution as two interrelated properties. This exercise illustrates further the strong relationship between bias and resolution.

Consider $\hat{\phi}_{BT}(\omega)$ as in (2.5.1), and without loss of generality assume $E\{\hat{r}(k)\} = r(k)$. (Generality is not lost because, if $E\{\hat{r}(k)\} = \alpha(k)r(k)$, then replacing $w(k)$ by $\alpha(k)w(k)$ and $\hat{r}(k)$ by $\hat{r}(k)/\alpha(k)$ results in an equivalent estimator with unbiased ACS estimates.) Find the weights $\{w(k)\}_{k=-M+1}^{M-1}$ that minimize the squared bias, as given by the error measure:

$$\epsilon = \frac{1}{2\pi}\int_{-\pi}^{\pi}\left[\phi(\omega) - E\{\hat{\phi}_{BT}(\omega)\}\right]^2 d\omega \tag{2.9.16}$$

In particular, show that the weight function that minimizes ϵ is the rectangular window. Recall that the rectangular window also has the narrowest main lobe, and hence the best resolution.

Exercise 2.16: A Property of the Bartlett Window

Let the window length, M, be given. Then, in the general case, the rectangular window can be expected to yield the windowed spectral estimate with the most favorable bias properties, owing to the fact that the sample covariance lags $\{\hat{r}(k)\}_{k=-(M-1)}^{M-1}$, appearing in (2.5.1), are left unchanged by this window (also see Exercise 2.15). The rectangular window, however, has the drawback that it is not positive definite and hence may produce negative spectral estimates. The Bartlett window, on the other hand, is positive definite and therefore yields a spectral estimate that is positive for all frequencies. Show that the latter window is the positive definite window which is closest to the rectangular one, in the sense of minimizing the following criterion:

$$\min_{\{w(k)\}} \sum_{k=0}^{M-1} |1 - w(k)| \qquad \text{subject to:}$$

$$\begin{aligned}&1)\ w(k) \equiv 0 \text{ for } |k| \geq M\\ &2)\ \{w(k)\}_{k=-\infty}^{\infty} \text{ is a positive definite sequence}\end{aligned} \tag{2.9.17}$$

Conclude that the Bartlett window is the positive definite window that distorts the sample covariances $\{\hat{r}(k)\}_{k=-(M-1)}^{M-1}$ *least* in the windowed spectral estimate formula. **Hint:** Any positive definite real window $\{w(k)\}_{k=-(M-1)}^{M-1}$ can be written as

$$w(k) = \sum_{i=0}^{M-1} b_i\, b_{i+k} \qquad (b_i = 0 \text{ for } i \geq M) \tag{2.9.18}$$

for some real–valued parameters $\{b_i\}_{i=0}^{M-1}$. Make use of the above parameterization of the set of positive definite windows to transform (2.9.17) into an optimization problem without constraints.

Computer Exercises

Tools for Periodogram Spectral Estimation:

The text web site `www.prenhall.com/~stoica` contains the following MATLAB functions for use in computing periodogram-based spectral estimates. In each case, `y` is the input data vector, `L` controls the frequency sample spacing of the output, and the output vector `phi`= $\phi(\omega_k)$ where $\omega_k = \frac{2\pi k}{L}$. MATLAB functions that generate the Correlogram, Blackman–Tukey, Windowed Periodogram, Bartlett, Welch, and Daniell spectral estimates are as follows:

- `phi = correlogramse(y,L)`
  Implements the correlogram spectral estimate in equation (2.2.2).
- `phi = btse(y,w,L)`
  Implements the Blackman–Tukey spectral estimate in equation (2.5.1); `w` is the vector $[w(0), \ldots, w(M-1)]^T$.
- `phi = periodogramse(y,v,L)`
  Implements the windowed periodogram spectral estimate in equation (2.6.24); `v` is a vector of window function elements $[v(1), \ldots, v(N)]^T$, and should be the same size as `y`. If `v` is a vector of ones, this function implements the unwindowed periodogram spectral estimate in equation (2.2.1).
- `phi = bartlettse(y,M,L)`
  Implements the Bartlett spectral estimate in equations (2.7.2) and (2.7.3); `M` is the size of each subsequence as in equation (2.7.2).
- `phi = welchse(y,v,K,L)`
  Implements the Welch spectral estimate in equation (2.7.8); `M` is the size of each subsequence, `v` is the window function $[v(1), \ldots, v(M)]^T$ applied to each subsequence, and `K` is the overlap parameter, as in equation (2.7.7).
- `phi = daniellse(y,J,Ntilde)`
  Implements the Daniell spectral estimate in equation (2.7.16); `J` and `Ntilde` correspond to J and $\tilde{N}$ there.

Exercise C2.17: Zero Padding Effects on Periodogram Estimators

In this exercise we study the effect zero padding has on the periodogram. Consider the sequence

$$y(t) = 10\sin(0.2 \cdot 2\pi t + \phi_1) + 5\sin((0.2 + 1/N)2\pi t + \phi_2) + e(t), \qquad (2.9.19)$$

where $t = 0, \ldots, N-1$, and $e(t)$ is white Gaussian noise with variance 1. Let $N = 64$ and $\phi_1 = \phi_2 = 0$.

From the results in Chapter 4, we find the spectrum of $y(t)$ to be

$$\begin{aligned}\phi(\omega) &= 50\pi\left[\delta(\omega - 0.2 \cdot 2\pi) + \delta(\omega + 0.2 \cdot 2\pi)\right] \\ &\quad + 12.5\pi\left[\delta(\omega - (0.2 + 1/N) \cdot 2\pi) + \delta(\omega + (0.2 + 1/N) \cdot 2\pi)\right] + 1\end{aligned}$$

Plot the periodogram for the sequence $\{y(t)\}$, and the sequence $\{y(t)\}$ zero padded with N, $3N$, $5N$, and $7N$ zeroes.

Explain the difference between the four periodograms. Why does the first periodogram not give a good description of the spectral content of the signal? Note that zero padding does not change the resolution of the estimator.

Exercise C2.18: Resolution and Leakage Properties of the Periodogram

We have seen from Section 2.4 that the expected value of the periodogram is the convolution of the true spectrum $\phi_y(\omega)$ with the Fourier transform of a Bartlett window, denoted $W_B(\omega)$ (see equation (2.4.15)). The shape and size of the $W_B(\omega)$ function determines the amount of *smearing* and *leakage* in the periodogram. Similarly, in Section 2.5 we introduced a windowed periodogram in (2.6.24) whose expected value is equal to the expected value of a corresponding Blackman–Tukey estimate with weights $w(k)$ given by (2.6.31). Different window functions than the rectangular window could be used in the periodogram estimate, giving rise to correspondingly different windows in the correlogram estimate. The choice of window affects the resolution and leakage properties of the periodogram (correlogram) spectral estimate.

Resolution Properties: The amount of smearing of the spectral estimate is determined by the width of the main lobe, and the amount of leakage is determined by the energy in the sidelobes. The amount of smearing is what limits the resolving power of the periodogram, and is studied empirically below.

We first study the resolution properties by considering a sequence made up of two sinusoids in noise, where the two sinusoidal frequencies are "close". Consider

$$y(t) = a_1 \sin(f_0 \cdot 2\pi t + \phi_1) + a_2 \sin((f_0 + \alpha/N)2\pi t + \phi_2) + e(t), \qquad (2.9.20)$$

where $e(t)$ is real–valued Gaussian white noise with zero mean and variance σ^2. We choose $f_0 = 0.2$ and $N = 256$, but the results are nearly independent of f_0 and N.

(a) Determine empirically the 3 dB width of the main lobe of $W_B(\omega)$ as a function of N, and verify equation (2.4.18). Also determine the peak sidelobe height (in dB) as a function of N. Note that the sidelobe level of a window function is generally independent of N. Verify this by examining plots of the magnitude of $W_B(\omega)$ for several values of N; try both linear and dB scales in your plots.

(b) Set $\sigma^2 = 0$ (this eliminates the statistical variation in the periodogram, so that the bias properties can be isolated and studied). Set $a_1 = a_2 = 1$ and

$\phi_1 = \phi_2 = 0$. Plot the (zero–padded) periodogram of $y(t)$ for various α and determine the resolution threshold (*i.e.*, the minimum value of α for which the two frequency components can be resolved). How does this value of α compare with the predicted resolution in Section 2.4?

(c) Repeat part (b) for a Hamming–windowed correlogram estimate.

(d) For reasonably high signal–to–noise ratio (SNR) values and reasonably close signal amplitudes, the resolution thresholds in parts (b) and (c) above are not very sensitive to variations in the signal amplitudes and frequency f_0. However, these thresholds are sensitive to the phases ϕ_1 and ϕ_2, especially if α is smaller than 1. Try two pairs (ϕ_1, ϕ_2) so that the two sinusoids are in phase and out of phase, respectively, at the center of the observation interval, and compare the resolution thresholds. Also, try different values of a_1, a_2, and σ^2 to verify that their values have relatively little effect on the resolution threshold.

Spectral Leakage: In this part we analyze the effects of leakage on the periodogram estimate. Leakage properties can be clearly seen when trying to estimate two sinusoidal terms that are well separated but have greatly differing amplitudes.

(a) Generate the sinusoidal sequence above for $\alpha = 4$, $\sigma^2 = 0$, and $\phi_1 = \phi_2 = 0$. Set $a_1 = 1$ and vary a_2 (choose $a_2 = 1, 0.1, 0.01$, and 0.001, for example). Compute the periodogram (using a rectangular data window), and comment on the ability to identify the second sinusoidal term from the spectral estimate.

(b) Repeat part (a) for $\alpha = 12$. Does the amplitude threshold for identifiability of the second sinusoidal term change?

(c) Explain your results in parts (a) and (b) by looking at the amplitude of the Bartlett window's Fourier transform at frequencies corresponding to α/N for $\alpha = 4$ and $\alpha = 12$.

(d) The Bartlett window (and many other windows) has the property that the leakage level depends on the distance between spectral components in the data, as seen in parts (a) and (b). For many practical applications it may be known what dynamic range the sinusoidal components in the data may have, and it is thus desirable to use a data window with a constant sidelobe level that can be chosen by the user. The Chebyshev window (or Taylor window) is a good choice for these applications, because the user can select the (constant) sidelobe level in the window design (see the MATLAB command `chebwin`).

Assume we know that the maximum dynamic range of sinusoidal components is 60 dB. Design a Chebyshev window $v(t)$ and corresponding Blackman–Tukey window $w(k)$ using (2.6.31) so that the two sinusoidal components of the data can be resolved for this dynamic range using (i) the Blackman–Tukey spectral estimator with window $w(k)$, and (ii) the windowed periodogram method with window $v(t)$. Plot the Fourier transform of the window and determine the spectral resolution of the window.

Test your window design by computing the Blackman–Tukey and windowed periodogram estimates for two sinusoids whose amplitudes differ by 50 dB in dynamic range, and whose frequency separation is the minimum value you predicted. Compare the resolution results with your predictions. Explain why the smaller amplitude sinusoid can be detected using one of the methods but not the other.

Exercise C2.19: Bias and Variance Properties of the Periodogram Spectral Estimate

In this exercise we verify the theoretical predictions about bias and variance properties of the periodogram spectral estimate. We use autoregressive moving average (ARMA) signals (see Chapter 3) as test signals.

Bias Properties — Resolution and Leakage: We consider a random process $y(t)$ generated by filtering white noise:

$$y(t) = H(q)e(t)$$

where $e(t)$ is zero mean Gaussian white noise with variance $\sigma^2 = 1$, and the filter $H(q)$ is given by:

$$H(q) = \sum_{k=1}^{2} A_k \left[\frac{1 - z_k q^{-1}}{1 - p_k q^{-1}} + \frac{1 - z_k^* q^{-1}}{1 - p_k^* q^{-1}} \right] \tag{2.9.21}$$

with

$$\begin{aligned} p_1 &= 0.99e^{i2\pi 0.3} & p_2 &= 0.99e^{i2\pi(0.3+\alpha)} \\ z_1 &= 0.95e^{i2\pi 0.3} & z_2 &= 0.95e^{i2\pi(0.3+\alpha)} \end{aligned} \tag{2.9.22}$$

We first let $A_1 = A_2 = 1$ and $\alpha = 0.05$.

(a) Plot the true spectrum $\phi(\omega)$. Using a sufficiently fine grid for ω so that approximation errors are small, plot the ACS using an inverse FFT of $\phi(\omega)$.

(b) For $N = 64$, plot the Fourier transform of the Bartlett window, and also plot the expected value of the periodogram estimate $\hat{\phi}_p(\omega)$ as given by equation (2.4.8). We see that for this example and data length, the main lobe width of the Bartlett window is wider than the widths of the spectral peaks in $\phi(\omega)$. Discuss how this relatively wide main lobe width affects the resolution properties of the estimator.

(c) Generate 50 realizations of $y(t)$, each of length $N = 64$ data points. You can generate the data by passing white noise through the filter $H(q)$ (see the

MATLAB commands `dimpulse` and `filter`); be sure to discard a sufficient number of initial filter output points to effectively remove the transient part of the filter output. Compute the periodogram spectral estimates for each data sequence; plot 10 spectral estimates overlayed on a single plot. Also plot the average of the 50 spectral estimates. Compare the average with the predicted expected value as found in part (a).

(d) The resolution of the spectral peaks in $\phi(\omega)$ will depend on their separation relative to the width of the Bartlett window. Generate 50 realizations of $y(t)$ for $N = 256$, and find the minimum value of α so that the spectral peaks can be resolved for most realizations. Compare your results with the predicted formula (2.4.18) for spectral resolution.

(e) Leakage from the Bartlett window will impact the ability to identify peaks of different amplitudes. To illustrate this, generate 50 realizations of $y(t)$ for $N = 64$, for both $\alpha = 4/N$ and $\alpha = 12/N$. For each value of α, set $A_1 = 1$, and vary A_2 to find the minimum amplitude for which the lower amplitude peak can reliably be identified from the spectral estimates for most realizations. Compare this value with the Bartlett window sidelobe level for $\omega = 2\pi\alpha$ and for the two values of α. Does the window sidelobe level accurately reflect the amplitude separation required to identify the second peak?

Variance Properties: In this part we will verify that the variance of the periodogram is almost independent of the data length, and compare the empirical variance with theoretical predictions. For this part, we consider a broadband signal $y(t)$ for which the Bartlett window smearing and leakage effects are small.

Consider the broadband ARMA process

$$y(t) = \frac{B_1(q)}{A_1(q)}\, e(t)$$

with

$$\begin{aligned} A_1(q) &= 1 - 1.3817q^{-1} + 1.5632q^{-2} - 0.8843q^{-3} + 0.4096q^{-4} \\ B_1(q) &= 1 + 0.3544q^{-1} + 0.3508q^{-2} + 0.1736q^{-3} + 0.2401q^{-4} \end{aligned}$$

(a) Plot the true spectrum $\phi(\omega)$.

(b) Generate 50 Monte–Carlo data realizations using different noise sequences, and compute the corresponding 50 periodogram spectral estimates. Plot the sample mean, the sample mean plus one sample standard deviation and sample mean minus one sample standard deviation spectral estimate curves. Do this for $N = 64$, 256, and 1024. Note that the variance does not decrease with N.

(c) Compare the sample variance to the predicted variance in equation (2.4.21). It may help to plot $\text{stdev}\{\hat{\phi}(\omega)\}/\phi(\omega)$ and determine to what degree this curve is approximately constant. Discuss your results.

Exercise C2.20: Refined Methods: Variance–Resolution Tradeoff

In this exercise we apply the Blackman–Tukey and Welch estimators to both a narrowband and broadband random process. We consider the same processes in Chapters 3 and 5 to facilitate comparison with the spectral estimation methods developed in those chapters.

Broadband ARMA Process: Generate realizations of the broadband autoregressive moving–average (ARMA) process

$$y(t) = \frac{B_1(q)}{A_1(q)} e(t)$$

with

$$\begin{aligned} A_1(q) &= 1 - 1.3817q^{-1} + 1.5632q^{-2} - 0.8843q^{-3} + 0.4096q^{-4} \\ B_1(q) &= 1 + 0.3544q^{-1} + 0.3508q^{-2} + 0.1736q^{-3} + 0.2401q^{-4} \end{aligned}$$

Choose the number of samples as $N = 256$.

(a) Generate 50 Monte–Carlo data realizations using different noise sequences, and compute the corresponding 50 spectral estimates using the following methods:

- The Blackman–Tukey spectral estimate using the Bartlett window $w_B(t)$. Try both $M = N/4$ and $M = N/16$.
- The Welch spectral estimate using the rectangular window $w_R(t)$, and using both $M = N/4$ and $M = N/16$ and overlap parameter $K = M/2$.

Plot the sample mean, the sample mean plus one sample standard deviation and sample mean minus one sample standard deviation spectral estimate curves. Compare with the periodogram results from Exercise C2.19, and with each other.

(b) Judging from the plots you have obtained, how has the variance decreased in the refined estimates? How does this variance decrease compare to the theoretical expectations?

(c) As discussed in the text, the value of M should be chosen to compromise between low "smearing" and low variance. For the Blackman–Tukey estimate, experiment with different values of M and different window functions to find a "best design" (in your judgment), and plot the corresponding spectral estimates.

Narrowband ARMA Process: Generate realizations of the narrowband ARMA process

$$y(t) = \frac{B_2(q)}{A_2(q)} e(t)$$

with

$$
\begin{aligned}
A_2(q) &= 1 - 1.6408q^{-1} + 2.2044q^{-2} - 1.4808q^{-3} + 0.8145q^{-4} \\
B_2(q) &= 1 + 1.5857q^{-1} + 0.9604q^{-2}
\end{aligned}
$$

and $N = 256$.

Repeat the experiments and comparisons in the broadband example for the narrowband process.

Exercise C2.21: Periodogram–Based Estimators applied to Measured Data

Consider the data sets in the files `sunspotdata.mat` and `lynxdata.mat`. These files can be obtained from the text web site `www.prenhall.com/~stoica`. Apply periodogram–based estimation techniques (possibly after some preprocessing; see the following) to estimate the spectral content of these data. Try to answer the following questions:

(a) Are there sinusoidal components (or periodic structure) in the data? If so, how many components and at what frequencies?

(b) Nonlinear transformations and linear or polynomial trend removal are often applied before spectral analysis of a time series. For the `lynx` data, compare your spectral analysis results from the original data, and the data transformed first by taking the logarithm of each sample and then by subtracting the sample mean of this logarithmic data. Does the logarithmic transformation make the data more sinusoidal in nature?

Chapter 3

PARAMETRIC METHODS FOR RATIONAL SPECTRA

3.1 Introduction

The principal difference between the spectral estimation methods of Chapter 2 and those in this chapter, is that in Chapter 2 we made no assumption on the studied signal (except for its stationarity). The *parametric* or *model–based methods* of spectral estimation assume that the signal satisfies a generating model with known functional form, and then proceed by estimating the parameters in the assumed model. The signal's spectral characteristics of interest are then derived from the estimated model. In those cases where the assumed model is a close approximation to the reality, it is no wonder that the parametric methods provide more accurate spectral estimates than the nonparametric techniques. The nonparametric approach to PSD estimation remains useful, though, in applications where there is little or no information about the signal in question.

Our discussion of parametric methods for spectral estimation is divided into two parts. In this chapter, we discuss parametric methods for rational spectra which form a dense set in the class of *continuous spectra* (see Section 3.2) [ANDERSON 1971; WEI 1990]. Our discussion on fitting rational spectral models to data series will be relatively brief. In particular, except for a discussion in Appendix B about the Cramér–Rao bound and the best accuracy achievable in this class of spectral models, we do not include detailed results on the statistical properties of the estimation methods discussed in the following sections since: (i) such results are readily available in the literature [KAY 1988; PRIESTLEY 1989; SÖDERSTRÖM AND STOICA 1989]; (ii) parametric methods provide consistent spectral estimates and hence (for large sample sizes, at least) the issue of statistical behavior is not so critical; and (iii) a detailed statistical analysis is beyond the scope of an introductory course. We also do not discuss the structure (or order) estimation problem. A review of methods that can be used to solve the structure estimation problem can be found in [CHOI 1992].

The second part of our discussion on parametric methods is contained in the next chapter where we consider *discrete spectra* such as those associated with sinusoidal signals buried in white noise. *Mixed spectra* (containing both continuous and discrete spectral components, such as in the case of sinusoidal signals corrupted by colored noise) are not covered explicitly in this text, but we remark that some methods in Chapter 4 can be extended to deal with such spectra as well.

3.2 Signals with Rational Spectra

A rational PSD is a rational function of $e^{-i\omega}$ (*i.e.*, the ratio of two polynomials in $e^{-i\omega}$):

$$\phi(\omega) = \frac{\sum_{k=-m}^{m} \gamma_k e^{-i\omega k}}{\sum_{k=-n}^{n} \rho_k e^{-i\omega k}} \tag{3.2.1}$$

where $\gamma_{-k} = \gamma_k^*$ and $\rho_{-k} = \rho_k^*$. The Weierstrass Theorem from calculus asserts that any continuous PSD can be approximated arbitrarily closely by a rational PSD of the form (3.2.1), provided the degrees m and n in (3.2.1) are chosen sufficiently large; that is, the rational PSDs form a *dense set in the class of all continuous spectra.* This observation partly motivates the significant interest in the model (3.2.1) for $\phi(\omega)$, among the researchers in the "spectral estimation community".

It is not difficult to show that, since $\phi(\omega) \geq 0$, the rational spectral density in (3.2.1) can be factored as follows:

$$\boxed{\phi(\omega) = \left| \frac{B(\omega)}{A(\omega)} \right|^2 \sigma^2} \tag{3.2.2}$$

where σ^2 is a positive scalar, and $A(\omega)$ and $B(\omega)$ are the polynomials

$$\begin{aligned} A(\omega) &= 1 + a_1 e^{-i\omega} + \ldots + a_n e^{-in\omega} \\ B(\omega) &= 1 + b_1 e^{-i\omega} + \ldots + b_m e^{-im\omega} \end{aligned} \tag{3.2.3}$$

The result (3.2.2) can similarly be expressed in the Z–domain. With the notation $\phi(z) = \sum_{k=-m}^{m} \gamma_k z^{-k} / \sum_{k=-n}^{n} \rho_k z^{-k}$, we can factor $\phi(z)$ as:

$$\phi(z) = \sigma^2 \frac{B(z)B^*(1/z^*)}{A(z)A^*(1/z^*)} \tag{3.2.4}$$

where, for example,

$$\begin{aligned} A(z) &= 1 + a_1 z^{-1} + \cdots + a_n z^{-n} \\ A^*(1/z^*) &= [A(1/z^*)]^* = 1 + a_1^* z + \cdots + a_n^* z^n \end{aligned}$$

Recall the notational convention in this text that we write, for example, $A(z)$ and $A(\omega)$ with the implicit understanding that when we convert from a function of z to a function of ω, we use the substitution $z = e^{i\omega}$.

We note that the zeroes and poles of $\phi(z)$ are in symmetric pairs about the unit circle; if $z_i = re^{i\theta}$ is a zero (pole) of $\phi(z)$, then $(1/z_i^*) = (1/r)e^{i\theta}$ is also a zero (pole) (see Exercise 1.3). Under the assumption that $\phi(z)$ has no pole with modulus equal to one, the region of convergence of $\phi(z)$ includes the unit circle $z = e^{i\omega}$. The result that (3.2.1) can be written as in (3.2.2) and (3.2.4) is called the *spectral factorization theorem* (see, *e.g.*, [SÖDERSTRÖM AND STOICA 1989; KAY 1988]).

The next point of interest is to compare (3.2.2) and (1.4.9). This comparison leads to the following result.

The arbitrary rational PSD in (3.2.2) can be associated with a signal obtained by filtering white noise of power σ^2 through the rational filter with transfer function $H(\omega) = B(\omega)/A(\omega)$.	(3.2.5)

The filtering referred to in (3.2.5) can be written in the time domain as

$$y(t) = \frac{B(q)}{A(q)} e(t) \tag{3.2.6}$$

or, alternatively,

$$A(q)y(t) = B(q)e(t) \tag{3.2.7}$$

where $y(t)$ is the filter output, and

q^{-1} = the unit delay operator $(q^{-k}y(t) = y(t-k))$
$e(t)$ = white noise of variance equal to σ^2

Hence, by means of the spectral factorization theorem, the parameterized model of $\phi(\omega)$ turned into a model of the signal itself. The spectral estimation problem can then be reduced to a problem of *signal modeling*. In the following sections, we present several methods for estimating the parameters in the signal model (3.2.7) and in two of its special cases corresponding to $m = 0$ and, respectively, $n = 0$.

A signal $y(t)$ satisfying the equation (3.2.6) is called an *autoregressive moving average* (ARMA or ARMA(n, m)) signal. If $m = 0$ in (3.2.6), then $y(t)$ is an *autoregressive* (AR or AR(n)) signal; and $y(t)$ is a *moving average* (MA or MA(m)) signal if $n = 0$. For easy reference, we summarize these naming conventions below.

$$\begin{array}{ll} \text{ARMA}: & A(q)y(t) = B(q)e(t) \\ \text{AR}: & A(q)y(t) = e(t) \\ \text{MA}: & y(t) = B(q)e(t) \end{array} \tag{3.2.8}$$

By assumption, $\phi(\omega)$ is finite for all ω values; as a result, $A(z)$ cannot have any zero exactly on the unit circle. Furthermore, since the poles and zeroes of $\phi(z)$ are in reciprocal pairs, as explained before, it is always possible to choose $A(q)$ to have all its zeroes strictly inside the unit disc. The corresponding model (3.2.6) is then said to be *stable*. If we assume, for simplicity, that $\phi(\omega)$ does not vanish at any ω then — similarly to the above — we can choose the polynomial $B(q)$ so that it has all zeroes inside the unit (open) disc. The corresponding model (3.2.6) is said to be *minimum phase* (see Exercise 3.1 for a motivation for the name minimum phase).

We remark that in the previous paragraph we actually provided a sketch of the proof of the spectral factorization theorem. That discussion also showed that the spectral factorization problem associated with a rational PSD has multiple solutions, with the stable and minimum phase ARMA model being only one of them. In the following, we will consider the problem of estimating the parameters in this particular ARMA equation. When the final goal is the estimation of $\phi(\omega)$, focusing on the stable and minimum phase ARMA model is no restriction.

3.3 Covariance Structure of ARMA Processes

In this section we derive an expression for the covariances of an ARMA process in terms of the parameters $\{a_i\}_{i=1}^n$, $\{b_i\}_{i=1}^m$, and σ^2. The expression provides a convenient method for estimating the ARMA parameters by replacing the true autocovariances with estimates obtained from data. Nearly all ARMA spectral estimation methods exploit this covariance structure either explicitly or implicitly, and thus it will be widely used in the remainder of the chapter.

Equation (3.2.7) can be written as

$$y(t) + \sum_{i=1}^{n} a_i y(t-i) = \sum_{j=0}^{m} b_j e(t-j), \qquad (b_0 = 1) \tag{3.3.1}$$

Multiplying (3.3.1) by $y^*(t-k)$ and taking expectation yields

$$r(k) + \sum_{i=1}^{n} a_i r(k-i) = \sum_{j=0}^{m} b_j E\left\{e(t-j)y^*(t-k)\right\} \tag{3.3.2}$$

Since the filter $H(q) = B(q)/A(q)$ is asymptotically stable and causal, we can write

$$H(q) = B(q)/A(q) = \sum_{k=0}^{\infty} h_k q^{-k}, \qquad (h_0 = 1)$$

which gives

$$y(t) = H(q)e(t) = \sum_{k=0}^{\infty} h_k e(t-k).$$

Then the term $E\left\{e(t-j)y^*(t-k)\right\}$ becomes

$$\begin{aligned} E\left\{e(t-j)y^*(t-k)\right\} &= E\left\{e(t-j)\sum_{s=0}^{\infty} h_s^* e^*(t-k-s)\right\} \\ &= \sigma^2 \sum_{s=0}^{\infty} h_s^* \delta_{j,k+s} = \sigma^2 h_{j-k}^* \end{aligned}$$

where we use the convention that $h_k = 0$ for $k < 0$. Thus, equation (3.3.2) becomes

$$r(k) + \sum_{i=1}^{n} a_i r(k-i) = \sigma^2 \sum_{j=0}^{m} b_j h_{j-k}^* \tag{3.3.3}$$

In general, h_k is a nonlinear function of the $\{a_i\}$ and $\{b_i\}$ coefficients. However, since $h_s = 0$ for $s < 0$, equation (3.3.3) for $k \geq m+1$ reduces to

$$\boxed{r(k) + \sum_{i=1}^{n} a_i r(k-i) = 0, \qquad \text{for } k > m} \tag{3.3.4}$$

Equation (3.3.4) is the basis for many estimators of the AR coefficients of AR(MA) processes, as we will see.

3.4 AR Signals

In the ARMA class, the *autoregressive or all–pole signals* constitute the type that is most frequently used in applications. The AR equation may model spectra with narrow peaks by placing zeroes of the A–polynomial in (3.2.2) (with $B(\omega) \equiv 1$) close to the unit circle. This is an important feature since narrowband spectra are quite common in practice. In addition, the estimation of parameters in AR signal models is a well–established topic; the estimates are found by solving a system of linear equations, and the stability of the estimated AR polynomial can be guaranteed.

We consider two methods for AR spectral estimation. The first is based directly on the linear relationship between the covariances and the AR parameters derived in equation (3.3.4); it is called the Yule–Walker method. The second method is based on a least squares solution of AR parameters using the time–domain equation $A(q)y(t) = e(t)$. This so–called least squares method is closely related to the problem of linear prediction, as we shall see.

3.4.1 Yule–Walker Method

In this section, we focus on a technique for estimating the AR parameters which is called the *Yule–Walker (YW) method* [YULE 1927; WALKER 1931]. For AR signals, $m = 0$ and $B(q) = 1$. Thus, equation (3.3.4) holds for $k > 0$. Also, we have from equation (3.3.3) that

$$r(0) + \sum_{i=1}^{n} a_i r(-i) = \sigma^2 \sum_{j=0}^{0} b_j h_j^* = \sigma^2 \tag{3.4.1}$$

Combining (3.4.1) and (3.3.4) for $k = 1, \ldots, n$ gives the following system of linear equations

$$\begin{bmatrix} r(0) & r(-1) & \dots & r(-n) \\ r(1) & r(0) & & \vdots \\ \vdots & & \ddots & r(-1) \\ r(n) & \dots & & r(0) \end{bmatrix} \begin{bmatrix} 1 \\ a_1 \\ \vdots \\ a_n \end{bmatrix} = \begin{bmatrix} \sigma^2 \\ 0 \\ \vdots \\ 0 \end{bmatrix} \tag{3.4.2}$$

The above equations are called the *Yule–Walker equations* or *Normal equations*, and form the basis of many AR estimation methods. If $\{r(k)\}_{k=0}^{n}$ were known, we could solve (3.4.2) for

$$\theta = [a_1, \dots, a_n]^T \tag{3.4.3}$$

by using all but the first row of (3.4.2):

$$\begin{bmatrix} r(1) \\ \vdots \\ r(n) \end{bmatrix} + \begin{bmatrix} r(0) & \cdots & r(-n+1) \\ \vdots & \ddots & \vdots \\ r(n-1) & \cdots & r(0) \end{bmatrix} \begin{bmatrix} a_1 \\ \vdots \\ a_n \end{bmatrix} = \begin{bmatrix} 0 \\ \vdots \\ 0 \end{bmatrix} \tag{3.4.4}$$

or, with obvious definitions,

$$r_n + R_n\theta = 0. \tag{3.4.5}$$

The solution is $\theta = -R_n^{-1}r_n$. Once θ is found, σ^2 can be obtained from the first row of (3.4.2) or, equivalently, from (3.4.1).

The Yule–Walker method for AR spectral estimation is based directly on (3.4.2). Given data $\{y(t)\}_{t=1}^{N}$, we first obtain sample covariances $\{\hat{r}(k)\}_{k=0}^{n}$ using the standard biased ACS estimator (2.2.4). We insert these ACS estimates in (3.4.2) and solve for $\hat{\theta}$ and $\hat{\sigma}^2$ as explained above in the known–covariance case.

Note that the covariance matrix in (3.4.2) can be shown to be positive definite for any n, and hence the solution to (3.4.2) is unique [Söderström and Stoica 1989]. When the covariances are replaced by standard biased ACS estimates, the matrix can be shown to be positive definite for any sample (not necessarily generated by an AR equation) that is not identically equal to zero; see the remark in the next section for a proof.

To explicitly stress the dependence of θ and σ^2 on the order n, we can write (3.4.2) as

$$\boxed{R_{n+1} \begin{bmatrix} 1 \\ \theta_n \end{bmatrix} = \begin{bmatrix} \sigma_n^2 \\ 0 \end{bmatrix}} \tag{3.4.6}$$

We will return to the above equation in Section 3.5.

3.4.2 Least Squares Method

The Yule–Walker method for estimating the AR parameters is based on equation (3.4.2) with the true covariance elements $\{r(k)\}$ replaced by the sample covariances $\{\hat{r}(k)\}$. In this section, we derive another type of AR estimator based on a least squares (LS) minimization criterion using the time–domain relation $A(q)y(t) = e(t)$. We develop the LS estimator by considering the closely related problem of *linear prediction*. We then interpret the LS method as a Yule–Walker-type method that uses a different estimate of R_{n+1} in equation (3.4.6).

We first relate the Yule–Walker equations to the linear prediction problem. Let $y(t)$ be an AR process of order n. Then $y(t)$ satisfies

$$\begin{aligned} e(t) &= y(t) + \sum_{i=1}^{n} a_i y(t-i) = y(t) + \varphi^T(t)\theta \\ &\triangleq y(t) + \hat{y}(t) \end{aligned} \tag{3.4.7}$$

where $\varphi(t) = [y(t-1), \ldots, y(t-n)]^T$. We interpret $\hat{y}(t)$ as a *linear prediction* of $y(t)$ from the n previous samples $y(t-1), \ldots, y(t-n)$, and we interpret $e(t)$ as the corresponding *prediction error*. See the Complement in Section 3.9.1 and also Exercises 3.3–3.5 for more discussion on this and other related linear prediction problems.

The vector θ that minimizes the prediction error variance $\sigma_n^2 \triangleq E\left\{|e(t)|^2\right\}$ is the AR coefficient vector in (3.4.6), as we will show. From (3.4.7) we have

$$\begin{aligned} \sigma_n^2 = E\left\{|e(t)|^2\right\} &= E\left\{\left[y^*(t) + \theta^*\varphi^c(t)\right]\left[y(t) + \varphi^T(t)\theta\right]\right\} \\ &= r(0) + r_n^*\theta + \theta^* r_n + \theta^* R_n \theta \end{aligned} \tag{3.4.8}$$

where r_n and R_n are defined in equations (3.4.4)–(3.4.5). The vector θ that minimizes (3.4.8) is given by (see Result R34 in Appendix A)

$$\theta = -R_n^{-1} r_n \tag{3.4.9}$$

with corresponding minimum prediction error

$$\sigma_n^2 = r(0) - r_n^* R_n^{-1} r_n \tag{3.4.10}$$

Equations (3.4.9) and (3.4.10) are exactly the Yule–Walker equations in (3.4.5) and (3.4.1) (or, equivalently, in (3.4.6)). Thus, we see that the Yule–Walker equations can be interpreted as the solution to the problem of finding the best linear predictor of $y(t)$ from its n most recent past samples. For this reason, AR modeling is sometimes referred to as *linear predictive modeling.*

The Least Squares AR estimation method is based on a finite–sample approximate solution of the above minimization problem. Given a finite set of measurements $\{y(t)\}_{t=1}^{N}$, we approximate the minimization of $E\left\{|e(t)|^2\right\}$ by the finite–sample cost function

$$f(\theta) = \sum_{t=N_1}^{N_2} |e(t)|^2 = \sum_{t=N_1}^{N_2} \left| y(t) + \sum_{i=1}^{n} a_i y(t-i) \right|^2$$

$$
= \left\| \begin{bmatrix} y(N_1) \\ y(N_1+1) \\ \vdots \\ y(N_2) \end{bmatrix} + \begin{bmatrix} y(N_1-1) & \cdots & y(N_1-n) \\ y(N_1) & \cdots & y(N_1+1-n) \\ \vdots & & \vdots \\ y(N_2-1) & \cdots & y(N_2-n) \end{bmatrix} \theta \right\|^2
$$

$$
\triangleq \|y + Y\theta\|^2 \tag{3.4.11}
$$

where we assume $y(t) = 0$ for $t < 1$ and $t > N$.

The vector θ that minimizes $f(\theta)$ is given by (see Result R32 in Appendix A)

$$
\boxed{\hat{\theta} = -(Y^*Y)^{-1}(Y^*y)} \tag{3.4.12}
$$

where, as seen from (3.4.11), the definitions of Y and y depend on the choice of (N_1, N_2) considered. For example, for $N_1 = 1$ and $N_2 = N + n$, Y and y are equal to:

$$
y = \begin{bmatrix} y(1) \\ y(2) \\ \vdots \\ \hline y(n+1) \\ y(n+2) \\ \vdots \\ y(N) \\ \hline 0 \\ 0 \\ \vdots \\ 0 \end{bmatrix}, \quad Y = \begin{bmatrix} 0 & 0 & \cdots & 0 \\ y(1) & 0 & & \vdots \\ \vdots & \ddots & \ddots & 0 \\ \hline y(n) & y(n-1) & \cdots & y(1) \\ y(n+1) & y(n) & \cdots & y(2) \\ \vdots & & & \vdots \\ y(N-1) & y(N-2) & \cdots & y(N-n) \\ \hline y(N) & y(N-1) & \cdots & y(N-n+1) \\ 0 & y(N) & & \\ \vdots & \ddots & \ddots & \vdots \\ 0 & \cdots & 0 & y(N) \end{bmatrix} \tag{3.4.13}
$$

Notice the Toeplitz structure of Y, and also that y matches this Toeplitz structure when it is appended to the left of Y; that is, $[y|Y]$ also shares this Toeplitz structure.

The two most common choices for N_1 and N_2 are:

- $N_1 = 1$, $N_2 = N + n$ (considered above). This choice yields the so-called *autocorrelation method.*

- $N_1 = n+1$, $N_2 = N$. This choice corresponds to removing the first n and last n rows of Y and y in equation (3.4.13), and hence eliminates all the arbitrary zero values there. The estimate (3.4.12) with this choice of (N_1, N_2) is often named the *covariance method.* We refer to this method as the *covariance LS method*, or the *LS method*.

Other choices for N_1 and N_2 have also been suggested. For example, the *prewindow method* uses $N_1 = 1$ and $N_2 = N$, and the *postwindow method* uses $N_1 = n + 1$ and $N_2 = N$.

The least squares methods can be interpreted as approximate solutions to the Yule–Walker equations in (3.4.4) by recognizing that Y^*Y and Y^*y are, to within a multiplicative constant, finite–sample estimates of R_n and r_n, respectively. In fact, it is easy to show that for the autocorrelation method, the elements of $(Y^*Y)/N$ and $(Y^*y)/N$ are *exactly* the biased ACS estimates (2.2.4) used in the Yule–Walker AR estimate. Writing $\hat{\theta}$ in (3.4.12) as

$$\hat{\theta} = -\left[\frac{1}{N}(Y^*Y)\right]^{-1}\left[\frac{1}{N}(Y^*y)\right]$$

we see as a consequence that

> The autocorrelation method of least squares AR estimation is equivalent to the Yule–Walker method.

Remark: We can now prove a claim made in the previous subsection that the matrix Y^*Y in (3.4.12), with Y given by (3.4.13), is positive definite for any sample $\{y(t)\}_{t=1}^{N}$ that is not identically equal to zero. To prove this claim it is necessary and sufficient to show that rank$(Y) = n$. If $y(1) \neq 0$, then clearly rank$(Y) = n$. If $y(1) = 0$ and $y(2) \neq 0$, then again we clearly have rank$(Y) = n$, and so on. ■

For the covariance AR estimator, $(Y^*Y)/(N-n)$ and $(Y^*y)/(N-n)$ are unbiased estimates of R_n and r_n in equations (3.4.4) and (3.4.5), and they do not use any measurement data outside the available interval $1 \leq t \leq N$. On the other hand, the matrix $(Y^*Y)/(N-n)$ is not Toeplitz, so the Levinson–Durbin or Delsarte–Genin algorithms in the next section cannot be used (although similar fast algorithms for the covariance method have been developed; see, *e.g.*, [MARPLE 1987]).

As N increases, the difference between the covariance matrix estimates used by the Yule–Walker and the least squares (LS) methods diminishes. Consequently, in large samples (*i.e.*, for $N \gg 1$), the YW and LS estimates of the AR parameters nearly coincide with one another.

For small or medium sample lengths, the Yule–Walker and covariance LS methods may behave differently. *First*, the estimated AR model obtained with the Yule–Walker method is always guaranteed to be *stable* (see, *e.g.*, [STOICA AND NEHORAI 1987] and Exercise 3.8), whereas the estimated LS model may be unstable. For applications in which one is interested in the AR model (and not just the AR spectral estimate), stability of the model is often an important requirement. It may, therefore, be thought that the potential instability of the AR model provided by the LS method is a significant drawback of this method. The case, however, is that estimated LS models which are unstable only appear infrequently and, moreover, when they do occur there are simple means to "stabilize" them (for instance, by reflecting the unstable poles inside the unit circle). Hence, to conclude this point, the lack of guaranteed stability is a drawback of the LS method, when compared with the Yule–Walker method, but often not a serious one.

Second, the LS model has been found to be more accurate than the Yule–Walker model in the sense that the estimated parameters of the former are on the average closer to the true values than those of the latter [MARPLE 1987; KAY 1988]. Since the finite–sample statistical analysis of these methods is underdeveloped, a theoretical explanation of this behavior is not possible at this time. Only heuristic explanations are available. One such explanation is that the assumption that $y(t) = 0$ outside the interval $1 \leq t \leq N$, and the corresponding zero elements in Y and y, result in bias in the Yule–Walker estimates of the AR parameters. When N is not much greater than n, this bias can be significant.

3.5 Order–Recursive Solutions to the Yule–Walker Equations

In applications, where *a priori* information about the true order n is usually lacking, AR models with different orders have to be tested and hence the Yule–Walker system of equations, (3.4.6), has to be solved for $n = 1$ up to $n = n_{\max}$ (some prespecified maximum order). By using a general solving method, this task requires $O(n_{\max}^4)$ flops. This may be a significant computational burden if $n_{\max}$ is large. This is, for example, the case in the applications dealing with narrowband signals, where values of 50 or even 100 for $n_{\max}$ are not uncommon. In such applications, it may be important to reduce the number of flops required to determine $\{\theta_n, \sigma_n^2\}$ in (3.4.6). In order to be able to do so, the special algebraic structure of (3.4.6) should be exploited, as explained next.

The matrix R_{n+1} in the Yule–Walker system of equations is highly structured: it is *Hermitian* and *Toeplitz.* The first algorithm which exploited this fact to determine $\{\theta_n, \sigma_n^2\}_{n=1}^{n_{\max}}$ in $n_{\max}^2$ flops was the *Levinson–Durbin algorithm* (LDA) [LEVINSON 1947; DURBIN 1960]. The number of flops required by the LDA is on the order of $n_{\max}$ times smaller than that required by a general linear equation solver to determine $(\theta_{n_{\max}}, \sigma_{n_{\max}}^2)$, and on the order of $n_{\max}^2$ times smaller than that required by a general linear equation solver to determine $\{\theta_n, \sigma_n^2\}_{n=1}^{n_{\max}}$. The LDA is discussed in Section 3.5.1. In Section 3.5.2 we present another algorithm, the *Delsarte–Genin algorithm* (DGA), also named the *split–Levinson algorithm,* which in the case of real–valued signals is about two times faster than the LDA [DELSARTE AND GENIN 1986].

Both the LDA and DGA solve, recursively in the order n, equation (3.4.6). The only requirement is that the matrix there be positive definite, Hermitian, and Toeplitz. Thus, the algorithms apply equally well to the Yule–Walker AR estimator (or, equivalently, the autocorrelation least squares AR method), in which the "true" ACS elements are replaced by estimates. Hence, to cover both cases simultaneously, in the following:

$$\boxed{\rho_k \text{ is used to represent either } r(k) \text{ or } \hat{r}(k)} \qquad (3.5.1)$$

By using the above convention, we have

$$R_{n+1} = \begin{bmatrix} \rho_0 & \rho_{-1} & \cdots & \rho_{-n} \\ \rho_1 & \rho_0 & & \vdots \\ \vdots & & \ddots & \rho_{-1} \\ \rho_n & \cdots & \rho_1 & \rho_0 \end{bmatrix} = \begin{bmatrix} \rho_0 & \rho_1^* & \cdots & \rho_n^* \\ \rho_1 & \rho_0 & & \vdots \\ \vdots & & \ddots & \rho_1^* \\ \rho_n & \cdots & \rho_1 & \rho_0 \end{bmatrix} \tag{3.5.2}$$

The following notational convention will also be frequently used in this section. For a vector $x = [x_1 \dots x_n]^T$, we define

$$\tilde{x} = [x_n^* \dots x_1^*]^T$$

An important property of any Hermitian Toeplitz matrix R is that

$$y = Rx \quad \Rightarrow \quad \tilde{y} = R\tilde{x} \tag{3.5.3}$$

The result (3.5.3) follows from the following calculation

$$\begin{aligned} \tilde{y}_i &= y_{n-i+1}^* = \sum_{k=1}^{n} R_{n-i+1,k}^* x_k^* \\ &= \sum_{k=1}^{n} \rho_{n-i+1-k}^* x_k^* = \sum_{p=1}^{n} \rho_{p-i}^* x_{n-p+1}^* = \sum_{p=1}^{n} R_{i,p}\tilde{x}_p \\ &= (R\tilde{x})_i \end{aligned}$$

where $R_{i,j}$ denotes the (i,j)th element of the matrix R.

3.5.1 Levinson–Durbin Algorithm

The basic idea of the LDA is to solve (3.4.6) *recursively in* n, starting from the solution for $n = 1$ (which is easily determined). By using (3.4.6) and the nested structure of the R matrix, we can write

$$R_{n+2}\begin{bmatrix} 1 \\ \theta_n \\ \hline 0 \end{bmatrix} = \left[\begin{array}{c|c} R_{n+1} & \begin{matrix}\rho_{n+1}^* \\ \tilde{r}_n\end{matrix} \\ \hline \begin{matrix}\rho_{n+1} & \tilde{r}_n^*\end{matrix} & \rho_0 \end{array}\right]\begin{bmatrix} 1 \\ \theta_n \\ \hline 0 \end{bmatrix} = \begin{bmatrix} \sigma_n^2 \\ 0 \\ \hline \alpha_n \end{bmatrix} \tag{3.5.4}$$

where

$$r_n = [\rho_1 \dots \rho_n]^T \tag{3.5.5}$$

$$\alpha_n = \rho_{n+1} + \tilde{r}_n^* \theta_n \tag{3.5.6}$$

Equation (3.5.4) would be the counterpart of (3.4.6) when n is increased by one, if α_n in (3.5.4) could be nulled. To do so, let

$$k_{n+1} = -\alpha_n / \sigma_n^2 \tag{3.5.7}$$

Table 3.1. The Levinson–Durbin Algorithm

Initialization:

$$\theta_1 = -\rho_1/\rho_0 = k_1 \qquad [1 \text{ flop}]$$

$$\sigma_1^2 = \rho_0 - |\rho_1|^2/\rho_0 \qquad [1 \text{ flop}]$$

For $n = 1, \ldots, n_{\max}$, do:

$$k_{n+1} = -\frac{\rho_{n+1} + \tilde{r}_n^* \theta_n}{\sigma_n^2} \qquad [n+1 \text{ flops}]$$

$$\sigma_{n+1}^2 = \sigma_n^2 (1 - |k_{n+1}|^2) \qquad [2 \text{ flops}]$$

$$\theta_{n+1} = \begin{bmatrix} \theta_n \\ 0 \end{bmatrix} + k_{n+1} \begin{bmatrix} \tilde{\theta}_n \\ 1 \end{bmatrix} \qquad [n \text{ flops}]$$

It follows from (3.5.3) and (3.5.4) that

$$R_{n+2} \left\{ \begin{bmatrix} 1 \\ \theta_n \\ 0 \end{bmatrix} + k_{n+1} \begin{bmatrix} 0 \\ \tilde{\theta}_n \\ 1 \end{bmatrix} \right\} = \begin{bmatrix} \sigma_n^2 \\ 0 \\ \alpha_n \end{bmatrix} + k_{n+1} \begin{bmatrix} \alpha_n^* \\ 0 \\ \sigma_n^2 \end{bmatrix} = \begin{bmatrix} \sigma_n^2 + k_{n+1}\alpha_n^* \\ 0 \end{bmatrix} \tag{3.5.8}$$

which has the same structure as

$$R_{n+2} \begin{bmatrix} 1 \\ \theta_{n+1} \end{bmatrix} = \begin{bmatrix} \sigma_{n+1}^2 \\ 0 \end{bmatrix} \tag{3.5.9}$$

Comparing (3.5.8) and (3.5.9) and making use of the fact that the solution to (3.4.6) is unique for any n, we reach the conclusion that

$$\theta_{n+1} = \begin{bmatrix} \theta_n \\ 0 \end{bmatrix} + k_{n+1} \begin{bmatrix} \tilde{\theta}_n \\ 1 \end{bmatrix} \tag{3.5.10}$$

and

$$\sigma_{n+1}^2 = \sigma_n^2 \left(1 - |k_{n+1}|^2\right) \tag{3.5.11}$$

constitute the solution to (3.4.6) for order $(n+1)$.

Equations (3.5.10) and (3.5.11) form the core of the LDA. The initialization of these recursive–in–n equations is straightforward. Table 3.1 summarizes the LDA in a form that should be convenient for machine coding. The LDA has many interesting properties and uses for which we refer to [SÖDERSTRÖM AND STOICA 1989; MARPLE 1987; KAY 1988]. The coefficients k_i in the LDA are often called the *reflection coefficients*; $-k_i$ are also called the *partial correlation (PARCOR) coefficients*. The motivation for the name "partial correlation coefficient" is developed in the Complement in Section 3.9.1

It can be seen from Table 3.1 that the LDA requires on the order of $2n$ flops to compute $\{\theta_{n+1}, \sigma_{n+1}^2\}$ from $\{\theta_n, \sigma_n^2\}$. Hence a total of about $n_{\max}^2$ flops is needed to compute all the solutions to the Yule–Walker system of equations, from $n = 1$ to $n = n_{\max}$. This confirms the claim that the LDA reduces the computational burden associated with a general solver by two orders of magnitude.

3.5.2 Delsarte–Genin Algorithm

In the *real data case* (*i.e.*, whenever $y(t)$ is real valued), the Delsarte–Genin algorithm (DGA), or the *split Levinson algorithm*, exploits some further structure of the Yule–Walker problem, which is not exploited by the LDA, to decrease even more the number of flops required to solve for $\{\theta_n, \sigma_n^2\}$ [DELSARTE AND GENIN 1986]. In the following, we present a derivation of the DGA which is simpler than the original derivation. As already stated, we assume that the covariance elements $\{\rho_k\}$ in the Yule–Walker equations are real valued.

Let Δ_n be defined by

$$R_{n+1}\Delta_n = \beta_n \begin{bmatrix} 1 \\ \vdots \\ 1 \end{bmatrix} \tag{3.5.12}$$

where the scalar β_n is unspecified for the moment. As the matrix R_{n+1} is positive definite, the $(n+1)$–vector Δ_n is uniquely defined by (3.5.12) (once β_n is specified; as a matter of fact, note that β_n only has a scaling effect on the components of Δ_n). It follows from (3.5.12) and (3.5.3) that Δ_n is a "symmetric vector", *i.e.*, it satisfies

$$\Delta_n = \tilde{\Delta}_n \tag{3.5.13}$$

The key idea of the DGA is to introduce such symmetric vectors into the computations involved by the LDA, as only half of the elements of these vectors need to be computed.

Next, note that by using the nested structure of R_{n+1} and the defining equation (3.5.12), we can write:

$$R_{n+1}\begin{bmatrix} 0 \\ \Delta_{n-1} \end{bmatrix} = \begin{bmatrix} \rho_0 & r_n^T \\ r_n & R_n \end{bmatrix}\begin{bmatrix} 0 \\ \Delta_{n-1} \end{bmatrix} = \begin{bmatrix} \gamma_{n-1} \\ \beta_{n-1} \\ \vdots \\ \beta_{n-1} \end{bmatrix} \tag{3.5.14}$$

where r_n is defined in (3.5.5) and

$$\gamma_{n-1} = r_n^T \Delta_{n-1} \tag{3.5.15}$$

The systems of equations (3.5.12) and (3.5.14) can be linearly combined into a system having the structure of (3.4.6). To do so, let

$$\lambda_n = \beta_n / \beta_{n-1} \tag{3.5.16}$$

Then, from (3.5.12), (3.5.14) and (3.5.16), we get

$$R_{n+1} \left\{ \Delta_n - \lambda_n \begin{bmatrix} 0 \\ \Delta_{n-1} \end{bmatrix} \right\} = \begin{bmatrix} \beta_n - \lambda_n \gamma_{n-1} \\ 0 \end{bmatrix} \tag{3.5.17}$$

It will be shown that β_n can always be chosen so as to make the first element of Δ_n equal to one,

$$(\Delta_n)_1 = 1 \tag{3.5.18}$$

In such a case, (3.5.17) has exactly the same structure as (3.4.6) and, as the solutions to these two systems of equations are unique, we are led to the following relations:

$$\begin{bmatrix} 1 \\ \theta_n \end{bmatrix} = \Delta_n - \lambda_n \begin{bmatrix} 0 \\ \Delta_{n-1} \end{bmatrix} \tag{3.5.19}$$

$$\sigma_n^2 = \beta_n - \lambda_n \gamma_{n-1} \tag{3.5.20}$$

Furthermore, since $(\Delta_n)_1 = 1$ and Δ_n is a symmetric vector, we must also have $(\Delta_n)_{n+1} = 1$. This observation, along with (3.5.19) and the fact that k_n is the last element of θ_n (see (3.5.10)), gives the following expression for k_n:

$$k_n = 1 - \lambda_n \tag{3.5.21}$$

The equations (3.5.19)–(3.5.21) express the LDA variables $\{\theta_n, \sigma_n^2, k_n\}$ as functions of $\{\Delta_n\}$ and $\{\beta_n\}$. It remains to derive recursive–in–n formulas for $\{\Delta_n\}$ and $\{\beta_n\}$, and also to prove that (3.5.18) really holds. This is done in the following.

Let $\{\beta_n\}$ be defined recursively by the following second–order difference equation:

$$\beta_n = 2\beta_{n-1} - \alpha_n \beta_{n-2} \tag{3.5.22}$$

where

$$\alpha_n = (\beta_{n-1} - \gamma_{n-1}) / (\beta_{n-2} - \gamma_{n-2}) \tag{3.5.23}$$

The initial values required to start the recursion (3.5.22) are: $\beta_0 = \rho_0$ and $\beta_1 = \rho_0 + \rho_1$. With this definition of $\{\beta_n\}$, we claim that the vectors $\{\Delta_n\}$ (as defined in (3.5.12)) satisfy (3.5.18) as well as the following second–order recursion:

$$\Delta_n = \begin{bmatrix} \Delta_{n-1} \\ 0 \end{bmatrix} + \begin{bmatrix} 0 \\ \Delta_{n-1} \end{bmatrix} - \alpha_n \begin{bmatrix} 0 \\ \Delta_{n-2} \\ 0 \end{bmatrix} \tag{3.5.24}$$

In order to prove the above claim, we first apply the result (3.5.3) to (3.5.14) to get

$$R_{n+1} \begin{bmatrix} \Delta_{n-1} \\ 0 \end{bmatrix} = \begin{bmatrix} \beta_{n-1} \\ \vdots \\ \beta_{n-1} \\ \gamma_{n-1} \end{bmatrix} \tag{3.5.25}$$

Next, we note that

$$R_{n+1}\begin{bmatrix}0\\ \Delta_{n-2}\\ 0\end{bmatrix} = \begin{bmatrix}\rho_0 & r_{n-1}^T & \rho_n\\ r_{n-1} & R_{n-1} & \tilde{r}_{n-1}\\ \rho_n & \tilde{r}_{n-1}^T & \rho_0\end{bmatrix}\begin{bmatrix}0\\ \Delta_{n-2}\\ 0\end{bmatrix} = \begin{bmatrix}\gamma_{n-2}\\ \beta_{n-2}\\ \vdots\\ \beta_{n-2}\\ \gamma_{n-2}\end{bmatrix} \tag{3.5.26}$$

The right–hand sides of equations (3.5.14), (3.5.25) and (3.5.26) can be linearly combined, as described below, to get the right–hand side of (3.5.12):

$$\begin{bmatrix}\gamma_{n-1}\\ \beta_{n-1}\\ \vdots\\ \beta_{n-1}\end{bmatrix} + \begin{bmatrix}\beta_{n-1}\\ \vdots\\ \beta_{n-1}\\ \gamma_{n-1}\end{bmatrix} - \alpha_n\begin{bmatrix}\gamma_{n-2}\\ \beta_{n-2}\\ \vdots\\ \beta_{n-2}\\ \gamma_{n-2}\end{bmatrix} = \beta_n\begin{bmatrix}1\\ \vdots\\ 1\end{bmatrix} \tag{3.5.27}$$

The equality in (3.5.27) follows from the defining equations of β_n and α_n. This observation, in conjunction with (3.5.14), (3.5.25) and (3.5.26), gives the following system of linear equations

$$R_{n+1}\left\{\begin{bmatrix}\Delta_{n-1}\\ 0\end{bmatrix} + \begin{bmatrix}0\\ \Delta_{n-1}\end{bmatrix} - \alpha_n\begin{bmatrix}0\\ \Delta_{n-2}\\ 0\end{bmatrix}\right\} = \beta_n\begin{bmatrix}1\\ \vdots\\ 1\end{bmatrix} \tag{3.5.28}$$

which has exactly the structure of (3.5.12). Since the solutions to (3.5.12) and (3.5.28) are unique, they must coincide and hence (3.5.24) follows.

Next, turn to the condition (3.5.18). From (3.5.24) we see that $(\Delta_n)_1 = (\Delta_{n-1})_1$. Hence, in order to prove that (3.5.18) holds, it suffices to show that $\Delta_1 = [1\ \ 1]^T$. The initial values $\beta_0 = \rho_0$ and $\beta_1 = \rho_0+\rho_1$ (purposely chosen for the sequence $\{\beta_n\}$), when inserted in (3.5.12), give $\Delta_0 = 1$ and $\Delta_1 = [1\ \ 1]^T$. With this observation, the proof of (3.5.18) and (3.5.24) is finished.

The DGA consists of the equations (3.5.16) and (3.5.19)–(3.5.24). These equations include second–order recursions and appear to be more complicated than the first–order recursive equations of the LDA. In reality, owing to the *symmetry of the* Δ_n *vectors*, the DGA is computationally more efficient than the LDA (see below). The DGA equations are summarized in Table 3.2, along with an approximate count of the number of flops required for implementation.

The DGA can be implemented in two principal modes, depending on the application at hand.

DGA — Mode 1: In most AR modeling exercises, we do not really need all $\{\theta_n\}_{n=1}^{n_{\max}}$. We do, however, need $\{\sigma_1^2, \sigma_2^2, \ldots\}$ for the purpose of order determination. Assume that we determined the AR order on the basis of the σ^2 sequence.

Table 3.2. The Delsarte–Genin Algorithm

DGA equations	Operation count no. of (×)	no. of (+)
Initialization:		
$\Delta_0 = 1,\ \beta_0 = \rho_0,\ \gamma_0 = \rho_1$	–	–
$\Delta_1 = [1\ \ 1]^T,\ \beta_1 = \rho_0 + \rho_1,\ \gamma_1 = \rho_1 + \rho_2$	–	2
For $n = 2, \ldots, n_{\max}$ do:		
A. $\alpha_n = (\beta_{n-1} - \gamma_{n-1})/(\beta_{n-2} - \gamma_{n-2})$	1	2
$\beta_n = 2\beta_{n-1} - \alpha_n \beta_{n-2}$	2	1
$\Delta_n = \begin{bmatrix} \Delta_{n-1} \\ 0 \end{bmatrix} + \begin{bmatrix} 0 \\ \Delta_{n-1} \end{bmatrix} - \alpha_n \begin{bmatrix} 0 \\ \Delta_{n-2} \\ 0 \end{bmatrix}$	$\sim n/2$	$\sim n$
$\gamma_n = r_{n+1}^T \Delta_n = (\rho_1 + \rho_{n+1}) + \Delta_{n,2}(\rho_2 + \rho_n) + \ldots$	$\sim n/2$	$\sim n$
B. $\lambda_n = \beta_n / \beta_{n-1}$	1	–
$\sigma_n^2 = \beta_n - \lambda_n \gamma_{n-1}$	1	1
$k_n = 1 - \lambda_n$	–	1
C. $\begin{bmatrix} 1 \\ \theta_n \end{bmatrix} = \Delta_n - \lambda_n \begin{bmatrix} 0 \\ \Delta_{n-1} \end{bmatrix}$	$\sim n/2$	$\sim n$

For simplicity, let this order be denoted by $n_{\max}$. Then the only θ vector to be computed is $\theta_{n_{\max}}$. We may also need to compute the $\{k_n\}$ sequence since this bears useful information about the stability of the determined AR model (see, *e.g.*, [SÖDERSTRÖM AND STOICA 1989; KAY 1988; THERRIEN 1992]).

In the modeling application outlined above, we need to iterate only the groups A and B of equations in the previous DGA summary. The matrix equation C is computed only for $n = n_{\max}$. This way of implementing the DGA requires the following number of multiplications and additions:

$$\text{no. of } (\times) \simeq n_{\max}^2/2 \qquad \text{no. of } (+) \simeq n_{\max}^2 \tag{3.5.29}$$

Recall that, for LDA, no. of $(\times)$ = no. of $(+) \simeq n_{\max}^2$. Thus, the DGA is approximately *two times faster* than the LDA (on computers for which multiplication is much more time consuming than addition). We may remark that in some parameter estimation applications, the equations in group B of the DGA can also be left out, but this will speed up the implementation of the DGA only slightly.

DGA — Mode 2: In other applications, we need all $\{\theta_n\}_{n=1}^{n_{\max}}$. An example of such an application is the Cholesky factorization of the inverse covariance matrix $R_{n_{\max}}^{-1}$ (see, *e.g.*, Exercise 3.7 and [SÖDERSTRÖM AND STOICA 1989]). In such a case, we need to iterate all equations in the DGA, which results in the following number of arithmetic operations:

$$\text{no. of } (\times) \simeq 0.75n_{\max}^2 \qquad \text{no. of } (+) \simeq 1.5n_{\max}^2 \tag{3.5.30}$$

This is still about 25% faster than the LDA (assuming, once again, that the computation time required for multiplication dominates the time corresponding to an addition).

In closing this section, we note that the computational comparisons between the DGA and the LDA above neglected terms on the order $O(n_{\max})$. This is acceptable if $n_{\max}$ is reasonably large (say, $n_{\max} \geq 10$). If $n_{\max}$ is small, then these comparisons are no longer valid and, in fact, LDA may be computationally more efficient than the DGA in such a case. In such low–dimensional applications, the LDA is therefore to be preferred to the DGA. Also recall that the LDA is the algorithm to use with complex–valued data, since the DGA does not appear to have a computationally efficient extension for complex–valued data.

3.6 MA Signals

According to the definition in (3.2.8), an *MA signal* is obtained by filtering white noise with an *all–zero filter*. Owing to this all–zero structure, it is not possible to use an MA equation to model a spectrum with sharp peaks unless the MA order is chosen "sufficiently large". This is to be contrasted to the ability of the AR (or "all–pole") equation to model narrowband spectra by using fairly low model orders (*cf.* the discussion in the previous sections). The MA model provides a good approximation for those spectra which are characterized by broad peaks and sharp nulls. Such spectra are encountered less frequently in applications than narrowband spectra, so there is a somewhat limited engineering interest in using the MA signal model for spectral estimation. Another reason for this limited interest is that the MA parameter estimation problem is basically a nonlinear one, and is significantly more difficult to solve than the AR parameter estimation problem. In any case, the types of difficulties we must face in MA and ARMA estimation problems are quite similar, and hence we may almost always prefer to use the more general ARMA model in lieu of the MA one. For these reasons, our discussion of the MA spectral estimation will be brief.

One method to estimate an MA spectrum consists of two steps: (i) Estimate the MA parameters $\{b_k\}_{k=1}^m$ and σ^2; and (ii) Insert the estimated parameters from the first step in the MA PSD formula (see (3.2.2)):

$$\phi(\omega) = \sigma^2 |B(\omega)|^2 \tag{3.6.1}$$

The difficulty with this approach lies in step (i) which is a nonlinear estimation problem. Approximate linear solutions to this problem do, however, exist. One of these approximate procedures, perhaps the most used method for MA parameter estimation, is based on a two–stage least squares methodology [DURBIN 1959]. It is called *Durbin's method* and will be described in the next section in the more general context of ARMA parameter estimation.

Another method to estimate an MA spectrum is based on the reparameterization of the PSD in terms of the covariance sequence. We see from (3.2.8) that for an MA of order m,

$$\boxed{r(k) = 0 \qquad \text{for } |k| > m} \tag{3.6.2}$$

Owing to this simple observation, the definition of the PSD as a function of $\{r(k)\}$ turns into a finite–dimensional spectral model:

$$\phi(\omega) = \sum_{k=-m}^{m} r(k)e^{-i\omega k} \tag{3.6.3}$$

Hence a simple estimator of MA PSD is obtained by inserting estimates of $\{r(k)\}_{k=0}^{m}$ in (3.6.3). If the standard sample covariances $\{\hat{r}(k)\}$ are used to estimate $\{r(k)\}$, then we obtain:

$$\boxed{\hat{\phi}(\omega) = \sum_{k=-m}^{m} \hat{r}(k)e^{-i\omega k}} \tag{3.6.4}$$

This spectral estimate is of the form of the Blackman–Tukey estimator (2.5.1). More precisely, (3.6.4) coincides with a Blackman–Tukey estimator using a rectangular window of length $2m+1$. This was not unexpected. If we impose the zero–bias restriction on the nonparametric approach to spectral estimation (to make the comparison with the parametric approach fair) then the Blackman–Tukey estimator with a rectangular window of length $2m+1$ implicitly assumes that the covariance lags outside the window interval are equal to zero. This is, however, precisely the assumption behind the MA signal model; see (3.6.2). Alternatively, if we make use of the assumption (3.6.2) in a Blackman–Tukey estimator, then we definitely end up with (3.6.4) as in such a case this is the spectral estimator in the Blackman–Tukey class with zero bias and "minimum" variance.

The analogy between the Blackman–Tukey and MA spectrum estimation methods makes it simpler to understand a problem associated with the MA spectral estimator (3.6.4). Owing to the (implicit) use of a rectangular window in (3.6.4), the so–obtained spectral estimate is not necessarily positive at all frequencies (see (2.5.5) and the discussion following that equation). Indeed, it was often noted in applications that (3.6.4) produces negative PSD estimates. In order to cure this deficiency of (3.6.4), we may use another lag window which is guaranteed to be positive semidefinite, in lieu of the rectangular one. This way of correcting $\hat{\phi}(\omega)$ in (3.6.4) is, of course, reminiscent of the Blackman–Tukey approach. It should be noted, however, that the so–corrected $\hat{\phi}(\omega)$ is no longer an unbiased estimator of the PSD of an MA(m) signal (see, *e.g.*, [Moses and Beex 1986] for details on this aspect).

3.7 ARMA Signals

Spectra with both sharp peaks and deep nulls cannot be modeled by either AR or MA equations of reasonably small orders. There are, of course, other instances of rational spectra that cannot be exactly described as AR or MA spectra. It is in these cases where the more general *ARMA model*, also called the *pole–zero model*, is valuable. However, the great initial promise of ARMA spectral estimation diminishes to some extent because there is yet no well–established algorithm, from both theoretical and practical standpoints, for ARMA parameter estimation. The "theoretically optimal ARMA estimators" are based on iterative procedures whose global convergence is not guaranteed. The "practical ARMA estimators", on the other hand, are computationally simple and often quite reliable, but their statistical accuracy may be poor in some cases. In the following, we describe two ARMA spectral estimation algorithms which have been used in applications with a reasonable degree of success.

3.7.1 Modified Yule–Walker Method

The modified Yule–Walker method is a two–stage procedure for estimating the ARMA spectral density. In the first stage we estimate the AR coefficients using equation (3.3.4). In the second stage, we use the AR coefficient and ACS estimates in equation (3.2.1) to estimate the γ_k coefficients. We describe the two steps below.

Writing equation (3.3.4) for $k = m+1, m+2, \ldots, m+M$ in a matrix form gives

$$\begin{bmatrix} r(m) & r(m-1) & \cdots & r(m-n+1) \\ r(m+1) & r(m) & & r(m-n+2) \\ \vdots & & \ddots & \vdots \\ r(m+M-1) & \cdots & \cdots & r(m-n+M) \end{bmatrix} \begin{bmatrix} a_1 \\ \vdots \\ a_n \end{bmatrix} = - \begin{bmatrix} r(m+1) \\ r(m+2) \\ \vdots \\ r(m+M) \end{bmatrix} \tag{3.7.1}$$

If we set $M = n$ in (3.7.1) we obtain a system of n equations in n unknowns. This constitutes a generalization of the Yule–Walker system of equations that holds in the AR case. Hence, these equations are said to form the *modified Yule–Walker* (MYW) system of equations [GERSH 1970; KINKEL, PERL, SCHARF, AND STUBBERUD 1979; BEEX AND SCHARF 1981; CADZOW 1982]. Replacing the theoretical covariances $\{r(k)\}$ by their sample estimates $\{\hat{r}(k)\}$ in these equations leads to:

$$\boxed{\begin{bmatrix} \hat{r}(m) & \cdots & \hat{r}(m-n+1) \\ \vdots & & \vdots \\ \hat{r}(m+n-1) & \cdots & \hat{r}(m) \end{bmatrix} \begin{bmatrix} \hat{a}_1 \\ \vdots \\ \hat{a}_n \end{bmatrix} = - \begin{bmatrix} \hat{r}(m+1) \\ \vdots \\ \hat{r}(m+n) \end{bmatrix}} \tag{3.7.2}$$

The above linear system can be solved for $\{\hat{a}_i\}$, which are called the *modified Yule–Walker estimates* of $\{a_i\}$. The square matrix in (3.7.2) can be shown to be nonsingular under mild conditions. Note that there exist fast algorithms of the Levinson type for solving *non–Hermitian* Toeplitz systems of equations of the form of (3.7.2); they require about twice the computational burden of the LDA algorithm (see [MARPLE 1987; KAY 1988; SÖDERSTRÖM AND STOICA 1989]).

The MYW AR estimate has reasonable accuracy if the zeroes of $B(z)$ in the ARMA model are well inside the unit circle. However, (3.7.2) may give very inaccurate estimates in those cases where the poles and zeroes of the ARMA model description are closely spaced together at positions near the unit circle. Such ARMA models, with nearly coinciding poles and zeroes of modulus close to one, correspond to narrowband signals. The covariance sequence of narrowband signals decays very slowly. Indeed, as we know, the more concentrated a signal is in frequency, usually the more expanded it is in time, and vice versa. This means that there is "information" in the higher–lag covariances of the signal that can be exploited to improve the accuracy of the AR coefficient estimates. We can exploit the additional information by choosing $M > n$ in equation (3.7.1) and solving the so–obtained overdetermined system of equations. If we replace the true covariances in (3.7.1) with $M > n$ by finite–sample estimates, there will in general be no exact solution. A most natural idea to overcome this problem is to solve the resultant equations

$$\hat{R}\hat{a} \simeq -\hat{r} \tag{3.7.3}$$

in a least squares (LS) or total least squares (TLS) sense (see Appendix A). Here, $\hat{R}$ and $\hat{r}$ represent the ACS matrix and vector in (3.7.1) with sample ACS estimates replacing the true ACS there. For instance, the (weighted) least squares solution to (3.7.3) is mathematically given by[1]

$$\hat{a} = -(\hat{R}^* W \hat{R})^{-1}(\hat{R}^* W \hat{r}) \tag{3.7.4}$$

where W is an $M \times M$ positive definite weighting matrix. The AR estimate derived from (3.7.3) with $M > n$ is called the *overdetermined modified YW estimate* [BEEX AND SCHARF 1981; CADZOW 1982].

Some notes on the choice between (3.7.2) and (3.7.3), and on the selection of M, are in order.

1. Choosing $M > n$ does not always improve the accuracy of the previous AR coefficient estimates. In fact, if the poles and zeroes are not close to the unit circle, choosing $M > n$ can make the accuracy *worse*. When the ACS decays slowly to zero, however, choosing $M > n$ generally improves the accuracy of $\hat{a}$ [CADZOW 1982; STOICA, FRIEDLANDER, AND SÖDERSTRÖM 1987B]. A qualitative explanation for this phenomenon can be seen by thinking of a finite–sample ACS estimate as being the sum of its "signal" component $r(k)$ and a "noise" component due to finite–sample estimation: $\hat{r}(k) = r(k) + n(k)$. If the ACS decays slowly to zero, the signal component is "large" compared to the noise component even for relatively large values of k, and including

[1]From a numerical viewpoint, equation (3.7.4) is not a particularly good way to solve (3.7.3). A more numerically sound approach is to use the QR decomposition; see Section A.8.2 for details.

$\hat{r}(k)$ in the estimation of $\hat{a}$ improves accuracy. If the noise component of $\hat{r}(k)$ dominates, including $\hat{r}(k)$ in the estimation of $\hat{a}$ may decrease the accuracy of $\hat{a}$.

2. The *statistical and numerical accuracies* of the solution $\{\hat{a}_i\}$ to (3.7.3) are quite interrelated. In more exact but still loose terms, it can be shown that the statistical accuracy of $\{\hat{a}_i\}$ is poor (good) if the condition number of the matrix $\hat{R}$ in (3.7.3) is large (small) (see [STOICA, FRIEDLANDER, AND SÖDERSTRÖM 1987B; SÖDERSTRÖM AND STOICA 1989] and also Appendix A). This observation suggests that M should be selected so as to make the matrix in (3.7.3) reasonably well–conditioned. In order to make a connection between this rule of thumb for selecting M and the previous explanation for the poor accuracy of (3.7.2) in the case of narrowband signals, note that for slowly decaying covariance sequences the columns of the matrix in (3.7.2) are nearly linearly dependent. Hence, the condition number of the covariance matrix may be quite high in such a case, and we may need to increase M in order to lower the condition number to a reasonable value.

3. The weighting matrix W in (3.7.4) can also be chosen to improve the accuracy of the AR coefficient estimates. A simple first choice is $W = I$, resulting in the regular (unweighted) least squares estimate. Some accuracy improvement can be obtained by choosing W to be diagonal with decreasing positive diagonal elements (to reflect the decreased confidence in higher ACS lag estimates). In addition, optimal weighting matrices have been derived (see [STOICA, FRIEDLANDER, AND SÖDERSTRÖM 1987A]); the optimal weight minimizes the covariance of $\hat{a}$ (for large N) over all choices of W. Unfortunately, the optimal weight depends on the (unknown) ARMA parameters. Thus, to use optimally weighted methods, a two–step "bootstrap" approach is used, in which a fixed W is first chosen and initial parameter estimates are obtained; these initial estimates are used to form an optimal W, and a second estimation gives the "optimal accuracy" AR coefficients. As a general rule, the performance gain in using optimal weighting is relatively small compared to the computational overhead required to compute the optimal weighting matrix. Most accuracy improvement can be realized by choosing $M > n$ and $W = I$ for many problems. We refer the reader to [STOICA, FRIEDLANDER, AND SÖDERSTRÖM 1987A; CADZOW 1982] for a discussion on the effect of W on the accuracy of $\hat{a}$ and on optimal weighting matrices.

Once the AR estimates are obtained, we turn to the problem of estimating the MA part of the ARMA spectrum. Let

$$\gamma_k = E\left\{[B(q)e(t)][B(q)e(t-k)]^*\right\} \tag{3.7.5}$$

denote the covariances of the MA part. Since the PSD of this part of the ARMA signal model is given by (see (3.6.1) and (3.6.3)):

$$\sigma^2|B(\omega)|^2 = \sum_{k=-m}^{m} \gamma_k e^{-i\omega k} \tag{3.7.6}$$

it suffices to estimate $\{\gamma_k\}$ in order to characterize the spectrum of the MA part. From (3.2.7) and (3.7.5), we obtain

$$\begin{aligned} \gamma_k &= E\{[A(q)y(t)][A(q)y(t-k)]^*\} \\ &= \sum_{j=0}^{n}\sum_{p=0}^{n} a_j a_p^* E\{y(t-j)y^*(t-k-p)\} \\ &= \sum_{j=0}^{n}\sum_{p=0}^{n} a_j a_p^* r(k+p-j) \qquad (a_0 \triangleq 1) \end{aligned} \tag{3.7.7}$$

for $k = 0, \ldots, m$. Inserting the previously calculated estimates of $\{a_k\}$ and $\{r_k\}$ in (3.7.7) leads to the following estimator of $\{\gamma_k\}$

$$\hat{\gamma}_k = \begin{cases} \sum_{j=0}^{n}\sum_{p=0}^{n} \hat{a}_j \hat{a}_p^* \hat{r}(k+p-j), & k = 0, \ldots, m \ (\hat{a}_0 \triangleq 1) \\ \hat{\gamma}_{-k}^*, & k = -1, \ldots, -m \end{cases} \tag{3.7.8}$$

Finally, the ARMA spectrum is estimated as follows:

$$\hat{\phi}(\omega) = \frac{\sum_{k=-m}^{m} \hat{\gamma}_k e^{-i\omega k}}{|\hat{A}(\omega)|^2} \tag{3.7.9}$$

The *MA estimate* used by the above ARMA spectral estimator is of the type (3.6.4) encountered in the MA context. Hence, the criticism of (3.6.4) in the previous section is still valid. In particular, the numerator in (3.7.9) is not guaranteed to be positive for all ω values, which may lead to negative ARMA spectral estimates (see, *e.g.*, [KINKEL, PERL, SCHARF, AND STUBBERUD 1979; MOSES AND BEEX 1986]).

Since (3.7.9) relies on the modified YW method of AR parameter estimation, we call (3.7.9) the *modified YW ARMA spectral estimator*. Refined versions of this ARMA spectral estimator, which improve the estimation accuracy if N is sufficiently large, were proposed in [STOICA AND NEHORAI 1986; STOICA, FRIEDLANDER, AND SÖDERSTRÖM 1987A; MOSES, ŠIMONYTĖ, STOICA, AND SÖDERSTRÖM 1994]. A related ARMA spectral estimation method is outlined in Exercise 3.13

3.7.2 Two–Stage Least Squares Method

If the noise sequence $\{e(t)\}$ were known, then the problem of estimating the parameters in the ARMA model (3.2.7) would have been a simple *input–output system identification* problem which could be solved by a diversity of means of which the most simple is the *least squares (LS) method.* In the LS method, we express equation (3.2.7) as

$$y(t) + \varphi^T(t)\theta = e(t) \tag{3.7.10}$$

where

$$\begin{aligned} \varphi^T(t) &= [y(t-1), \ldots, y(t-n) | -e(t-1), \ldots, -e(t-m)] \\ \theta &= [a_1, \ldots, a_n | b_1, \ldots, b_m]^T \end{aligned}$$

Writing (3.7.10) in matrix form for $t = L+1, \ldots, N$ (for some $L > \max(m,n)$) gives

$$z + Z\theta = e \tag{3.7.11}$$

where

$$Z = \left[\begin{array}{ccc|ccc} y(L) & \ldots & y(L-n+1) & -e(L) & \ldots & -e(L-m+1) \\ y(L+1) & \ldots & y(L-n+2) & -e(L+1) & \ldots & -e(L-m+2) \\ \vdots & & \vdots & \vdots & & \vdots \\ y(N-1) & \ldots & y(N-n) & -e(N-1) & \ldots & -e(N-m) \end{array} \right] \tag{3.7.12}$$

$$z = [y(L+1), y(L+2), \ldots, y(N)]^T \tag{3.7.13}$$

$$e = [e(L+1), e(L+2), \ldots, e(N)]^T \tag{3.7.14}$$

Assume we know Z; then we could solve for θ in (3.7.11) by minimizing $\|e\|^2$. This leads to a least squares estimate similar to the AR LS estimate introduced in Section 3.4.2 (see also Result R32 in Appendix A):

$$\hat{\theta} = -(Z^*Z)^{-1}(Z^*z) \tag{3.7.15}$$

Of course, the $\{e(t)\}$ in Z are not known. However, they may be estimated as described next.

Since the ARMA model (3.2.7) is *minimum phase*, by assumption, it can alternatively be written as an infinite–order AR equation:

$$(1 + \alpha_1 q^{-1} + \alpha_2 q^{-2} + \ldots) y(t) = e(t) \tag{3.7.16}$$

where the coefficients $\{\alpha_k\}$ of $1 + \alpha_1 q^{-1} + \alpha_2 q^{-2} + \cdots \triangleq A(q)/B(q)$ converge to zero as k increases. An idea to estimate $\{e(t)\}$ is to first determine the AR parameters $\{\alpha_k\}$ in (3.7.16) and next obtain $\{e(t)\}$ by filtering $\{y(t)\}$ as in (3.7.16). Of course, we cannot estimate an infinite number of (independent) parameters from a finite number of samples. In practice, the AR equation must be approximated by one of order K (say). The parameters in the *truncated AR model* of $y(t)$ can be estimated by using either the YW or the LS procedure in Section 3.4.

Table 3.3. Two–Stage Least Squares ARMA Method

Step 1. Estimate the parameters $\{\alpha_k\}$ in an AR(K) model of $y(t)$ by the YW or covariance LS method. Let $\{\hat{\alpha}_k\}_{k=1}^{K}$ denote the estimated parameters.

Obtain an estimate of the noise sequence $\{e(t)\}$ by

$$\hat{e}(t) = y(t) + \sum_{k=1}^{K} \hat{\alpha}_k y(t-k)$$

for $t = K+1, \ldots, N$.

Step 2. Replace $e(t)$ in (3.7.12) by $\hat{e}(t)$ determined in Step 1. Obtain $\hat{\theta}$ from (3.7.15) with $L = K + m$. Estimate

$$\hat{\sigma}^2 = \frac{1}{N-L}\tilde{e}^*\tilde{e}$$

where $\tilde{e} = Z\hat{\theta} + z$ is the LS error from (3.7.11).

Insert $\{\hat{\theta}, \hat{\sigma}^2\}$ into the PSD expression (3.2.2) to estimate the ARMA spectrum.

The above discussion leads to the two–stage LS algorithm summarized in Table 3.3. The two–stage LS parameter estimator is also discussed, for example, in [Mayne and Firoozan 1982; Söderström and Stoica 1989]. The spectral estimate is guaranteed to be positive for all frequencies by construction. Owing to the practical requirement to truncate the AR model (3.7.16), the two–stage LS estimate is biased. The bias can be made small by choosing K sufficiently large; however, K should not be too large with respect to N or the accuracy of $\hat{\theta}$ in Step 2 will decrease. The difficult case for this method is apparently that of ARMA signals with *zeroes close to the unit circle.* In such a case, it may be necessary to select a very large value of K in order to keep the approximation (bias) errors in Step 1 at a reasonable level. The computational burden of Step 1 may then become prohibitively large. It should be noted, however, that the case of ARMA signals with zeroes near the unit circle is a difficult one for all known ARMA estimation methods [Kay 1988; Marple 1987; Söderström and Stoica 1989].

Finally, we remark that the two–stage LS algorithm may be modified to estimate the parameters in MA models, simply by skipping over the estimation of AR parameters in Step 2. The so–obtained method was for the first time suggested in [Durbin 1959], and is often called *Durbin's Method.*

3.8 Multivariate ARMA Signals

The multivariate analog of the ARMA signal in equation (3.2.7) is:

$$A(q)y(t) = B(q)e(t) \tag{3.8.1}$$

where $y(t)$ and $e(t)$ are $ny \times 1$ vectors, and $A(q)$ and $B(q)$ are $ny \times ny$ matrix polynomials in the unit delay operator. The task of estimating the matrix coefficients, $\{A_i, B_j\}$ say, of the AR and MA polynomials in (3.8.1) is much more complicated than in the scalar case for at least one reason: The representation of $y(t)$ in (3.8.1), with all elements in $\{A_i, B_j\}$ assumed to be unknown, may well be *nonunique* even though the orders of $A(q)$ and $B(q)$ may have been chosen correctly. More precisely, assume that we are given the spectral density matrix of an ARMA signal $y(t)$ along with the (minimal) orders of the AR and MA polynomials in its ARMA equation. If all elements of $\{A_i, B_j\}$ are considered to be unknown, then, *unlike in the scalar case*, the previous information may not be sufficient to determine the matrix coefficients $\{A_i, B_j\}$ uniquely (see, *e.g.*, [HANNAN AND DEISTLER 1988] and also Exercise 3.15). The lack of uniqueness of the representation may lead to a *numerically ill-conditioned parameter estimation* problem. For instance, this would be the case with the multivariate analog of the *modified Yule–Walker* method discussed in Section 3.7.1.

Apparently the only possible cure to the aforementioned problem consists of using a *canonical parameterization* for the AR and MA coefficients. Basically this amounts to setting some of the elements of $\{A_i, B_j\}$ to known values, such as 0 or 1, hence reducing the number of unknowns. The problem, however, is that to know which elements should be set to 0 or 1 in a specific case, we need to know *ny indices* (called "structure indices") which are usually difficult to determine in practice [KAILATH 1980; HANNAN AND DEISTLER 1988]. The difficulty in obtaining those indices has hampered the use of canonical parameterizations in applications. For this reason we do not go into any detail of the canonical forms for ARMA signals. The nonuniqueness of the fully parameterized ARMA equation will, however, receive further attention in the next subsection.

Concerning the other approach to ARMA parameter estimation discussed in Section 3.7.2, namely *the two–stage least squares method*, it is worth noting that it *can be extended to the multivariate case in a straightforward manner.* In particular there is *no* need for using a canonical parameterization in either step of the extended method (see, *e.g.*, [SÖDERSTRÖM AND STOICA 1989]). Working the details of the extension is left as an interesting exercise to the reader. We stress that the two–stage LS approach is perhaps the only real competitor to the subspace ARMA parameter estimation method described in the next subsections.

3.8.1 ARMA State–Space Equations

The difference equation representation in (3.8.1) can be transformed into the following *state–space representation*, and vice versa (see, *e.g.*, [AOKI 1987; KAILATH

1980]):

$$\boxed{\begin{aligned} x(t+1) &= Ax(t) + Be(t) \qquad (n \times 1) \\ y(t) &= Cx(t) + e(t) \qquad (ny \times 1) \end{aligned}} \tag{3.8.2}$$

Thereafter, $x(t)$ is the state vector of dimension n; A, B, and C are matrices of appropriate dimensions (with A having all eigenvalues inside the unit circle); and $e(t)$ is white noise with zero mean and covariance matrix denoted by Q:

$$E\{e(t)\} = 0 \tag{3.8.3}$$

$$E\{e(t)e^*(s)\} = Q\delta_{t,s} \tag{3.8.4}$$

where Q is positive definite by assumption.

The *transfer filter* corresponding to (3.8.2), also called *the ARMA shaping filter*, is readily seen to be:

$$H(q) = q^{-1}C(I - Aq^{-1})^{-1}B + I \tag{3.8.5}$$

By paralleling the calculation leading to (1.4.9), it is then possible to show that the *ARMA power spectral density (PSD) matrix* is given by:

$$\boxed{\phi(\omega) = H(\omega)QH^*(\omega)} \tag{3.8.6}$$

(The derivation of (3.8.6) is left as an exercise to the reader.)

In the next subsections, we will introduce a methodology for estimating the matrices A, B, C, and Q of the state–space equation (3.8.2), and hence the ARMA's power spectral density (via (3.8.5) and (3.8.6)). In this subsection, we derive a number of results that prepare the discussion in the next subsections.

Let

$$R_k = E\{y(t)y^*(t-k)\} \tag{3.8.7}$$

$$P = E\{x(t)x^*(t)\} \tag{3.8.8}$$

Observe that, for $k \geq 1$,

$$\begin{aligned} R_k &= E\{[Cx(t+k) + e(t+k)][x^*(t)C^* + e^*(t)]\} \\ &= CE\{x(t+k)x^*(t)\}C^* + CE\{x(t+k)e^*(t)\} \end{aligned} \tag{3.8.9}$$

From equation (3.8.2), we obtain (by induction):

$$x(t+k) = A^k x(t) + \sum_{\ell=0}^{k-1} A^{k-\ell-1}\, Be(t+\ell) \tag{3.8.10}$$

which implies that

$$E\{x(t+k)x^*(t)\} = A^k P \tag{3.8.11}$$

and

$$E\{x(t+k)e^*(t)\} = A^{k-1}BQ \tag{3.8.12}$$

Inserting (3.8.11) and (3.8.12) into (3.8.9) yields:

$$\boxed{R_k = CA^{k-1}D \qquad (\text{for } k \geq 1)} \tag{3.8.13}$$

where

$$\boxed{D = APC^* + BQ} \tag{3.8.14}$$

From the first equation in (3.8.2), we also readily obtain

$$P = APA^* + BQB^* \tag{3.8.15}$$

and from the second equation,

$$R_0 = CPC^* + Q \tag{3.8.16}$$

It follows from (3.8.14) and (3.8.16) that

$$\boxed{B = (D - APC^*)Q^{-1}} \tag{3.8.17}$$

and, respectively,

$$\boxed{Q = R_0 - CPC^*} \tag{3.8.18}$$

Finally, inserting (3.8.17) and (3.8.18) into (3.8.15) gives the following *Riccati equation* for P:

$$\boxed{P = APA^* + (D - APC^*)(R_0 - CPC^*)^{-1}(D - APC^*)^*} \tag{3.8.19}$$

The above results lead to a number of interesting observations.

The (Non)Uniqueness Issue: It is well known that a linear nonsingular transformation of the state vector in (3.8.2) leaves the transfer function matrix associated with (3.8.2) unchanged. To be more precise, let the new state vector be given by:

$$\tilde{x}(t) = Tx(t), \qquad (|T| \neq 0) \tag{3.8.20}$$

It can be verified that the state–space equations in $\tilde{x}(t)$, corresponding to (3.8.2), are:

$$\begin{aligned} \tilde{x}(t+1) &= \tilde{A}\tilde{x}(t) + \tilde{B}e(t) \\ y(t) &= \tilde{C}\tilde{x}(t) + e(t) \end{aligned} \tag{3.8.21}$$

where

$$\tilde{A} = TAT^{-1}; \quad \tilde{B} = TB; \quad \tilde{C} = CT^{-1} \tag{3.8.22}$$

As $\{y(t)\}$ and $\{e(t)\}$ in (3.8.21) are the same as in (3.8.2), the transfer function $H(q)$ from $e(t)$ to $y(t)$ must be the same for both (3.8.2) and (3.8.21). (Verifying this by direct calculation is left to the reader.) The consequence is that there exists an *infinite number* of triples (A, B, C) (with *all* matrix elements assumed unknown) that lead to the same ARMA transfer function, and hence the same ARMA covariance sequence and PSD matrix. For the transfer function matrix, the nonuniqueness induced by the *similarity transformation* (3.8.22) is the only type possible (as we know from the deterministic system theory, *e.g.*, [KAILATH 1980]). For the covariance sequence and the PSD, however, other types of nonuniqueness are also possible (see, *e.g.*, [FAURRE 1976] and [SÖDERSTRÖM AND STOICA 1989, Problem 6.3]).

Most ARMA estimation methods require the use of a uniquely parameterized representation. The previous discussion has clearly shown that letting all elements of A, B, C, and Q be unknown does not lead to such a unique representation. The latter representation is obtained only if a canonical form is used. As already explained, the ARMA parameter estimation methods relying on canonical parameterizations are impractical. The subspace–based estimation approach discussed in the next subsection circumvents the canonical parameterization requirement in an interesting way: The nonuniqueness of the ARMA representation with A, B, C, and Q fully parameterized is reduced to the nonuniqueness of a certain decomposition of covariance matrices; then by choosing a specific decomposition, a triplet (A, B, C) is isolated and determined in a numerically well–posed manner.

The Minimality Issue: Let, for some integer–valued m,

$$\mathcal{O} = \begin{bmatrix} C \\ CA \\ \vdots \\ CA^{m-1} \end{bmatrix} \tag{3.8.23}$$

and

$$\mathcal{C}^* = [D \;\; AD \;\; \cdots \;\; A^{m-1}D] \tag{3.8.24}$$

The similarity between the above matrices and the *observability and controllability matrices*, respectively, from the theory of deterministic state–space equations is evident. In fact, it follows from the aforementioned theory and from (3.8.13) that the triplet (A, D, C) is a *minimal representation* (i.e., one with the minimum possible dimension n) *of the covariance sequence* $\{R_k\}$ if and only if (see, *e.g.*, [KAILATH 1980; HANNAN AND DEISTLER 1988]):

$$\boxed{\operatorname{rank}(\mathcal{O}) = \operatorname{rank}(\mathcal{C}) = n \qquad (\text{for } m \geq n)} \tag{3.8.25}$$

As shown previously, the other matrices P, Q, and B of the state–space equation (3.8.2) can be obtained from A, C, and D (see equations (3.8.19), (3.8.18), and (3.8.17), respectively). It follows that the state–space equation (3.8.2) is a minimal representation of the ARMA covariance sequence $\{R_k\}$ if and only if the condition (3.8.25) is satisfied. In what follows, we assume that the "minimality condition" (3.8.25) holds true.

3.8.2 Subspace Parameter Estimation — Theoretical Aspects

We begin with showing how A, C, and D can be obtained from a sequence of theoretical ARMA covariances. Let

$$R = \begin{bmatrix} R_1 & R_2 & \cdots & R_m \\ R_2 & R_3 & \cdots & R_{m+1} \\ \vdots & \vdots & & \vdots \\ R_m & R_{m+1} & \cdots & R_{2m-1} \end{bmatrix}$$

$$= E\left\{ \begin{bmatrix} y(t) \\ \vdots \\ y(t+m-1) \end{bmatrix} [y^*(t-1) \cdots y^*(t-m)] \right\} \tag{3.8.26}$$

denote the *block–Hankel matrix* of covariances. (The name given to (3.8.26) is due to its special structure: the submatrices on its block antidiagonals are identical. Such a matrix is a block extension to the standard Hankel matrix; see Definition D14 in Appendix A.) According to (3.8.13), we can factor R as follows:

$$R = \begin{bmatrix} C \\ CA \\ \vdots \\ CA^{m-1} \end{bmatrix} [D \; AD \; \cdots A^{m-1} \; D] = \mathcal{O}\mathcal{C}^* \tag{3.8.27}$$

It follows from (3.8.25) and (3.8.27) that (see Result R4 in Appendix A):

$$\operatorname{rank}(R) = n \qquad (\text{for } m \geq n) \tag{3.8.28}$$

Hence, n could in principle be obtained as the rank of R. To determine A, C, and D let us consider the singular value decomposition (SVD) of R (see Appendix A):

$$R = U\Sigma V^* \tag{3.8.29}$$

where Σ is a nonsingular $n \times n$ diagonal matrix, and

$$U^*U = V^*V = I \qquad (n \times n)$$

By comparing (3.8.27) and (3.8.29), we obtain

$$\mathcal{O} = U\Sigma^{1/2}T \quad \text{for some nonsingular transformation matrix } T \tag{3.8.30}$$

because the columns of both $\mathcal{O}$ and $U\Sigma^{1/2}$ are bases of the range space of R. Henceforth, $\Sigma^{1/2}$ denotes a square root of Σ (that is, $\Sigma^{1/2}\Sigma^{1/2} = \Sigma$). By inserting (3.8.30) in the equation $\mathcal{O}\mathcal{C}^* = U\Sigma V^*$, we also obtain:

$$\mathcal{C} = V\Sigma^{1/2}(T^{-1})^* \tag{3.8.31}$$

Next, observe that

$$\mathcal{O}T^{-1} = \begin{bmatrix} (CT^{-1}) \\ (CT^{-1})(TAT^{-1}) \\ \vdots \\ (CT^{-1})(TAT^{-1})^{m-1} \end{bmatrix} \tag{3.8.32}$$

and

$$T\mathcal{C}^* = [(TD)\cdots(TAT^{-1})^{m-1}(TD)] \tag{3.8.33}$$

This implies that *by identifying $\mathcal{O}$ and $\mathcal{C}$ with the matrices made from all possible bases of the range spaces of R and R^*, respectively, we obtain the set of similarity–equivalent triples* (A, D, C). Hence, picking up a certain basis yields a specific triple (A, D, C) in the aforementioned set. *This is how the subspace approach to ARMA state–space parameter estimation circumvents the nonuniqueness problem associated with a fully parameterized model.*

In view of the previous discussion we can, for instance, set $T = I$ in (3.8.30) and (3.8.31) and obtain C as the first ny rows of $U\Sigma^{1/2}$ and D as the first ny columns of $\Sigma^{1/2}V^*$. Then, A may be obtained as the solution to the linear system of equations

$$(\bar{U}\Sigma^{1/2})A = \underline{U}\Sigma^{1/2} \tag{3.8.34}$$

where $\bar{U}$ and $\underline{U}$ are the matrices made from the first and, respectively, the last $(m-1)$ block rows of U. Once A, C, and D have been determined, P is obtained by solving the Riccati equation (3.8.19) and then Q and B are derived from (3.8.18) and (3.8.17). Algorithms for solving the Riccati equation are presented, for instance, in [VAN OVERSCHEE AND DE MOOR 1996] and the references therein.

A modification of the above procedure that does not change the solution obtained in the theoretical case but which appears to have *beneficial effects on the parameter estimates obtained from finite samples* is as follows. Let us denote the two vectors appearing in (3.8.26) by the following symbols:

$$f(t) = [y^T(t)\cdots y^T(t+m-1)]^T \tag{3.8.35}$$

$$p(t) = [y^T(t-1)\cdots y^T(t-m)]^T \tag{3.8.36}$$

Let

$$R_{fp} = E\{f(t)p^*(t)\} \tag{3.8.37}$$

and let R_{ff} and R_{pp} be similarly defined. Redefine the matrix in (3.8.26) as

$$R = R_{ff}^{-1/2}R_{fp}R_{pp}^{-1/2} \tag{3.8.38}$$

where $R_{ff}^{-1/2}$ and $R_{pp}^{-1/2}$ are the Hermitian square roots of R_{ff}^{-1} and R_{pp}^{-1} (see Appendix A). A heuristic explanation why the previous modification should lead to better parameter estimates in finite samples is as follows. The matrix R in (3.8.26) is equal to R_{fp}, whereas the R in (3.8.38) can be written as $R_{\tilde{f}\tilde{p}}$ where both $\tilde{f}(t) = R_{ff}^{-1/2} f(t)$ and $\tilde{p}(t) = R_{pp}^{-1/2} p(t)$ have unity covariance matrices. Owing to the latter property the cross–covariance matrix $R_{\tilde{f}\tilde{p}}$ and its singular elements are usually estimated more accurately from finite samples than are R_{fp} and its singular elements. This fact should eventually lead to better parameter estimates.

By making use of the factorization (3.8.27) of R_{fp} along with the formula (3.8.38) for the matrix R, we can write:

$$\begin{aligned} R &= R_{ff}^{-1/2} R_{fp} R_{pp}^{-1/2} = \\ &= R_{ff}^{-1/2} \mathcal{O} \mathcal{C}^* R_{pp}^{-1/2} = U \Sigma V^* \end{aligned} \tag{3.8.39}$$

where $U\Sigma V^*$ is now the SVD of R in (3.8.38). Identifying $R_{ff}^{-1/2}\mathcal{O}$ with $U\Sigma^{1/2}$ and $R_{pp}^{-1/2}\mathcal{C}$ with $V\Sigma^{1/2}$, we obtain

$$\mathcal{O} = R_{ff}^{1/2} U \Sigma^{1/2} \tag{3.8.40}$$

$$\mathcal{C} = R_{pp}^{1/2} V \Sigma^{1/2} \tag{3.8.41}$$

The matrices A, C, and D can be determined from these equations as previously described. Then we can derive P, Q, and B as has also been indicated before.

3.8.3 Subspace Parameter Estimation — Implementation Aspects

Let $\hat{R}_{fp}$ be the sample estimate of R_{fp}, for example,

$$\hat{R}_{fp} = \frac{1}{N} \sum_{t=m+1}^{N-m+1} f(t) p^*(t) \tag{3.8.42}$$

and let $\hat{R}_{ff}$ etc be similarly defined. Compute $\hat{R}$ as

$$\hat{R} = \hat{R}_{ff}^{-1/2} \hat{R}_{fp} \hat{R}_{pp}^{-1/2} \tag{3.8.43}$$

and its SVD. Estimate n as the "practical rank" of $\hat{R}$:

$$\boxed{\hat{n} = \text{p-rank}(\hat{R})} \tag{3.8.44}$$

(*i.e.*, the number of singular values of $\hat{R}$ which are significantly larger than the remaining ones; statistical tests for deciding whether a singular value of a given sample covariance matrix is significantly different from zero are discussed in, *e.g.*,

[FUCHS 1987].) Let $\hat{U}$, $\hat{\Sigma}$ and $\hat{V}$ denote the matrices made from the $\hat{n}$ principal singular elements of $\hat{R}$, corresponding to the matrices U, Σ and V in (3.8.39). Take

$$\boxed{\begin{aligned} \hat{C} &= \text{the first } ny \text{ rows of } \hat{R}_{ff}^{1/2}\hat{U}\hat{\Sigma}^{1/2} \\ \hat{D} &= \text{the first } ny \text{ columns of } \hat{\Sigma}^{1/2}\hat{V}^*\hat{R}_{pp}^{1/2} \end{aligned}} \tag{3.8.45}$$

Next, let

$$\begin{aligned} \bar{\Gamma} \text{ and } \underline{\Gamma} = {} & \text{the matrices made from the first and, respectively, last} \\ & (m-1) \text{ block rows of } \hat{R}_{ff}^{1/2}\hat{U}\hat{\Sigma}^{1/2}. \end{aligned} \tag{3.8.46}$$

Estimate A as

$$\boxed{\hat{A} = \text{the LS or TLS solution to } \bar{\Gamma}A \simeq \underline{\Gamma}} \tag{3.8.47}$$

Finally, estimate P as

$$\boxed{\begin{aligned} \hat{P} = {} & \text{the positive definite solution, if any, of the Riccati equation} \\ & \text{(3.8.19) with } A,\ C,\ D \text{ and } R_0 \text{ replaced by their estimates} \end{aligned}} \tag{3.8.48}$$

and Q and B as:

$$\boxed{\begin{aligned} \hat{Q} &= \hat{R}_0 - \hat{C}\hat{P}\hat{C}^* \\ \hat{B} &= (\hat{D} - \hat{A}\hat{P}\hat{C}^*)\hat{Q}^{-1} \end{aligned}} \tag{3.8.49}$$

In some cases, the previous procedure cannot be completed because the Riccati equation has no positive definite solution or even no solution at all. (In the case of a real–valued ARMA signal, for instance, that equation may have no real–valued solution.) In such cases, we can *approximately* determine P as follows. (Note that only the estimation of P has to be modified; all the other parameter estimates can be obtained as described above.)

A straightforward calculation making use of (3.8.11) and (3.8.12) yields:

$$\begin{aligned} E\{x(t)y^*(t-k)\} &= A^k PC^* + A^{k-1}BQ \\ &= A^{k-1}D \qquad (\text{for } k \geq 1) \end{aligned} \tag{3.8.50}$$

Hence,

$$C^* = E\{x(t)p^*(t)\} \tag{3.8.51}$$

Let

$$\psi = C^* R_{pp}^{-1} \tag{3.8.52}$$

and define $\epsilon(t)$ via the equation:

$$x(t) = \psi p(t) + \epsilon(t) \tag{3.8.53}$$

It is not difficult to verify that $\epsilon(t)$ is uncorrelated with $p(t)$. Indeed,

$$E\{\epsilon(t)p^*(t)\} = E\{[x(t) - \psi p(t)]p^*(t)\} = \mathcal{C}^* - \psi R_{pp} = 0 \tag{3.8.54}$$

This implies that *the first term in (3.8.53) is the least squares approximation of $x(t)$ based on the past signal values in $p(t)$* (see, *e.g.*, [Söderström and Stoica 1989] and Appendix A). It follows from this observation that $\psi p(t)$ approaches $x(t)$ as m increases. Hence,

$$\psi R_{pp}\psi^* = \mathcal{C}^* R_{pp}^{-1}\mathcal{C} \to P \qquad (\text{as } m \to \infty) \tag{3.8.55}$$

However, in view of (3.8.41),

$$\mathcal{C}^* R_{pp}^{-1}\mathcal{C} = \Sigma \tag{3.8.56}$$

The conclusion is that, provided m is chosen large enough, we can approximate P as

$$\boxed{\tilde{P} = \hat{\Sigma}, \qquad \text{for } m \gg 1} \tag{3.8.57}$$

This is the alternative estimate of P which can be used in lieu of (3.8.48) whenever the latter estimation procedure fails. The estimate $\tilde{P}$ approaches the true value P as N tends to infinity *provided* m is also increased without bound at an appropriate rate. However, if (3.8.57) is used with too small a value of m the estimate of P so obtained may be heavily biased.

The reader interested in more aspects on the subspace approach to parameter estimation for rational models should consult [Aoki 1987; van Overschee and de Moor 1996; Rao and Arun 1992; Viberg 1995] and the references therein.

3.9 Complements

3.9.1 The Partial Autocorrelation Sequence

The sequence $\{k_j\}$ computed in equation (3.5.7) of the LDA has an interesting statistical interpretation, as explained next. The covariance lag ρ_j "measures" the degree of correlation between the data samples $y(t)$ and $y(t-j)$ (in the chapter ρ_j was equal to either $r(j)$ or $\hat{r}(j)$; here $\rho_j = r(j)$). The normalized covariance sequence $\{\rho_j/\rho_0\}$ is often called the *autocorrelation function*. Now, $y(t)$ and $y(t-j)$ are related to one another not only "directly" but also through the intermediate samples:

$$[y(t-1)\ldots y(t-j+1)]^T \triangleq \varphi(t)$$

Let $\epsilon_f(t)$ and $\epsilon_b(t-j)$ denote the errors of the LS linear predictions of $y(t)$ and $y(t-j)$, respectively, based on $\varphi(t)$ above; in particular, $\epsilon_f(t)$ and $\epsilon_b(t-j)$ must then be uncorrelated with $\varphi(t)$: $E\{\epsilon_f(t)\varphi^*(t)\} = E\{\epsilon_b(t-j)\varphi^*(t)\} = 0$. (Note that

$\epsilon_f(t)$ and $\epsilon_b(t-j)$ are termed forward and backward prediction errors respectively; see also Exercises 3.3 and 3.4.) Then show that

$$k_j = -\frac{E\left\{\epsilon_f(t)\epsilon_b^*(t-j)\right\}}{\left[E\left\{|\epsilon_f(t)|^2\right\}E\left\{|\epsilon_b(t-j)|^2\right\}\right]^{1/2}} \tag{3.9.1}$$

Hence, k_j is the negative of the so–called *partial correlation* (PARCOR) coefficient of $\{y(t)\}$ (which measures the "partial correlation" between $y(t)$ and $y(t-j)$ after the correlation due to the intermediate values $y(t-1), \ldots, y(t-j+1)$ has been eliminated).

Solution: Let

$$\epsilon_f(t) = y(t) + \varphi^T(t)\theta \tag{3.9.2}$$

where, similarly to (3.4.9),

$$\theta = -\{E\left\{\varphi^c(t)\varphi^T(t)\right\}\}^{-1}\, E\left\{\varphi^c(t)\, y(t)\right\} \triangleq -R^{-1}r$$

It is readily verified (by making use of the previous definition for θ) that:

$$E\left\{\varphi^c(t)\epsilon_f(t)\right\} = 0$$

which shows that $\epsilon_f(t)$, as defined above, is indeed the error of the linear *forward* LS prediction of $y(t)$, based on $\varphi(t)$.

Similarly, define the following linear *backward* LS prediction error:

$$\epsilon_b(t-j) = y(t-j) + \varphi^T(t)\alpha$$

where

$$\alpha = -\{E\left\{\varphi^c(t)\varphi^T(t)\right\}\}^{-1}E\left\{\varphi^c(t)y(t-j)\right\} = -R^{-1}\tilde{r} = \tilde{\theta}$$

The last equality above follows from (3.5.3). We thus have

$$E\left\{\varphi^c(t)\epsilon_b(t-j)\right\} = 0$$

as required.

Next, some simple calculations give:

$$\begin{aligned} E\left\{|\epsilon_f(t)|^2\right\} &= E\left\{y^*(t)[y(t) + \varphi^T(t)\theta]\right\} \\ &= \rho_0 + [\rho_1^* \cdots \rho_{j-1}^*]\theta = \sigma_{j-1}^2 \end{aligned}$$

$$\begin{aligned} E\left\{|\epsilon_b(t-j)|^2\right\} &= E\left\{y^*(t-j)[y(t-j) + \varphi^T(t)\alpha]\right\} \\ &= \rho_0 + [\rho_{j-1} \cdots \rho_1]\tilde{\theta} = \sigma_{j-1}^2 \end{aligned}$$

and

$$\begin{aligned} E\left\{\epsilon_f(t)\epsilon_b^*(t-j)\right\} &= E\left\{[y(t) + \varphi^T(t)\theta]y^*(t-j)\right\} \\ &= \rho_j + [\rho_{j-1} \cdots \rho_1]\theta = \alpha_{j-1} \end{aligned}$$

(*cf.* (3.4.1) and (3.5.6)). By using the previous equations in (3.9.1), we obtain

$$k_j = -\alpha_{j-1}/\sigma_{j-1}^2$$

which coincides with (3.5.7).

3.9.2 Some Properties of Covariance Extensions

Assume we are given a finite sequence $\{r(k)\}_{k=-(m-1)}^{m-1}$ with $r(-k) = r^*(k)$, and such that R_m in equation (3.4.6) is positive definite. We wish to explore whether or not the finite sequence can be extended to an infinite sequence that is a valid ACS.

(a) Show that there is an infinite set of values of $r(m)$ such that $R_{m+1} > 0$. Show that this set is a disk in the complex plane defined by $|r(m) - c_m| < b_m$; find c_m and b_m.

(b) Argue inductively to deduce that there is an infinite number of possible covariance extensions to $\{r(k)\}_{k=-(m-1)}^{m-1}$ so that the extension $\{r(k)\}_{k=-\infty}^{\infty}$ is a valid ACS.

(c) If $r(k)$ is chosen to be the center of its admissible region (that is, $r(k) = c_k$ for $k = m, m+1, m+2, \ldots$) then the extension is called the Maximum Entropy extension [BURG 1975]. (The name maximum entropy arises because the so-obtained spectrum has maximum entropy rate $\int_{-\pi}^{\pi} \ln \phi(\omega) d\omega$ under the Gaussian assumption [BURG 1975]; the entropy rate is closely related to the numerator in the spectral flatness measure introduced in Exercise 3.6). Show that for this choice, the reflection coefficients $k_m, k_{m+1}, \ldots$ are all zero, and thus the infinite ACS corresponds to an AR process of order less than or equal to $(m-1)$.

Solution: Using the result of Exercise 3.7, we have

$$|R_{m+1}| = \sigma_m^2 |R_m| \tag{3.9.3}$$

From the Levinson–Durbin algorithm,

$$\sigma_m^2 = \sigma_{m-1}^2 \left[1 - |k_m|^2\right] = \sigma_{m-1}^2 \left[1 - \frac{|r(m) + \tilde{r}_{m-1}^* \theta_{m-1}|^2}{\sigma_{m-1}^4}\right] \tag{3.9.4}$$

Combining (3.9.3) and (3.9.4) gives

$$|R_{m+1}| = |R_m| \cdot \sigma_{m-1}^2 \left[1 - \frac{|r(m) + \tilde{r}_{m-1}^* \theta_{m-1}|^2}{\sigma_{m-1}^4}\right] \tag{3.9.5}$$

which shows that $|R_{m+1}|$ is quadratic in $r(m)$. Since $\sigma_{m-1}^2 > 0$ and R_m is positive definite, it follows that

$$|R_{m+1}| > 0 \text{ if and only if } |r(m) + \tilde{r}_{m-1}^* \theta_{m-1}|^2 < \sigma_{m-1}^4 \tag{3.9.6}$$

The above region is an open disk in the complex plane; the values of c_m and b_m in the problem statement are immediate from (3.9.6). This solves part (a).

To show part (b), note that if $R_p > 0$ and we choose $r(p)$ inside the disk $|r(p) + \tilde{r}_{p-1}^{*}\theta_{p-1}|^2 < \sigma_{p-1}^4$, then $|R_{p+1}| > 0$. This implies $\sigma_p^2 > 0$, and the admissible disk for $r(p+1)$ has nonzero radius, so there are an infinite number of possible choices for $r(p+1)$ such that $|R_{p+2}| > 0$. Arguing inductively in this way for $p = m, m+1, \ldots$ shows that there are an infinite number of covariance extensions.

For part (c), if we choose $r(p) = -\tilde{r}_{p-1}^{*}\theta_{p-1}$ for $p = m, m+1, \ldots$ (*i.e.*, $r(p)$ is chosen to be at the center of each disk in (3.9.6)), then from (3.9.4) we see that the reflection coefficient $k_p = 0$. Thus, from the Levinson–Durbin algorithm (see equation (3.5.10)) we have

$$\theta_p = \begin{bmatrix} \theta_{p-1} \\ 0 \end{bmatrix} \tag{3.9.7}$$

and

$$\sigma_p^2 = \sigma_{p-1}^2 \tag{3.9.8}$$

Arguing inductively again, we find that $k_p = 0$, $\theta_p = \begin{bmatrix} \theta_{m-1} \\ 0 \end{bmatrix}$, and $\sigma_p^2 = \sigma_{m-1}^2$ for $p = m, m+1, \ldots$. Thus, the Maximum Entropy extension corresponds to an AR process of order less than or equal to $(m-1)$.

3.9.3 The Burg Method for AR Parameter Estimation

Burg [Burg 1975] developed a method for AR parameter estimation that is based on forward and backward prediction errors, and on direct estimation of the reflection coefficients in equation (3.9.1). In this complement, we develop the Burg estimator and discuss some of its properties.

Assume we have data measurements $\{y(t)\}$ for $t = 1, 2, \ldots, N$. Similarly to Section 3.9.1, we define the forward and backward prediction errors for a pth–order model as:

$$\hat{e}_{f,p}(t) = y(t) + \sum_{i=1}^{p} \hat{a}_{p,i} y(t-i), \qquad t = p+1, \ldots, N \tag{3.9.9}$$

$$\hat{e}_{b,p}(t) = y(t-p) + \sum_{i=1}^{p} \hat{a}_{p,i}^{*} y(t-p+i), \qquad t = p+1, \ldots, N \tag{3.9.10}$$

(we have shifted the time index on the definition of $e_b(t)$ from that in equation (3.9.2) to reflect that $\hat{e}_{b,p}(t)$ is computed using data up to time t; also, the fact that the coefficients in (3.9.10) are given by $\{\hat{a}_{p,i}^{*}\}$ follows from the Complement in Section 3.9.1). We use hats to denote estimated quantities, and we explicitly denote the order p in both the prediction error sequences and the AR coefficients. The AR parameters are related to the reflection coefficient $\hat{k}_p$ by (see (3.5.10))

$$\hat{a}_{p,i} = \begin{cases} \hat{a}_{p-1,i} + \hat{k}_p \hat{a}_{p-1,p-i}^{*}, & i = 1, \ldots, p-1 \\ \hat{k}_p, & i = p \end{cases} \tag{3.9.11}$$

Burg's method considers the recursive–in–order estimation of $\hat{k}_p$ *given that the AR coefficients for order* $p-1$ *have been computed.* In particular, Burg's method finds $\hat{k}_p$ to minimize the arithmetic mean of the forward and backward prediction error variance estimates:

$$\min_{\hat{k}_p} \frac{1}{2}\left[\hat{\rho}_f(p) + \hat{\rho}_b(p)\right] \tag{3.9.12}$$

where

$$\hat{\rho}_f(p) = \frac{1}{N-p}\sum_{t=p+1}^{N} |\hat{e}_{f,p}(t)|^2$$

$$\hat{\rho}_b(p) = \frac{1}{N-p}\sum_{t=p+1}^{N} |\hat{e}_{b,p}(t)|^2$$

and where $\{\hat{a}_{p-1,i}\}_{i=1}^{p-1}$ are assumed to be known from the recursion at the previous order.

Show that the prediction errors satisfy the following recursive–in–order expressions

$$\hat{e}_{f,p}(t) = \hat{e}_{f,p-1}(t) + \hat{k}_p \hat{e}_{b,p-1}(t-1) \tag{3.9.13}$$

$$\hat{e}_{b,p}(t) = \hat{e}_{b,p-1}(t-1) + \hat{k}_p^* \hat{e}_{f,p-1}(t) \tag{3.9.14}$$

and use these expressions to develop a recursive–in–order algorithm for estimating the AR coefficients. Show that the reflection coefficient that minimizes (3.9.12) is given by

$$\hat{k}_p = \frac{-2\sum_{t=p+1}^{N} \hat{e}_{f,p-1}(t)\hat{e}_{b,p-1}^*(t-1)}{\sum_{t=p+1}^{N}\left[|\hat{e}_{f,p-1}(t)|^2 + |\hat{e}_{b,p-1}(t-1)|^2\right]} \tag{3.9.15}$$

Finally, show that the resulting AR model is stable.

Solution: Equation (3.9.13) follows directly from (3.9.9)–(3.9.11) as

$$\begin{aligned}
\hat{e}_{f,p}(t) &= y(t) + \sum_{i=1}^{p-1}\left(\hat{a}_{p-1,i} + \hat{k}_p \hat{a}_{p-1,p-i}^*\right) y(t-i) + \hat{k}_p y(t-p) \\
&= \left[y(t) + \sum_{i=1}^{p-1} \hat{a}_{p-1,i} y(t-i)\right] + \hat{k}_p \left[y(t-p) + \sum_{i=1}^{p-1} \hat{a}_{p-1,i}^* y(t-p+i)\right] \\
&= \hat{e}_{f,p-1}(t) + \hat{k}_p \hat{e}_{b,p-1}(t-1)
\end{aligned}$$

Similarly,

$$\begin{aligned}
\hat{e}_{b,p}(t) &= y(t-p) + \sum_{i=1}^{p-1}[\hat{a}_{p-1,i}^* + \hat{k}_p^* \hat{a}_{p-1,p-i}] y(t-p+i) + \hat{k}_p^* y(t) \\
&= \hat{e}_{b,p-1}(t-1) + \hat{k}_p^* \hat{e}_{f,p-1}(t)
\end{aligned}$$

which shows (3.9.14).

To show (3.9.15), we first recognize that the quantity to be minimized in (3.9.12) is quadratic in $\hat{k}_p$ since

$$\begin{aligned}
\frac{1}{2}\left[\hat{\rho}_f(p) + \hat{\rho}_b(p)\right] &= \frac{1}{2(N-p)} \sum_{t=p+1}^{N} \left\{ \left| \hat{e}_{f,p-1}(t) + \hat{k}_p \hat{e}_{b,p-1}(t-1) \right|^2 \right. \\
&\qquad \left. + \left| \hat{e}_{b,p-1}(t-1) + \hat{k}_p^* \hat{e}_{f,p-1}(t) \right|^2 \right\} \\
&= \frac{1}{2(N-p)} \sum_{t=p+1}^{N} \left\{ \left[|\hat{e}_{f,p-1}(t)|^2 + |\hat{e}_{b,p-1}(t-1)|^2 \right] \left[1 + |\hat{k}_p|^2 \right] \right. \\
&\qquad + 2\hat{e}_{f,p-1}(t)\hat{e}^*_{b,p-1}(t-1)\hat{k}_p^* \\
&\qquad \left. + 2\hat{e}^*_{f,p-1}(t)\hat{e}_{b,p-1}(t-1)\hat{k}_p \right\}
\end{aligned}$$

Using Result R34 in Appendix A, we immediately find that the $\hat{k}_p$ that minimizes the above quantity is given by (3.9.15).

A recursive–in–order algorithm for estimating the AR parameters is as follows:

Initialization: $\hat{e}_{f,0}(t) = \hat{e}_{b,0}(t) = y(t)$.

For $p = 1, \ldots, n$,

(a) Compute $\hat{e}_{f,p-1}(t)$ and $\hat{e}_{b,p-1}(t)$ for $t = p+1, \ldots, N$ from (3.9.13) and (3.9.14).

(b) Compute $\hat{k}_p$ from (3.9.15).

(c) Compute $\hat{a}_{p,i}$ for $i = 1, \ldots, p$ from (3.9.11).

Then $\hat{\theta} = [\hat{a}_{p,1}, \ldots, \hat{a}_{p,p}]^T$ is the vector of AR coefficient estimates.

Finally, to show that the resulting AR model is stable, we need to show that $|\hat{k}_p| \leq 1$ for $p = 1, \ldots, n$ (see Exercise 3.9). To do so, we express $\hat{k}_p$ as

$$\hat{k}_p = \frac{-2c^* d}{c^* c + d^* d} \tag{3.9.16}$$

where

$$\begin{aligned}
c &= [\hat{e}_{b,p-1}(p), \ldots, \hat{e}_{b,p-1}(N-1)]^T \\
d &= [\hat{e}_{f,p-1}(p+1), \ldots, \hat{e}_{f,p-1}(N)]^T
\end{aligned}$$

Then

$$\begin{aligned}
0 \leq \|c - e^{i\alpha} d\|^2 &= c^* c + d^* d - 2\,\mathrm{Re}\left\{e^{i\alpha} c^* d\right\} \quad \text{for every } \alpha \in [-\pi, \pi] \\
&\Longrightarrow 2\,\mathrm{Re}\left\{e^{i\alpha} c^* d\right\} \leq c^* c + d^* d \quad \text{for every } \alpha \in [-\pi, \pi] \\
&\Longrightarrow 2|c^* d| \leq c^* c + d^* d \Longrightarrow |\hat{k}_p| \leq 1
\end{aligned}$$

The Burg algorithm is computationally simple, and is amenable to both order–recursive and time–recursive solutions. In addition, the Burg AR model estimate is guaranteed to be stable. On the other hand, the Burg method is suboptimal in that it estimates the n reflection coefficients by decoupling an n–dimensional minimization problem into the n one–dimensional minimizations in (3.9.12). This is in contrast to the Least Squares AR method in Section 3.4.2, in which the AR coefficients are found by an n–dimensional minimization. For large N, the two algorithms give very similar performance; for short or medium data lengths, the Burg algorithm usually behaves somewhere between the LS method and the Yule–Walker method.

3.9.4 The Gohberg–Semencul Formula

The Hermitian Toeplitz matrix R_{n+1} in (3.4.6) is highly structured. In particular, it is completely defined by its first column (or row). As shown in Section 3.5, exploitation of the special algebraic structure of (3.4.6) makes it possible to solve this system of equations very efficiently. In this complement we show that the Toeplitz structure of R_{n+1} may also be exploited to derive a *closed–form* expression for the inverse of this matrix. This expression is what is usually called the *Gohberg–Semencul (GS) formula* (or the Gohberg–Semencul–Heining formula, in recognition of the contribution also made by Heining to its discovery) [Söderström and Stoica 1989; Iohvidov 1982; Böttcher and Silbermann 1983]. As will be seen, an interesting consequence of the GS formula is the fact that, even if R_{n+1}^{-1} is *not* Toeplitz in general, it is still completely determined by its first column. Observe from (3.4.6) that the first column of R_{n+1}^{-1} is given by $[1 \;\; \theta]^T/\sigma^2$.

Solution: In what follows, we drop the subscript n of θ for notational convenience.

The derivation of the GS formula requires some preparations. First, note that the following nested structures of R_{n+1},

$$R_{n+1} = \begin{bmatrix} \rho_0 & r_n^* \\ r_n & R_n \end{bmatrix} = \begin{bmatrix} R_n & \tilde{r}_n \\ \tilde{r}_n^* & \rho_0 \end{bmatrix}$$

along with (3.4.6) and the result (3.5.3), imply that

$$\theta = -R_n^{-1} r_n, \qquad \tilde{\theta} = -R_n^{-1} \tilde{r}_n$$

$$\sigma_n^2 = \rho_0 - r_n^* R_n^{-1} r_n = \rho_0 - \tilde{r}_n^* R_n^{-1} \tilde{r}_n$$

Next, make use of the above equations and a standard formula for the inverse of a partitioned matrix (see Result R26 in Appendix A) to write

$$R_{n+1}^{-1} = \begin{bmatrix} 0 & 0 \\ 0 & R_n^{-1} \end{bmatrix} + \begin{bmatrix} 1 \\ \theta \end{bmatrix} [1 \;\; \theta^*]/\sigma_n^2 \tag{3.9.17}$$

$$= \begin{bmatrix} R_n^{-1} & 0 \\ 0 & 0 \end{bmatrix} + \begin{bmatrix} \tilde{\theta} \\ 1 \end{bmatrix} [\tilde{\theta}^* \;\; 1]/\sigma_n^2 \tag{3.9.18}$$

Finally, introduce the following $(n+1) \times (n+1)$ matrix

$$Z = \begin{bmatrix} 0 & \cdots & & 0 \\ 1 & \ddots & & \vdots \\ & \ddots & & \\ 0 & & 1 & 0 \end{bmatrix} = \begin{bmatrix} 0 & \cdots & 0 \\ & & \vdots \\ I_{n\times n} & & \\ & & 0 \end{bmatrix}$$

and observe that multiplication by Z of a vector or a matrix has the effects indicated below.

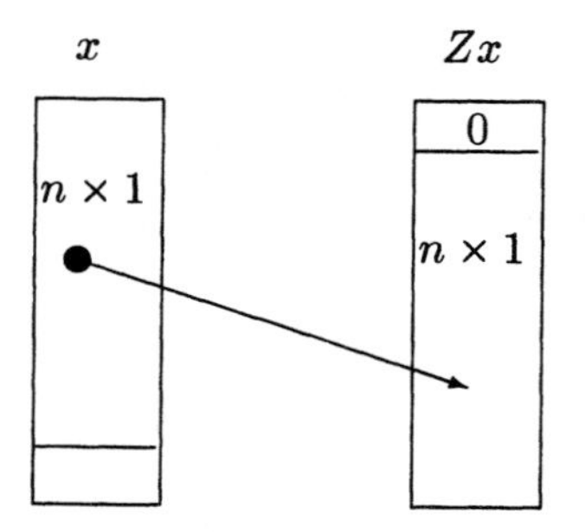

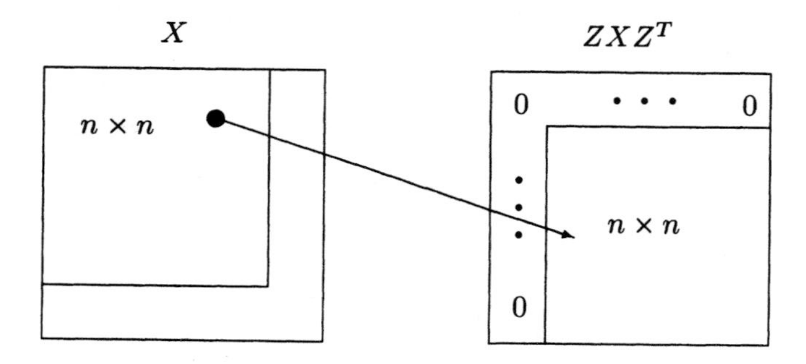

Owing to these effects of the linear transformation by Z, this matrix is called a *shift* or *displacement operator.*

We are now prepared to present a simple derivation of the GS formula. The basic idea of this derivation is to eliminate R_n^{-1} from the expressions for R_{n+1}^{-1} in (3.9.17) and (3.9.18) by making use of the above displacement properties of Z. Hence, using the expression (3.9.17) for R_{n+1}^{-1}, and its "dual" (3.9.18) for calculating $ZR_{n+1}^{-1}Z^T$, gives

$$R_{n+1}^{-1} - ZR_{n+1}^{-1}Z^T = \frac{1}{\sigma_n^2} \left\{ \begin{bmatrix} 1 \\ a_1 \\ \vdots \\ a_n \end{bmatrix} [1 \;\; a_1^* \dots a_n^*] - \begin{bmatrix} 0 \\ a_n^* \\ \vdots \\ a_1^* \end{bmatrix} [0 \;\; a_n \dots a_1] \right\} \tag{3.9.19}$$

Premultiplying and postmultiplying (3.9.19) by Z and Z^T, respectively, and then continuing to do so with the resulting equations, we obtain

$$ZR_{n+1}^{-1}Z^T - Z^2R_{n+1}^{-1}Z^{2^T} = \frac{1}{\sigma_n^2}\left\{\begin{bmatrix} 0 \\ 1 \\ a_1 \\ \vdots \\ a_{n-1} \end{bmatrix}[0 \;\; 1 \;\; a_1^* \ldots a_{n-1}^*] - \begin{bmatrix} 0 \\ 0 \\ a_n^* \\ \vdots \\ a_2^* \end{bmatrix}[0 \;\; 0 \;\; a_n \ldots a_2]\right\} \tag{3.9.20}$$

$$\vdots$$

$$Z^nR_{n+1}^{-1}Z^{n^T} - 0 = \frac{1}{\sigma_n^2}\left\{\begin{bmatrix} 0 \\ \vdots \\ 0 \\ 1 \end{bmatrix}[0 \ldots 0 \;\; 1]\right\} \tag{3.9.21}$$

In (3.9.21), use was made of the fact that Z is a *nilpotent matrix of order* $n+1$, in the sense that:

$$Z^{n+1} = 0$$

(which can be readily verified). Now, by simply summing up the above equations (3.9.19)–(3.9.21), we derive the following expression for R_{n+1}^{-1}:

$$\boxed{R_{n+1}^{-1} = \frac{1}{\sigma_n^2}\left\{\begin{bmatrix} 1 & & & 0 \\ a_1 & \ddots & & \\ \vdots & \ddots & \ddots & \\ a_n & \ldots & a_1 & 1 \end{bmatrix}\begin{bmatrix} 1 & a_1^* & \ldots & a_n^* \\ & \ddots & \ddots & \vdots \\ & & \ddots & a_1^* \\ 0 & & & 1 \end{bmatrix} - \begin{bmatrix} 0 & & & 0 \\ a_n^* & \ddots & & \\ \vdots & \ddots & \ddots & \\ a_1^* & \ldots & a_n^* & 0 \end{bmatrix}\begin{bmatrix} 0 & a_n & \ldots & a_1 \\ & \ddots & \ddots & \vdots \\ & & \ddots & a_n \\ 0 & & & 0 \end{bmatrix}\right\}} \tag{3.9.22}$$

which is the GS formula. Note from (3.9.22) that R_{n+1}^{-1} is, indeed, completely determined by its first column, as was claimed before.

The GS formula is inherently related to the Yule–Walker method of AR modeling, and this was one of the reasons for including it in this book. The GS formula is also useful in studying other spectral estimators, such as the Capon method, which is discussed in Chapter 5. The hope that the curious reader who studies this part will become interested in the fascinating topic of Toeplitz matrices and allied subjects was another reason for its inclusion. In particular, it is indeed fascinating to be able to derive an analytical formula for the inverse of a given matrix, as was shown above to be the case for Toeplitz matrices. The basic ideas of the previous derivation may be extended to more general matrices. Let us explain this briefly. For a given matrix X, the rank of $X - ZXZ^T$ is called the *displacement rank* of X under Z. As can be seen from (3.9.19), the inverse of a Hermitian Toeplitz matrix has a displacement rank equal to two. Now, assume we are given a (structured) matrix X for which we are able to find a nilpotent matrix Y such that X^{-1} has a *low* displacement rank under Y; the matrix Y does not need to have the previous form of Z. Then, paralleling the calculations in (3.9.19)–(3.9.22), we might be able to derive a simple "closed–form" expression for X^{-1}. See [FRIEDLANDER, MORF, KAILATH, AND LJUNG 1979] for more details on the topic of this complement.

3.10 Exercises

Exercise 3.1: The Minimum Phase Property

As stated in the text, a polynomial $A(z)$ is said to be minimum phase if all its zeroes are inside the unit circle. In this exercise, we motivate the name minimum phase. Specifically, we will show that if $A(z) = 1 + a_1 z^{-1} + \cdots + a_n z^{-n}$ has all its zeroes inside the unit circle, and $B(z)$ is any other polynomial in z^{-1} with $|B(\omega)| = |A(\omega)|$ and $B(0) = A(0)$ (where $B(\omega) \triangleq B(z)|_{z=e^{i\omega}}$), then the phase *lag* of $B(\omega)$ (which is given by $-\arg B(\omega)$) is greater than or equal to the phase lag of $A(\omega)$:

$$-\arg B(\omega) \geq -\arg A(\omega)$$

Since we can factor $A(z)$ as

$$A(z) = \prod_{k=1}^{n} (1 - \alpha_k z^{-1})$$

and $\arg A(\omega) = \sum_{k=1}^{n} \arg\left(1 - \alpha_k e^{-i\omega}\right)$, we begin by proving the minimum phase property for first–order polynomials. Let

$$\begin{aligned} C(z) &= 1 - \alpha z^{-1}, \qquad \alpha \triangleq re^{i\theta}, \quad r < 1 \\ D(z) &= z^{-1} - \alpha^* = C(z)\frac{z^{-1} - \alpha^*}{1 - \alpha z^{-1}}\frac{1-\alpha}{1-\alpha^*} \triangleq C(z)E(z) \end{aligned} \tag{3.10.1}$$

(a) Show that the zero of $D(z)$ is outside the unit circle, and that $|D(\omega)| = |C(\omega)|$.

(b) Show that

$$-\arg E(\omega) = \omega + 2\tan^{-1}\left[\frac{r\sin(\omega - \theta)}{1 - r\cos(\omega - \theta)}\right] - 2\tan^{-1}\left[\frac{r\sin(\theta)}{1 - r\cos(\theta)}\right]$$

Also, show that the above function is increasing and that $E(0) = 0$.

(c) Conclude that $-\arg D(\omega) \geq -\arg C(\omega)$ which justifies the name minimum phase for $C(z)$ in the first–order case.

(d) Generalize the first–order results proved in parts (a)–(c) to polynomials $A(z)$ and $B(z)$ of arbitrary order.

Exercise 3.2: Generating the ACS from ARMA Parameters

In this chapter we developed equations expressing the ARMA coefficients $\{\sigma^2, a_i, b_j\}$ in terms of the ACS $\{r(k)\}_{k=-\infty}^{\infty}$. Find the inverse map; that is, given $\sigma^2, a_1, \ldots, a_n, b_1 \ldots, b_m$, find equations to determine $\{r(k)\}_{k=-\infty}^{\infty}$.

Exercise 3.3: Relationship between AR Modeling and Forward Linear Prediction

Suppose we have a zero mean stationary process $\{y(t)\}$ (not necessarily AR) with ACS $\{r(k)\}_{k=-\infty}^{\infty}$. We wish to predict $y(t)$ by a linear combination of its n past values; that is, the predicted value is given by

$$\hat{y}_f(t) = \sum_{k=1}^{n} (-a_k) y(t-k)$$

We define the forward prediction error as

$$e_f(t) = y(t) - \hat{y}_f(t) = \sum_{k=0}^{n} a_k y(t-k)$$

with $a_0 = 1$. Show that the vector $\theta_f = [a_1 \ldots a_n]^T$ of prediction coefficients that minimizes the prediction error variance $\sigma_f^2 \triangleq E\{|e_f(t)|^2\}$ is the solution to (3.4.2). Show also that $\sigma_f^2 = \sigma_n^2$, *i.e.*, that σ_n^2 in (3.4.2) is the prediction error variance.

Furthermore, show that if $\{y(t)\}$ is an AR(p) process with $p \leq n$, then the prediction error is white noise, and that

$$\boxed{k_j = 0 \qquad \text{for } j > p}$$

where k_j is the jth reflection coefficient defined in (3.5.7). Show that, as a consequence, $a_{p+1}, \ldots, a_n = 0$. **Hint:** The calculations performed in Section 3.4.2 and in the Complement in Section 3.9.2 will be useful in solving this problem.

Exercise 3.4: Relationship between AR Modeling and Backward Linear Prediction

Consider the signal $\{y(t)\}$ as in Exercise 3.3. This time, we will consider backward prediction; that is, we will predict $y(t)$ from its n immediate future values:

$$\hat{y}_b(t) = \sum_{k=1}^{n} (-b_k) y(t+k)$$

with corresponding backward prediction error $e_b(t) = y(t) - \hat{y}_b(t)$. Such backward prediction is useful in applications where noncausal processing is permitted; for example, when the data has been prerecorded and is stored in memory or on a tape and we want to make inferences on samples that precede the observed ones. Find an expression similar to (3.4.2) for the backward prediction coefficient vector $\theta_b = [b_1 \dots b_n]^T$. Find a relationship between the θ_b and the corresponding forward prediction coefficient vector θ_f. Relate the forward and backward prediction error variances.

Exercise 3.5: Prediction Filters and Smoothing Filters

The smoothing filter is a practically useful variation on the theme of linear prediction. A result of Exercises 3.3 and 3.4 should be that for the forward and backward prediction filters

$$A(q) = 1 + \sum_{k=1}^{n} a_k q^{-k} \quad \text{and} \quad B(q) = 1 + \sum_{k=1}^{n} b_k q^{-k},$$

the prediction coefficients satisfy $a_k = b_k^*$, and the prediction error variances are equal.

Now consider the *smoothing filter*

$$e_s(t) = \sum_{k=1}^{m} c_k y(t-k) + y(t) + \sum_{k=1}^{m} d_k y(t+k).$$

(a) Derive a system of linear equations, similar to the forward and backward linear prediction equations, that relate the smoothing filter coefficients, the smoothing prediction error variance $\sigma_s^2 = E\left\{|e_s(t)|^2\right\}$, and the ACS of $y(t)$.

(b) For $n = 2m$, provide an example of a zero–mean stationary random process for which the minimum smoothing prediction error variance is *greater* than the minimum forward prediction error variance. Also provide a second example where the minimum smoothing filter prediction error variance is less than the corresponding minimum forward prediction error variance.

(c) Assume $m = n$, but now constrain the smoothing prediction coefficients to be complex–conjugate symmetric: $c_k = d_k^*$ for $k = 1, \dots, m$. In this case the two prediction filters and the smoothing filter have the same number of degrees of freedom. Prove that the minimum smoothing prediction error variance is less than or equal to the minimum (forward or backward) prediction error variance. **Hint:** Show that the unconstrained minimum smoothing error variance solution (where we do not impose the constraint $c_k = d_k^*$) satisfies $c_k = d_k^*$ anyway.

Exercise 3.6: Relationship between Minimum Prediction Error and Spectral Flatness

Consider a random process $\{y(t)\}$ with ACS $\{r(k)\}$ ($y(t)$ is not necessarily an AR process). We find an AR(n) model for $y(t)$ by solving (3.4.6) for σ_n^2 and θ_n. These parameters generate an AR PSD model:

$$\phi_{AR}(\omega) = \frac{\sigma_n^2}{|A(\omega)|^2}$$

whose inverse Fourier transform we denote by $\{r_{AR}(k)\}_{k=-\infty}^{\infty}$. In this exercise we explore the relationship between $\{r(k)\}$ and $\{r_{AR}(k)\}$, and between $\phi_y(\omega)$ and $\phi_{AR}(\omega)$.

(a) Verify that the AR model has the property that

$$r_{AR}(k) = r(k), \qquad k = 0, \ldots, n.$$

(b) We have seen from Exercise 3.3 that the AR model minimizes the nth–order forward prediction error variance; that is, the variance of

$$e(t) = y(t) + a_1 y(t-1) + \ldots + a_n y(t-n).$$

For the special case that $\{y(t)\}$ is AR of order n or less, we also know that $\{e(t)\}$ is white noise, so $\phi_e(\omega)$ is flat. We will extend this last property by showing that, for general $\{y(t)\}$, $\phi_e(\omega)$ is maximally flat in the sense that the AR model maximizes the *spectral flatness* measure given by

$$f_e = \frac{\exp\left[\frac{1}{2\pi}\int_{-\pi}^{\pi} \ln \phi_e(\omega) d\omega\right]}{\frac{1}{2\pi}\int_{-\pi}^{\pi} \phi_e(\omega)\, d\omega} \tag{3.10.2}$$

where

$$\phi_e(\omega) = |A(\omega)|^2\, \phi_y(\omega) = \sigma_n^2 \frac{\phi_y(\omega)}{\phi_{AR}(\omega)}.$$

Show that the measure f_e has the following "desirable" properties of a spectral flatness measure:

(i) f_e is unchanged if $\phi_e(\omega)$ is multiplied by a constant.

(ii) $0 \leq f_e \leq 1$.

(iii) $f_e = 1$ if and only if $\phi_e(\omega) = \text{constant}$.

Hint: Use the fact that if $A(\omega)$ is minimum phase then

$$\frac{1}{2\pi}\int_{-\pi}^{\pi} \ln |A(\omega)|^2\, d\omega = 0 \tag{3.10.3}$$

(The above result can be proven using the Cauchy integral formula). Since the solution to (3.4.2) is guaranteed to give a minimum phase $A(\omega)$ (see Exercise 3.8), show that (3.10.3) implies

$$f_e = f_y \frac{r_y(0)}{r_e(0)} \tag{3.10.4}$$

and thus that minimizing $r_e(0)$ maximizes f_e.

Exercise 3.7: Diagonalization of the Covariance Matrix

Show that R_{n+1} in equation (3.5.2) satisfies

$$L^* R_{n+1} L = D$$

where

$$L = \begin{bmatrix} 1 & 0 & \dots & 0 & 0 \\ & 1 & & \vdots & \vdots \\ & & \ddots & 0 & \\ & & & 1 & 0 \\ \theta_n & \theta_{n-1} & & \theta_1 & 1 \end{bmatrix} \quad \text{and} \quad D = \text{diag}\ [\sigma_n^2\ \sigma_{n-1}^2 \cdots \sigma_0^2]$$

and where θ_k and σ_k^2 are defined in (3.4.6). Use this property to show that

$$|R_{n+1}| = \prod_{k=0}^{n} \sigma_k^2$$

Exercise 3.8: Stability of Yule–Walker AR Models

Assume that the matrix R_{n+1} in equation (3.4.6) is positive definite. (This can be achieved by using the sample covariances in (2.2.4) to build R_{n+1}, as explained in Section 2.2.) Then show that the AR model obtained from the Yule–Walker equations (3.4.6) is stable in the sense that the polynomial $A(z)$ has all its zeroes strictly inside the unit circle. (Most of the available proofs for this property are discussed in [STOICA AND NEHORAI 1987]).

Exercise 3.9: Three Equivalent Representations for AR Processes

In this chapter we have considered three ways to parameterize an AR(n), but we have not explicitly shown when they are equivalent. Show that, for a nondegenerate AR(n) process (*i.e.*, one for which R_{n+1} is positive definite), the following three parameterizations are equivalent:

(R) $r(0), \dots, r(n)$ such that R_{n+1} is positive definite.

(K) $r(0), k_1, \dots, k_n$ such that $r(0) > 0$ and $|k_i| < 1$ for $i = 1, \dots, n$.

(A) $\sigma_n^2, a_1, \ldots, a_n$ such that $\sigma_n^2 > 0$ and all the zeroes of $A(z)$ are inside the unit circle.

Find the mapping from each parameterization to the others (some of these have already been derived in the text and in the previous exercises).

Exercise 3.10: Recurrence Properties of Reflection Coefficient Sequence for an MA Model

For an AR process of order n, the reflection coefficients satisfy $k_i = 0$ for $i > n$ (see Exercise 3.3), and the ACS satisfies the linear recurrence relationship $A(q)r(k) = 0$ for $k > 0$. Since an MA process of order m has the property that $r(i) = 0$ for $i > m$, we might wonder if a recurrence relationship holds for the reflection coefficients corresponding to a MA process. We will investigate this "conjecture" for a simple case.

Consider an MA process of order 1 with parameter b_1. Show that $|R_n|$ satisfies the relationship

$$|R_n| = r(0)|R_{n-1}| - |r(1)|^2|R_{n-2}|, \qquad n \geq 2$$

Show that $k_n = (-r(1))^n/|R_n|$ and that the reflection coefficient sequence satisfies the recurrence relationship:

$$\frac{1}{k_n} = -\frac{r(0)}{r(1)}\frac{1}{k_{n-1}} - \frac{r^*(1)}{r(1)}\frac{1}{k_{n-2}} \tag{3.10.5}$$

with appropriate initial conditions (state them). Show that the solution to (3.10.5) for $|b_1| < 1$ is

$$k_n = \frac{(1 - |b_1|^2)(-b_1)^n}{1 - |b_1|^{2n+2}} \tag{3.10.6}$$

This sequence decays exponentially to zero. When $b_1 = -1$, show that $k_n = 1/n$.

It has been shown that for large n, $B(q)k_n \simeq 0$, where $\simeq 0$ means that the residue is small compared to the k_n terms [Georgiou 1987]. This result holds even for MA processes of order higher than 1. Unfortunately, the result is of little practical use as a means of estimating the b_k coefficients since for large n the k_n values are (very) small.

Exercise 3.11: Asymptotic Variance of the ARMA Spectral Estimator

Consider the ARMA spectral estimator (3.2.2) with any consistent estimate of σ^2 and $\{a_i, b_j\}$. For simplicity, assume that the ARMA parameters are real; however, the result holds for complex ARMA processes as well. Show that the asymptotic (for large data sets) variance of this spectral estimator can be written in the form

$$E\left\{[\hat{\phi}(\omega) - \phi(\omega)]^2\right\} = C(\omega)\phi^2(\omega) \tag{3.10.7}$$

where $C(\omega) = \varphi^T(\omega)P\varphi(\omega)$. Here, P is the covariance matrix of the estimate of the parameter vector $[\sigma^2, a^T, b^T]^T$ and the vector $\varphi(\omega)$ has an expression that is to be found. Deduce that (3.10.7) has the same form as the asymptotic variance of the periodogram spectral estimator *but* with the essential difference that in the ARMA estimator case $C(\omega)$ goes to zero as the number of data samples processed increases (and that $C(\omega)$ in (3.10.7) is a function of ω). **Hint:** Use a Taylor series expansion of $\hat{\phi}(\omega)$ as a function of the estimated parameters $\{\hat{\sigma}^2, \hat{a}_i, \hat{b}_j\}$ (see, *e.g.*, Appendix B).

Exercise 3.12: Filtering Interpretation of Numerator Estimators in ARMA Estimation

An alternative method for estimating the MA part of an ARMA PSD is as follows. Assume we have estimated the AR coefficients (*e.g.*, from equation (3.7.2) or (3.7.4)). We filter $y(t)$ by $\hat{A}(q)$ to form $f(t)$:

$$f(t) = y(t) + \sum_{i=1}^{n} \hat{a}_i y(t-i), \quad t = n+1, \ldots, N.$$

Then estimate the ARMA PSD as

$$\hat{\phi}(\omega) = \frac{\sum_{k=-m}^{m} \hat{r}_f(k) e^{-i\omega k}}{|\hat{A}(\omega)|^2}$$

where $\hat{r}_f(k)$ are the standard ACS estimates for $f(t)$. Show that the above estimator is exactly equal to (3.7.8) and (3.7.9) when the MYW method with $M = n$ is used to estimate the AR parameters.

Exercise 3.13: An Alternative Expression for ARMA Power Spectral Density

Consider an ARMA(n, m) process. Show that

$$\phi(z) = \sigma^2 \frac{B(z)B^*(1/z^*)}{A(z)A^*(1/z^*)}$$

can be written as

$$\phi(z) = \frac{C(z)}{A(z)} + \frac{C^*(1/z^*)}{A^*(1/z^*)} \tag{3.10.8}$$

where

$$C(z) = \sum_{k=0}^{\max(m,n)} c_k z^{-k}$$

Show that the polynomial $C(z)$ satisfying (3.10.8) is unique, and find an expression for c_k in terms of $\{a_i\}$ and $\{r(k)\}$.

Equation (3.10.8) motivates an alternative estimation procedure to that in equations (3.7.8) and (3.7.9) for ARMA spectral estimation. In the alternative approach, we first estimate the AR coefficients $\{\hat{a}_i\}_{i=1}^{n}$ using, *e.g.*, equation (3.7.2). We then estimate the c_k coefficients using the formula found in this exercise, and finally insert the estimates $\hat{a}_k$ and $\hat{c}_k$ into the right–hand side of (3.10.8) to obtain a spectral estimate. Prove that this alternative estimator is equivalent to that in (3.7.8)–(3.7.9) under certain conditions, and find conditions on $\{\hat{a}_k\}$ so that they are equivalent. Also, compare (3.7.9) and (3.10.8) for ARMA(n, m) spectral estimation when $m < n$.

Exercise 3.14: Padé Approximation

A minimum phase (or causally invertible) ARMA(n, m) model $B(q)/A(q)$ can be equivalently represented as an AR(∞) model $1/C(q)$. The approximation of a ratio of polynomials by a polynomial of higher order was considered by Padé in the late 1800s. One possible application of the Padé approximation is to obtain an ARMA spectral model by first estimating the coefficients of a high–order AR model, then solving for a (low–order) ARMA model from the estimated AR coefficients. In this exercise we investigate the model relationships and some consequences of truncating the AR model polynomial coefficients.

Define:

$$\begin{aligned} A(q) &= 1 + a_1 q^{-1} + \cdots + a_n q^{-n} \\ B(q) &= 1 + b_1 q^{-1} + \cdots + b_m q^{-m} \\ C(q) &= 1 + c_1 q^{-1} + c_2 q^{-2} + \cdots \end{aligned}$$

(a) Show that

$$c_k = \begin{cases} 1, & k = 0 \\ a_k - \sum_{i=1}^{m} b_i c_{k-i}, & 1 \le k \le n \\ -\sum_{i=1}^{m} b_i c_{k-i}, & k > n \end{cases}$$

where we assume any polynomial coefficient is equal to zero outside its defined range.

(b) Using the equations above, derive a procedure for computing the a_i and b_j parameters from a given set of $\{c_k\}_{k=0}^{m+n}$ parameters. Assume m and n are known.

(c) The above equations give an exact representation using an infinite–order AR polynomial. In the Padé method, an *approximation* to $B(q)/A(q) = 1/C(q)$ is obtained by truncating (setting to zero) the c_k coefficients for $k > m + n$.

Suppose a stable minimum phase ARMA(n, m) filter is approximated by an AR$(m + n)$ filter using the Padé approximation. Give an example to show that the resulting AR approximation is not necessarily stable.

(d) Suppose a stable AR$(m+n)$ filter is approximated by a ratio $B_m(q)/A_n(q)$ as in part (b). Give an example to show that the resulting ARMA approximation is not necessarily stable.

Exercise 3.15: (Non)Uniqueness of Fully Parameterized ARMA Equations

The shaping filter (or transfer function) of the ARMA equation (3.8.1) is given by the following *matrix fraction*:

$$H(z) = A^{-1}(z)B(z), \qquad (ny \times ny) \tag{3.10.9}$$

where z is a dummy variable, and

$$A(z) = I + A_1 z^{-1} + \cdots + A_p z^{-p}$$

$$B(z) = I + B_1 z^{-1} + \cdots + B_p z^{-p}$$

(if the AR and MA orders, n and m, are different, then p above is equal to $\max(m, n)$). Assume that $A(z)$ and $B(z)$ are "fully parameterized" in the sense that all elements of the matrix coefficients $\{A_i,\ B_j\}$ are unknown.

The matrix fraction description (MFD) (3.10.9) of the ARMA shaping filter is unique if and only if there exist *no* matrix polynomials $\tilde{A}(z)$ and $\tilde{B}(z)$ of degree p and *no* matrix polynomial $L(z) \neq I$ such that

$$\tilde{A}(z) = L(z)A(z) \qquad \tilde{B}(z) = L(z)B(z) \tag{3.10.10}$$

This can be verified by making use of (3.10.9); see, *e.g.*, [KAILATH 1980].

Show that the above uniqueness condition is satisfied for the fully parameterized MFD if and only if

$$\boxed{\operatorname{rank}[A_p \ B_p] = ny} \tag{3.10.11}$$

Comment on the character of this condition: is it restrictive or not?

Computer Exercises

Tools for AR, MA, and ARMA Spectral Estimation:

The text web site `www.prenhall.com/~stoica` contains the following MATLAB functions for use in computing AR, MA, and ARMA spectral estimates. In each case, `y` is the input data vector, `n` is the desired AR order, and `m` is the desired MA order (if applicable). The outputs are `a`, the vector $[\hat{a}_1, \ldots, \hat{a}_n]^T$ of estimated AR parameters, `b`, the vector $[\hat{b}_1, \ldots, \hat{b}_m]^T$ of MA parameters (if applicable), and `sig2`, the noise variance estimate $\hat{\sigma}^2$.

- `[a,sig2]=yulewalker(y,n)`
  The Yule–Walker AR method given by equation (3.4.2).

- `[a,sig2]=lsar(y,n)`
  The covariance Least Squares AR method given by equation (3.4.12).

- `[a,gamma]=mywarma(y,n,m,M)`
  The modified Yule–Walker based ARMA spectral estimate given by equation (3.7.9), where the AR coefficients are estimated from the overdetermined set of equations (3.7.4) with $W = I$. Here, `M` is the number of Yule-Walker equations used in (3.7.4) and `gamma` is the vector $[\hat{\gamma}_0, \ldots, \hat{\gamma}_m]^T$.

- `[a,b,sig2]=lsarma(y,n,m,K)`
  The two–stage Least Squares ARMA method given in Section 3.7.2; `K` is the number of AR parameters to estimate in Step 1 of that algorithm.

Exercise C3.16: Comparison of AR, ARMA and Periodogram Methods for ARMA Signals

In this exercise we examine the properties of parametric methods for PSD estimation. We will use two ARMA signals, one broadband and one narrowband, to illustrate the performance of these parametric methods.

Broadband ARMA Process: Generate realizations of the broadband ARMA process

$$y(t) = \frac{B_1(q)}{A_1(q)} e(t)$$

with $\sigma^2 = 1$ and

$$\begin{aligned} A_1(q) &= 1 - 1.3817q^{-1} + 1.5632q^{-2} - 0.8843q^{-3} + 0.4096q^{-4} \\ B_1(q) &= 1 + 0.3544q^{-1} + 0.3508q^{-2} + 0.1736q^{-3} + 0.2401q^{-4} \end{aligned}$$

Choose the number of samples as $N = 256$.

(a) Estimate the PSD of the realizations by using the four AR and ARMA estimators described above. Use AR(4), AR(8), ARMA(4,4), and ARMA(8,8); for the MYW algorithm, use both $M = n$ and $M = 2n$; for the LS AR(MA) algorithms, use $K = 2n$. Illustrate the performance by plotting ten overlaid estimates of the PSD. Also, plot the true PSD on the same diagram.

In addition, plot pole or pole–zero estimates for the various methods. (For the MYW method, the zeroes can be found by spectral factorization of the numerator; comment on the difficulties you encounter, if any.)

(b) Compare the two AR algorithms. How are they different in performance?

(c) Compare the two ARMA algorithms. How does M impact performance of the MYW algorithm? How do the accuracies of the respective pole and zero estimates compare?

(d) Use an ARMA(4,4) model for the LS ARMA algorithm, and estimate the PSD of the realizations for K = 4, 8, 12, and 16. How does K impact performance of the algorithm?

(e) Compare the lower–order estimates with the higher–order estimates. In what way(s) does increasing the model order improve or degrade estimation performance?

(f) Compare the AR to the ARMA estimates. How does the AR(8) model perform with respect to the ARMA(4,4) model and the ARMA(8,8) model?

(g) Compare your results with those using the periodogram method on the same process (from Exercise C2.19 in Chapter 2). Comment on the difference between the methods with respect to variance, bias, and any other relevant properties of the estimators you notice.

Narrowband ARMA Process: Generate realizations of the narrowband ARMA process

$$y(t) = \frac{B_2(q)}{A_2(q)} e(t)$$

with $\sigma^2 = 1$ and

$$\begin{aligned} A_2(q) &= 1 - 1.6408q^{-1} + 2.2044q^{-2} - 1.4808q^{-3} + 0.8145q^{-4} \\ B_2(q) &= 1 + 1.5857q^{-1} + 0.9604q^{-2} \end{aligned}$$

(a) Repeat the experiments and comparisons in the broadband example for the narrowband process; this time, use the following model orders: AR(4), AR(8), AR(12), AR(16), ARMA(4,2), ARMA(8,4), and ARMA(12,6).

(b) Study qualitatively how the algorithm performances differ for narrowband and broadband data. Comment separately on performance near the spectral peaks and near the spectral valleys.

Exercise C3.17: AR and ARMA Estimators for Line Spectral Estimation

The ARMA methods can also be used to estimate line spectra (estimation of line spectra by other methods is the topic of Chapter 4). In this application, AR(MA) techniques are often said to provide ***super–resolution*** capabilities because they are able to resolve sinusoids too closely spaced in frequency to be resolved by periodogram–based methods.

We again consider the four AR and ARMA estimators described above.

(a) Generate realizations of the signal

$$y(t) = 10\sin(0.24\pi t + \varphi_1) + 5\sin(0.26\pi t + \varphi_2) + e(t), \qquad t = 1, \ldots, N$$

where $e(t)$ is (real) white Gaussian noise with variance σ^2, and where φ_1, φ_2 are independent random variables each uniformly distributed on $[0, 2\pi]$. From the results in Chapter 4, we find the spectrum of $y(t)$ to be

$$\begin{aligned} \phi(\omega) = {} & 50\pi\left[\delta(\omega - 0.24\pi) + \delta(\omega + 0.24\pi)\right] \\ & +12.5\pi\left[\delta(\omega - 0.26\pi) + \delta(\omega + 0.26\pi)\right] + \sigma^2 \end{aligned}$$

(b) Compute the "true" AR polynomial (using the true ACS sequence; see equation (4.1.6)) using the Yule–Walker equations for both AR(4), AR(12), ARMA(4,4) and ARMA(12,12) models when $\sigma^2 = 1$. This experiment corresponds to estimates obtained as $N \to \infty$. Plot $1/|A(\omega)|^2$ for each case, and find the roots of $A(z)$. Which method(s) are able to resolve the two sinusoids? How does the choice of M or K in the ARMA methods affect resolution or accuracy of the frequency estimates?

(c) Consider now $N = 64$, and set $\sigma^2 = 0$; this corresponds to the finite data length but infinite SNR case. Compute estimated AR polynomials using the four spectral estimators and the AR and ARMA model orders described above; for the MYW technique consider both $M = n$ and $M = 2n$, and for the LS ARMA technique use both $K = n$ and $K = 2n$. Plot $1/|A(\omega)|^2$, overlayed, for 50 different Monte–Carlo simulations (using different values of φ_1 and φ_2 for each). Also plot the zeroes of $A(z)$, overlayed, for these 50 simulations. Which method(s) are reliably able to resolve the sinusoids? Explain why. Note that as $\sigma^2 \to 0$, $y(t)$ corresponds to a (limiting) AR(4) process.

(d) Obtain spectral estimates ($\hat{\sigma}^2|\hat{B}(\omega)|^2/|\hat{A}(\omega)|^2$ for the ARMA estimators and $\hat{\sigma}^2/|\hat{A}(\omega)|^2$ for the AR estimators) for the four methods when $N = 64$ and $\sigma^2 = 1$. Plot ten overlayed spectral estimates and overlayed polynomial zeroes of the $\hat{A}(z)$ estimates. Experiment with different AR and ARMA model orders to see if the true frequencies are estimated more accurately; note also the appearance and severity of "spurious" sinusoids in the estimates for higher model orders. Which method(s) give reliable "super–resolution" estimation of the sinusoids? How does the model order influence the resolution properties? Which method appears to have the best resolution?

You may want to experiment further by changing the SNR and the relative amplitudes of the sinusoids to gain a better understanding of the relative differences between the methods. Also, experiment with different model orders and parameters K and M to understand their impact on estimation accuracy.

(e) Compare the estimation results with periodogram–based estimates obtained from the same signals. Discuss differences in resolution, bias, and variance of the techniques.

Exercise C3.18: AR and ARMA Estimators applied to Measured Data

Consider the data sets in the files `sunspotdata.mat` and `lynxdata.mat`. These files can be obtained from the text web site `www.prenhall.com/~stoica`. Apply your favorite AR and ARMA estimator(s) (for the `lynx` data, use both the original data and the logarithmically transformed data as in Exercise C2.21) to estimate the spectral content of these data. You will also need to determine appropriate model orders m and n. As in Exercise C2.21, try to answer the following questions: Are there sinusoidal components (or periodic structure) in the data? If so, how many components and at what frequencies? Discuss the relative strengths and weaknesses of parametric and nonparametric estimators for understanding the spectral content of these data. In particular, discuss how a combination of the two techniques can be used to estimate the spectral and periodic structure of the data.

Chapter 4

PARAMETRIC METHODS FOR LINE SPECTRA

4.1 Introduction

In several applications, particularly in communications, radar, sonar, geophysical seismology and so forth, the signals dealt with can be well described by the following *sinusoidal model*:

$$y(t) = x(t) + e(t) \quad ; \quad x(t) = \sum_{k=1}^{n} \alpha_k e^{i(\omega_k t + \varphi_k)} \tag{4.1.1}$$

where $x(t)$ denotes the noise–free complex–valued sinusoidal signal; $\{\alpha_k\}$, $\{\omega_k\}$, $\{\varphi_k\}$ are its *amplitudes, (angular) frequencies* and *initial phases*, respectively; and $e(t)$ is an additive observation noise. The complex–valued form (4.1.1), of course, is not encountered in practice as it stands; practical signals are real valued. However, as already mentioned in Chapter 1, in many applications both the *in–phase and quadrature components* of the studied signal are available. (See Chapter 6 for more details on this aspect.) In the case of a (real–valued) sinusoidal signal, this means that both the sine and the corresponding cosine components are available. These two components may be processed by arranging them in a two–dimensional vector signal or a complex–valued signal of the form of (4.1.1). Since the complex–valued description (4.1.1) of the in–phase and quadrature components of a sinusoidal signal is the most convenient one from a mathematical standpoint, we focus on it in this chapter.

The noise $\{e(t)\}$ in (4.1.1) is usually assumed to be (complex–valued) *circular white noise* as defined in (2.4.19). We also make the white noise assumption in this chapter. We may argue in the following way that the white noise assumption is not particularly restrictive. Let the continuous–time counterpart of the noise in (4.1.1) be correlated, but assume that the "correlation time" of the continuous–time noise is less than half of the shortest period of the sine wave components in the continuous–time counterpart of $x(t)$ in (4.1.1). If this mild condition is satisfied, then choosing the sampling period larger than the noise correlation time (yet smaller than half the shortest sinusoidal signal period, to avoid aliasing) results in a white discrete-time noise sequence $\{e(t)\}$. If the correlation condition above is not satisfied, but we know the shape of the noise spectrum, we can filter $y(t)$ by a linear *whitening filter* which makes the noise component at the filter output white; the sinusoidal components remain sinusoidal with the same frequencies, and with amplitudes and phases altered in a known way.

If the noise process is not white and has unknown spectral shape, then accurate frequency estimates can still be found if we estimate the sinusoids using the nonlinear least squares (NLS) method in Section 4.3 (see [STOICA AND NEHORAI 1989B], for example). Indeed, the properties of the NLS estimates in the colored and unknown noise case are quite similar to those for the white noise case, only with the sinusoidal signal amplitudes "adjusted" to give corresponding local SNRs — the signal–to–noise power ratio at each frequency ω_k. This amplitude adjustment is the same amplitude adjustment realized by the whitening filter approach. It is important to note that these comments only apply if the NLS method is used. The other estimation methods in this chapter (*e.g.*, the subspace–based methods) depend on the assumption that the noise is white, and may be adversely affected if the noise is not white (or is not prewhitened).

Concerning the signal in (4.1.1), we assume that $\omega_k \in [-\pi, \pi]$ and that $\alpha_k > 0$. We need to specify the sign of $\{\alpha_k\}$; otherwise we are left with a phase ambiguity. More precisely, without the condition $\alpha_k > 0$ in (4.1.1), both $\{\alpha_k, \omega_k, \varphi_k\}$ and $\{-\alpha_k, \omega_k, \varphi_k + \pi\}$ give the same signal $\{x(t)\}$, so the parameterization is not unique. As to the initial phases $\{\varphi_k\}$ in (4.1.1), one could assume that they are fixed (nonrandom) constants, which would result in $\{x(t)\}$ being a deterministic signal. In most applications, however, $\{\varphi_k\}$ are *nuisance parameters* and it is more convenient to assume that they are random variables. Note that if we try to mimic the conditions of a previous experiment as much as possible, we will usually be unable to ensure the same initial phases of the sine waves in the observed sinusoidal signal (this will be particularly true for received signals). Since there is usually no reason to believe that a specific set of initial phases is more likely than another one, or that two different initial phases are interrelated, we make the following assumption:

The initial phases $\{\varphi_k\}$ are independent random variables uniformly distributed on $[-\pi, \pi]$	(4.1.2)

The covariance function and the PSD of the noisy sinusoidal signal $\{y(t)\}$ can be calculated in a straightforward manner under the assumptions made above. By using (4.1.2), we get

$$E\left\{e^{i\varphi_p} e^{-i\varphi_j}\right\} = 1 \quad \text{for} \quad p = j$$

and for $p \neq j$

$$\begin{aligned} E\left\{e^{i\varphi_p} e^{-i\varphi_j}\right\} &= E\left\{e^{i\varphi_p}\right\} E\left\{e^{-i\varphi_j}\right\} \\ &= \left[\frac{1}{2\pi}\int_{-\pi}^{\pi} e^{i\varphi} d\varphi\right] \left[\frac{1}{2\pi}\int_{-\pi}^{\pi} e^{-i\varphi} d\varphi\right] = 0 \end{aligned}$$

Thus,

$$E\left\{e^{i\varphi_p}e^{-i\varphi_j}\right\} = \delta_{p,j} \tag{4.1.3}$$

Let

$$x_p(t) = \alpha_p e^{i(\omega_p t + \varphi_p)} \tag{4.1.4}$$

denote the pth sine wave in (4.1.1). It follows from (4.1.3) that

$$E\left\{x_p(t)x_j^*(t-k)\right\} = \alpha_p^2 e^{i\omega_p k}\delta_{p,j} \tag{4.1.5}$$

which, in turn, gives

$$\boxed{r(k) = E\left\{y(t)y^*(t-k)\right\} = \sum_{p=1}^{n} \alpha_p^2 e^{i\omega_p k} + \sigma^2 \delta_{k,0}} \tag{4.1.6}$$

and the derivation of the covariance function of $y(t)$ is completed. The PSD of $y(t)$ is given by the DTFT of $\{r(k)\}$ in (4.1.6), which is

$$\boxed{\phi(\omega) = 2\pi \sum_{p=1}^{n} \alpha_p^2 \delta(\omega - \omega_p) + \sigma^2} \tag{4.1.7}$$

where $\delta(\omega - \omega_p)$ is the Dirac impulse (or Dirac delta "function") which, by definition, has the property that

$$\int_{-\pi}^{\pi} F(\omega)\delta(\omega - \omega_p)d\omega = F(\omega_p) \tag{4.1.8}$$

for any function $F(\omega)$ that is continuous at ω_p. The expression (4.1.7) for $\phi(\omega)$ may be verified by inserting it in the inverse transform formula (1.3.8) and checking that the result is the covariance function. Doing so, we obtain

$$\frac{1}{2\pi}\int_{-\pi}^{\pi} [2\pi \sum_{p=1}^{n} \alpha_p^2 \delta(\omega - \omega_p) + \sigma^2] e^{i\omega k} d\omega = \sum_{p=1}^{n} \alpha_p^2 e^{i\omega_p k} + \sigma^2 \delta_{k,o} = r(k) \tag{4.1.9}$$

which is the desired result.

The PSD (4.1.7) is depicted in Figure 4.1. It consists of a "floor" of constant level equal to the noise power σ^2, along with n vertical lines (or impulses) located at the sinusoidal frequencies $\{\omega_k\}$ and having zero support but nonzero areas equal to 2π times the sine wave powers $\{\alpha_k^2\}$. Owing to its appearance, as exhibited in Figure 4.1, $\phi(\omega)$ in (4.1.7) is called a *line*, or *discrete, spectrum*.

It is evident from the previous discussion that a spectral analysis based on the parametric PSD model (4.1.7) reduces to the problem of estimating the parameters of the signal in (4.1.1). In most applications, such as those listed at the beginning of this chapter, the parameters of major interest are the locations of the spectral lines, namely the sinusoidal frequencies. In the following sections, we present a number of

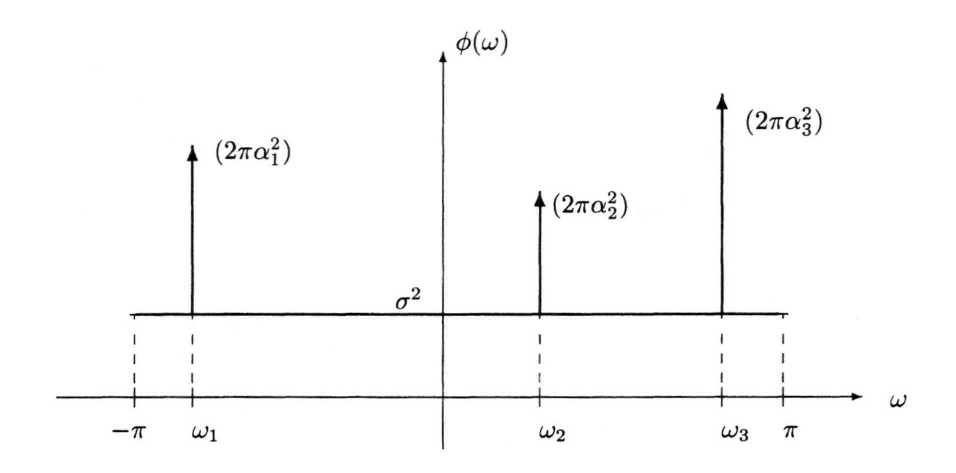

Figure 4.1. The PSD of a complex sinusoidal signal in additive white noise.

methods for *spectral line analysis.* We focus on the problem of *frequency estimation* meaning determination of $\{\omega_k\}_{k=1}^n$ from a set of observations $\{y(t)\}_{t=1}^N$. Once the frequencies have been determined, estimation of the other signal parameters (or PSD parameters) becomes a simple *linear regression problem.* More precisely, for given $\{\omega_k\}$ the observations $y(t)$ can be written as a linear regression function whose coefficients are equal to the remaining unknowns $\{\alpha_k e^{i\varphi_k} \triangleq \beta_k\}$:

$$y(t) = \sum_{k=1}^{n} \beta_k e^{i\omega_k t} + e(t) \tag{4.1.10}$$

If desired, $\{\beta_k\}$ (and hence $\{\alpha_k\}$, $\{\varphi_k\}$) in (4.1.10) can be obtained by a least squares method (as in equation (4.3.8) below). Alternatively, one may determine the signal powers $\{\alpha_k^2\}$ — for given $\{\omega_k\}$ — from the sample version of (4.1.6):

$$\hat{r}(k) = \sum_{p=1}^{n} \alpha_p^2 e^{i\omega_p k} + \text{residuals} \qquad \text{for } k \geq 1 \tag{4.1.11}$$

where the residuals arise from finite–sample estimation of $r(k)$; this is, once more, a linear regression with $\{\alpha_p^2\}$ as unknown coefficients. The solution to either linear regression problem is straightforward and is discussed in Section A.8 of Appendix A.

The methods for frequency estimation that will be described in the following sections are sometimes called ***high–resolution*** (or, even, ***super–resolution***) techniques. This is due to their ability to resolve spectral lines separated in frequency $f = \omega/2\pi$

by less than $1/N$ cycles per sampling interval, which is the resolution limit for the classical periodogram–based methods. All of the high–resolution methods to be discussed in the following provide *consistent estimates* of $\{\omega_k\}$ under the assumptions we made. Their consistency will surface in the following discussion in an obvious manner and hence we do not need to pay special attention to this aspect. Nor do we discuss in detail other statistical properties of the frequency estimates obtained by these high–resolution methods, though in Appendix B we review the Cramér–Rao bound and the best accuracy that can be achieved by such methods. For derivations and discussions of the statistical properties not addressed in this text, we refer the interested reader to [STOICA, SÖDERSTRÖM, AND TI 1989; STOICA AND SÖDERSTRÖM 1991; STOICA, MOSES, FRIEDLANDER, AND SÖDERSTRÖM 1989; STOICA AND NEHORAI 1989B]. Let us briefly summarize the conclusions of these analyses: All the high–resolution methods presented in the following provide very accurate frequency estimates, with only small differences in their statistical performances. Furthermore, the computational burdens associated with these methods are rather similar. Hence, selecting one of the high–resolution methods for frequency estimation is essentially a "matter of taste" even though we will identify some advantages of one of these methods, named ESPRIT, over the others.

We should point out that the comparison in the previous paragraph between the high–resolution methods and the periodogram–based techniques is unfair in the sense that periodogram–based methods do not assume any knowledge about the data, whereas high–resolution methods exploit an exact description of the studied signal. Owing to the additional information assumed, a parametric method should be expected to offer better resolution than the nonparametric method of the periodogram. On the other hand, when no two spectral lines in the spectrum are separated by less than $1/N$, the *unmodified periodogram* turns out to be an excellent frequency estimator which may beat any of the high–resolution methods (as we shall see). One may ask why the *unmodified* periodogram is preferred over the many windowed or smoothed periodogram techniques to which we paid so much attention in Chapter 2. The explanation actually follows from the discussion in that chapter. The unmodified periodogram can be viewed as a Blackman–Tukey "windowed" estimator with a rectangular window of maximum length equal to $2N + 1$. Of all window sequences, this is exactly the one which has the narrowest main lobe and hence the one which affords the maximum spectral resolution, a desirable property for high-resolution spectral line scenarios. It should be noted, however, that if the sinusoidal components in the signal are not too closely spaced in frequency, but their amplitudes differ significantly from one another, then a mildly windowed periodogram (to avoid leakage) may perform better than the unwindowed periodogram. (in the unwindowed periodogram, the weaker sinusoids may be obscured by the leakage from the stronger ones, and hence they may not be visible in a plot of the estimated spectrum).

In order to simplify the following discussion we assume that the number of sinusoidal components, n, in (4.1.1) is known. When n is unknown, which may well be the case in many applications, it can be determined from the available data as described for example in [FUCHS 1988; KAY 1988; MARPLE 1987; PROAKIS, RADER, LING, AND NIKIAS 1992; SÖDERSTRÖM AND STOICA 1989]. The problem of determining n is called the *order estimation* problem.

4.2 Models of Sinusoidal Signals in Noise

The frequency estimation methods presented in this chapter rely on three different models for the noisy sinusoidal signal (4.1.1). This section introduces the three models of (4.1.1).

4.2.1 Nonlinear Regression Model

The nonlinear regression model is given by (4.1.1). Note that $\{\omega_k\}$ enter in a nonlinear fashion in (4.1.1), hence the name "nonlinear regression" given to this type of model for $\{y(t)\}$. The other two models for $\{y(t)\}$, to be discussed in the following, are derived from (4.1.1); they are descriptions of the data that are not as complete as (4.1.1). However, they preserve the information required to determine the frequencies $\{\omega_k\}$ which, as already stated, are the parameters of major interest. Hence, in some sense, these two models are more appropriate for frequency estimation since they do not include some of the *nuisance parameters* which appear in (4.1.1).

4.2.2 ARMA Model

It can be readily verified that

$$(1 - e^{i\omega_k} q^{-1}) x_k(t) \equiv 0 \tag{4.2.1}$$

where q^{-1} denotes the unit delay (or shift) operator introduced in Chapter 1. Hence, $(1 - e^{i\omega_k} q^{-1})$ is an *annihilating filter* for the kth component in $x(t)$. By using this simple observation, we obtain the following *homogeneous AR* equation for $\{x(t)\}$

$$\boxed{A(q)x(t) = 0} \tag{4.2.2}$$

and the following *ARMA model* for the noisy data $\{y(t)\}$:

$$\boxed{\begin{aligned} A(q)y(t) &= A(q)e(t) \\ A(q) &= \prod_{k=1}^{n} (1 - e^{i\omega_k} q^{-1}) \end{aligned}} \tag{4.2.3}$$

It may be a useful exercise to derive equation (4.2.2) in a different way. The PSD of $x(t)$ consists of n spectral lines located at $\{\omega_k\}_{k=1}^{n}$. It should then be clear, in view of the relation (1.4.9) governing the transfer of a PSD through a linear system, that any filter which has zeroes at frequencies $\{\omega_k\}$ is an annihilating filter for $x(t)$.

The polynomial $A(q)$ in (4.2.3) is the simplest kind of such an annihilating filter. This polynomial bears *complete information* about $\{\omega_k\}$ and hence the problem of estimating the frequencies can be reduced to that of determining $A(q)$.

We remark that the ARMA model (4.2.3) has a very special form (a reason for which it is sometimes called a "degenerate" ARMA). All its poles and zeroes are located exactly on the unit circle. Furthermore, its AR and MA parts are identical. It might be tempting to cancel the common poles and zeroes in (4.2.3). However, such an operation leads to the wrong conclusion that $y(t) = e(t)$ and, therefore, should be invalid. Let us explain briefly why cancelation in (4.2.3) is not allowed. The ARMA equation description of a signal $y(t)$ is *asymptotically* equivalent to the associated transfer function description (in the sense that both give the same signal sequence, for $t \to \infty$) if and only if the poles are situated strictly inside the unit circle. If there are poles on the unit circle, then the equivalence between these two descriptions ceases. In particular, the solution of an ARMA equation with poles on the unit circle strongly depends on the initial conditions, whereas the transfer function description does not include a dependence on initial values.

4.2.3 Covariance Matrix Model

A notation that will often be used in the following is:

$$\boxed{\begin{aligned} a(\omega) &\triangleq [1 \;\; e^{-i\omega} \ldots e^{-i(m-1)\omega}]^T \qquad (m \times 1) \\ A &= [a(\omega_1) \ldots a(\omega_n)] \qquad (m \times n) \end{aligned}} \tag{4.2.4}$$

In (4.2.4), m is a positive integer which is not yet specified. Note that the matrix A introduced above is a Vandermonde matrix which enjoys the following rank property (see Result R24 in Appendix A):

$$\boxed{\operatorname{rank}(A) = n \quad \text{if} \quad m \geq n \;\; \text{and} \;\; \omega_k \neq \omega_p \;\; \text{for} \;\; k \neq p} \tag{4.2.5}$$

By making use of the previous notation, along with (4.1.1) and (4.1.4), we can write

$$\boxed{\begin{aligned} \tilde{y}(t) &\triangleq \begin{bmatrix} y(t) \\ y(t-1) \\ \vdots \\ y(t-m+1) \end{bmatrix} = A\tilde{x}(t) + \tilde{e}(t) \\ \tilde{x}(t) &= [x_1(t) \ldots x_n(t)]^T \\ \tilde{e}(t) &= [e(t) \ldots e(t-m+1)]^T \end{aligned}} \tag{4.2.6}$$

The following expression for the covariance matrix of $\tilde{y}(t)$ can be readily derived from (4.1.5) and (4.2.6)

$$R \triangleq E\{\tilde{y}(t)\tilde{y}^*(t)\} = APA^* + \sigma^2 I \quad ; \quad P = \begin{bmatrix} \alpha_1^2 & & 0 \\ & \ddots & \\ 0 & & \alpha_n^2 \end{bmatrix} \tag{4.2.7}$$

The above equation constitutes the covariance matrix model of the data. As we will show later, the *eigenstructure* of R contains complete information on the frequencies $\{\omega_k\}$, and this is exactly where the usefulness of (4.2.7) lies.

From equations (4.2.6) and (4.1.5), we also derive for later use the following result:

$$\begin{aligned} \Gamma &\triangleq E\left\{ \begin{bmatrix} y(t-L-1) \\ \vdots \\ y(t-L-M) \end{bmatrix} [y^*(t) \ldots y^*(t-L)] \right\} \\ &= E\left\{A_M \tilde{x}(t-L-1)\tilde{x}^*(t) A_{L+1}^*\right\} \\ &= A_M P_{L+1} A_{L+1}^* \qquad (L, M \geq 1) \end{aligned} \tag{4.2.8}$$

where A_K stands for A in (4.2.4) with $m = K$, and

$$P_K = \begin{bmatrix} \alpha_1^2 e^{-i\omega_1 K} & & 0 \\ & \ddots & \\ 0 & & \alpha_n^2 e^{-i\omega_n K} \end{bmatrix}$$

As we explain in detail later, the *null space* of the matrix Γ (with $L, M \geq n$) gives complete information on the frequencies $\{\omega_k\}$.

4.3 Nonlinear Least Squares Method

An intuitively appealing approach to spectral line analysis, based on the *nonlinear regression model* (4.1.1), consists of determining the unknown parameters as the minimizers of the following criterion:

$$f(\omega, \alpha, \varphi) = \sum_{t=1}^{N} \left| y(t) - \sum_{k=1}^{n} \alpha_k e^{i(\omega_k t + \varphi_k)} \right|^2 \tag{4.3.1}$$

where ω is the vector of frequencies ω_k, and similarly for α and φ. The sinusoidal model determined as above has the smallest "sum of squares" distance to the observed data $\{y(t)\}_{t=1}^N$. Since f is a nonlinear function of its arguments $\{\omega, \varphi, \alpha\}$, the method which obtains parameter estimates by minimizing (4.3.1) is called the *nonlinear method of least squares (LS)*. When the (white) noise $e(t)$ is Gaussian distributed, the minimization of (4.3.1) can also be interpreted as the *method of maximum likelihood* (see Appendix B); in that case, minimization of (4.3.1) can be shown to provide the parameter values which are most likely to "explain" the observed data sequence (see [SÖDERSTRÖM AND STOICA 1989; KAY 1988; MARPLE 1987]).

The criterion in (4.3.1) depends on both $\{\alpha_k\}$ and $\{\varphi_k\}$ as well as on $\{\omega_k\}$. However, it can be *concentrated with respect to the nuisance parameters* $\{\alpha_k, \varphi_k\}$, as explained next. By making use of the following notation,

$$\beta_k = \alpha_k e^{i\varphi_k} \tag{4.3.2}$$

$$\beta = [\beta_1 \dots \beta_n]^T \tag{4.3.3}$$

$$Y = [y(1) \dots y(N)]^T \tag{4.3.4}$$

$$B = \begin{bmatrix} e^{i\omega_1} & \cdots & e^{i\omega_n} \\ \vdots & & \vdots \\ e^{iN\omega_1} & \cdots & e^{iN\omega_n} \end{bmatrix} \tag{4.3.5}$$

we can write the function f in (4.3.1) as

$$f = (Y - B\beta)^*(Y - B\beta) \tag{4.3.6}$$

The Vandermonde matrix B in (4.3.5) (which resembles the matrix A defined in (4.2.4)) has full column rank equal to n under the weak condition that $N \geq n$; in this case, $(B^*B)^{-1}$ exists. By using this observation, we can put (4.3.6) in the more convenient form:

$$\begin{aligned} f = {} & [\beta - (B^*B)^{-1}B^*Y]^*[B^*B][\beta - (B^*B)^{-1}B^*Y] \\ & + Y^*Y - Y^*B(B^*B)^{-1}B^*Y \end{aligned} \tag{4.3.7}$$

For any choice of $\omega = [\omega_1, \dots, \omega_n]^T$ in B (which is such that $\omega_k \neq \omega_p$ for $k \neq p$), we can choose β to make the first term of f zero; thus, we see that the vectors β and ω which minimize f are given by

$$\boxed{\begin{aligned} \hat{\omega} &= \arg\max_\omega [Y^*B(B^*B)^{-1}B^*Y] \\ \hat{\beta} &= (B^*B)^{-1}B^*Y|_{\omega=\hat{\omega}} \end{aligned}} \tag{4.3.8}$$

It can be shown that, as N tends to infinity, $\hat{\omega}$ obtained as above converges to ω (*i.e.*, $\hat{\omega}$ is a consistent estimate) and, in addition, the estimation errors $\{\hat{\omega}_k - \omega_k\}$ have the following (asymptotic) covariance matrix:

$$\boxed{\operatorname{Cov}(\hat{\omega}) = \frac{6\sigma^2}{N^3} \begin{bmatrix} 1/\alpha_1^2 & & 0 \\ & \ddots & \\ 0 & & 1/\alpha_n^2 \end{bmatrix}} \tag{4.3.9}$$

(see [Stoica and Nehorai 1989a; Stoica, Moses, Friedlander, and Söderström 1989]). In the case of Gaussian noise, the matrix in (4.3.9) can also be shown to equal the *Cramér–Rao limit matrix* which gives a lower bound on the covariance matrix of any unbiased estimator of ω (see Appendix B). Hence, under the Gaussian hypothesis the nonlinear LS method provides the most accurate (*i.e.*, minimum variance) frequency estimates in a fairly general class of estimators. As a matter of fact, the variance of $\{\hat{\omega}_k\}$ (as given by (4.3.9)) may take quite small values for reasonably large sample lengths N and signal–to–noise ratios $\text{SNR}_k = \alpha_k^2/\sigma^2$. For example, for $N = 300$ and $\text{SNR}_k = 30\text{dB}$ it follows from (4.3.9) that we may expect frequency estimation errors on the order of 10^{-5}, which is comparable with the roundoff errors in a 32–bit fixed–point processor.

The nonlinear least squares (NLS) method has another advantage that sets it apart from the subspace-based approaches that are discussed in the remainder of the chapter. The NLS method does not critically depend on the assumption that the noise process is white. If the noise process is not white, the NLS still gives consistent frequency estimates. In fact, the asymptotic covariance of the frequency estimates is diagonal and $\text{var}(\hat{\omega}_k) = 6/(N^3\text{SNR}_k)$, where $\text{SNR}_k = \alpha_k^2/\phi_n(\omega_k)$ (here $\phi_n(\omega)$ is the noise PSD) is the "local" signal-to-noise ratio of the sinusoid at frequency ω_k (see [Stoica and Nehorai 1989b], for example). Interestingly enough, the nonlinear LS method remains the most accurate method (if the data length is large) even in those cases where the (Gaussian) noise is colored [Stoica and Nehorai 1989b]. This fact spurred a renewed interest in the nonlinear LS approach and in reliable algorithms for performing the minimization required in (4.3.1) (see, *e.g.*, [Hwang and Chen 1993; Ying, Potter, and Moses 1994; Li and Stoica 1996b; Umesh and Tufts 1996]).

Unfortunately, the good statistical performance associated with the nonlinear LS method of frequency estimation is difficult to achieve, for the following reason. The function (4.3.8) has a *complicated multimodal shape* with a *very sharp global maximum* corresponding to $\hat{\omega}$ [Stoica, Moses, Friedlander, and Söderström 1989]. Hence, finding $\hat{\omega}$ by a search algorithm requires very accurate initialization. Initialization procedures that provide fairly accurate approximations of the maximizer of (4.3.8) have been proposed in [Kumaresan, Scharf, and Shaw 1986], [Bresler and Macovski 1986], [Ziskind and Wax 1988]. However, there is no available method which is guaranteed to provide frequency estimates within the attraction domain of the global maximum $\hat{\omega}$ of (4.3.8). As a consequence, a search algorithm may well fail to converge to $\hat{\omega}$, or may even diverge.

The kind of difficulties indicated above, that must be faced when using the nonlinear LS method in applications, limits the practical interest in this approach to frequency estimation. There are, however, some instances when the nonlinear LS approach may be turned into a practical frequency estimation method. Consider, first, the case of a single sine wave ($n = 1$). A straightforward calculation shows that, in such a case, the first equation in (4.3.8) can be rewritten in the following form:

$$\hat{\omega} = \arg\max_{\omega} \hat{\phi}_p(\omega) \tag{4.3.10}$$

where $\hat{\phi}_p(\omega)$ is the periodogram (see (2.2.1))

$$\hat{\phi}_p(\omega) = \frac{1}{N}\left|\sum_{t=1}^{N} y(t)e^{-i\omega t}\right|^2 \tag{4.3.11}$$

Hence, the nonlinear LS estimate of the frequency of a single sine wave buried in observation noise is precisely given by the highest peak of the unmodified periodogram.

Note that the above result is only approximately true (for $N \gg 1$) in the case of *real–valued* sinusoidal signals, a fact which lends additional support to the claim made in Chapter 1 that the analysis of the case of real–valued signals faces additional complications not encountered in the complex–valued case. Each real–valued sinusoid can be written as a sum of two complex exponentials, and the treatment of the real case with $n = 1$ is similar to the complex case with $n > 1$ presented below.

Next, consider the case of multiple sine waves ($n > 1$). The key condition that makes it possible to treat this case in a manner similar to the one above, is that the minimum frequency separation between the sine waves in the studied signal is larger than the periodogram's resolution limit:

$$\boxed{\Delta\omega = \inf_{k \neq p} |\omega_k - \omega_p| > 2\pi/N} \tag{4.3.12}$$

Since the estimation errors $\{\hat{\omega}_k - \omega_k\}$ from the nonlinear LS estimates are of order $O(1/N^{3/2})$ (because $\text{cov}(\hat{\omega}) = O(1/N^3)$; see (4.3.9)), equation (4.3.12) implies a similar inequality for the nonlinear LS frequency estimates $\{\hat{\omega}_k\}$: $\Delta\hat{\omega} > 2\pi/N$. It should then be possible to *resolve all n sine waves* in the noisy signal and to obtain *reasonable approximations* $\{\tilde{\omega}_k\}$ to $\{\hat{\omega}_k\}$ by evaluating the function in (4.3.8) at the points of a grid corresponding to the sampling of each frequency variable as in the FFT:

$$\omega_k = \frac{2\pi}{N}j \qquad j = 0, \ldots, N-1 \qquad (k = 1, \ldots, n) \tag{4.3.13}$$

Of course, a direct application of such a grid method for the approximate maximization of (4.3.8) would be computationally burdensome for large values of n or N. However, it can be greatly simplified as described in the following.

The p, k element of the matrix B^*B occurring in (4.3.8), when evaluated *at the points of the grid* (4.3.13), is given by

$$[B^*B]_{p,k} = N \qquad \text{for } \; p = k \tag{4.3.14}$$

and

$$[B^*B]_{p,k} = \sum_{t=1}^{N} e^{i(\omega_k - \omega_p)t} = e^{i(\omega_k-\omega_p)}\frac{e^{iN(\omega_k-\omega_p)} - 1}{e^{i(\omega_k-\omega_p)} - 1}$$

$$= 0 \qquad \text{for } p \neq k \tag{4.3.15}$$

which implies that the function to be minimized in (4.3.8) has, in such a case, the following form:

$$\sum_{k=1}^{n} \frac{1}{N} \left| \sum_{t=1}^{N} y(t) e^{-i\omega_k t} \right|^2 \tag{4.3.16}$$

The previous additive decomposition in n functions of $\omega_1, \ldots, \omega_n$ (respectively) leads to the conclusion that $\{\tilde{\omega}_k\}$ (which, by definition, maximize (4.3.16) at the points of the grid (4.3.13)) are given by the n largest peaks of the periodogram. To show this, let us write the function in (4.3.16) as

$$g(\omega_1, \ldots, \omega_n) = \sum_{k=1}^{n} \hat{\phi}_p(\omega_k)$$

where $\hat{\phi}_p(\omega)$ is once again the periodogram. Observe that

$$\frac{\partial g(\omega_1, \ldots, \omega_n)}{\partial \omega_k} = \hat{\phi}'_p(\omega_k)$$

and

$$\frac{\partial^2 g(\omega_1, \ldots, \omega_n)}{\partial \omega_k \partial \omega_j} = \hat{\phi}''_p(\omega_k)\delta_{k,j}$$

Hence, the maximum points of (4.3.16) satisfy

$$\hat{\phi}'_p(\omega_k) = 0 \quad \text{and} \quad \hat{\phi}''_p(\omega_k) < 0 \quad \text{for } k = 1, \ldots, n$$

It follows that the set of maximizers of (4.3.16) is given by all possible combinations of n elements from the periodogram's peak locations. Now, recall the assumption made that $\{\omega_k\}$, and hence their estimates $\{\hat{\omega}_k\}$, are *distinct.* Under this assumption the highest maximum of $g(\omega_1, \ldots, \omega_n)$ is given by the locations of the n largest peaks of $\hat{\phi}_p(\omega)$, which is the desired result.

The above findings are summarized as:

> Under the condition (4.3.12), the unmodified periodogram resolves all the n sine waves present in the noisy signal. Furthermore, the locations $\{\tilde{\omega}_k\}$ of the n largest peaks in the periodogram provide $O(1/N)$ approximations to the nonlinear LS frequency estimates $\{\hat{\omega}_k\}$. In the case of $n = 1$, we have $\tilde{\omega}_1 = \hat{\omega}_1$ exactly. (4.3.17)

The fact that the differences $\{\tilde{\omega}_k - \hat{\omega}_k\}$ are $O(1/N)$ means, of course, that the computationally convenient estimates $\{\tilde{\omega}_k\}$ (derived from the periodogram) will generally have an inflated variance compared to $\{\hat{\omega}_k\}$. However, $\{\tilde{\omega}_k\}$ can at least be used as initial values in a numerical implementation of the nonlinear LS estimator. In any case, the above discussion indicates that, under (4.3.12), the periodogram performs quite well as a frequency estimator (which actually is the task for which it was introduced by Schuster nearly a century ago!).

In the following sections, we present several "high–resolution" methods for frequency estimation, which exploit the *covariance matrix models.* More precisely, all of these methods derive frequency estimates by exploiting the properties of the eigendecomposition of data covariance matrices and, in particular, the subspaces associated with those matrices. For this reason, these methods are sometimes referred to by the generic name of *subspace methods.* However, in spite of their common subspace theme, the methods are quite different, and we will treat them in separate sections below. The main features of these methods can be summarized as follows: (i) Their statistical performance is close to the ultimate performance corresponding to the nonlinear LS method (and given by the Cramér–Rao lower bound, (4.3.9)); (ii) Unlike the nonlinear LS method, these methods are not based on multidimensional search procedures; and (iii) They do not depend on a "resolution condition", such as (4.3.12), which means that they may generally have a lower resolution threshold than that of the periodogram. The chief drawback of these methods, as compared with the nonlinear LS method, is that their performance significantly degrades if the measurement noise in (4.1.1) cannot be assumed to be white.

4.4 High–Order Yule–Walker Method

The high–order Yule–Walker (HOYW) method of frequency estimation can be derived from the *ARMA model* of the sinusoidal data, (4.2.3), similarly to its counterpart in the rational PSD case (see Section 3.7 and [Cadzow 1982; Stoica, Söderström, and Ti 1989; Stoica, Moses, Söderström, and Li 1991]). Actually, the HOYW method is based on an ARMA model of an order L *higher* than the minimal order n, for a reason that will be explained shortly.

If the polynomial $A(q)$ in (4.2.3) is multiplied by any other polynomial $\bar{A}(q)$, say of degree equal to $L-n$, then we obtain a higher–order ARMA representation of our sinusoidal data, given by

$$y(t) + b_1 y(t-1) + \ldots + b_L y(t-L) = e(t) + b_1 e(t-1) + \ldots + b_L e(t-L) \quad (4.4.1)$$

or

$$B(q)y(t) = B(q)e(t)$$

where

$$B(q) = 1 + \sum_{k=1}^{L} b_k q^{-k} \triangleq A(q)\bar{A}(q) \quad (4.4.2)$$

Equation (4.4.1) can be rewritten in the following more condensed form (with obvious notation):

$$[y(t) \;\; y(t-1) \ldots y(t-L)] \begin{bmatrix} 1 \\ b \end{bmatrix} = e(t) + \ldots + b_L e(t-L) \tag{4.4.3}$$

Premultiplying (4.4.3) by $[y^*(t-L-1) \ldots y^*(t-L-M)]^T$ and taking the expectation leads to

$$\Gamma^c \begin{bmatrix} 1 \\ b \end{bmatrix} = 0 \tag{4.4.4}$$

where the matrix Γ is defined in (4.2.8) and M is a positive integer which is yet to be specified. In order to obtain (4.4.4) as indicated above, we made use of the fact that $E\{y^*(t-k)e(t)\} = 0$ for $k > 0$.

The similarity of (4.4.4) with the Yule–Walker system of equations encountered in Chapter 3 (see equation (3.7.1)) is more readily seen if (4.4.4) is rewritten in the following more detailed form:

$$\begin{bmatrix} r(L) & \ldots & r(1) \\ \vdots & & \vdots \\ r(L+M-1) & \ldots & r(M) \end{bmatrix} b = - \begin{bmatrix} r(L+1) \\ \vdots \\ r(L+M) \end{bmatrix} \tag{4.4.5}$$

Owing to this analogy, the set of equations (4.4.5) associated with the noisy sinusoidal signal $\{y(t)\}$ is said to form a HOYW system.

The HOYW matrix equation (4.4.4) can also be obtained directly from (4.2.8). For any $L \geq n$ and any polynomial $\bar{A}(q)$ (used in the defining equation, (4.4.2), for b), the elements of the vector

$$A_{L+1}^T \begin{bmatrix} 1 \\ b \end{bmatrix} \tag{4.4.6}$$

are equal to zero. Indeed, the kth row of (4.4.6) is

$$\begin{aligned} [1 \;\; e^{-i\omega_k} \ldots e^{-iL\omega_k}] \begin{bmatrix} 1 \\ b \end{bmatrix} &= 1 + \sum_{p=1}^{L} b_p e^{-i\omega_k p} \\ &= A(\omega_k)\bar{A}(\omega_k) = 0, \quad k = 1, \ldots, n \end{aligned} \tag{4.4.7}$$

(since $A(\omega_k) = 0$, *cf.* (4.2.3)). It follows from (4.2.8) and (4.4.7) that the vector $\begin{bmatrix} 1 \\ b \end{bmatrix}$ lies in the *null space* of Γ^c (see Definition D2 in Appendix A),

$$\Gamma^c \begin{bmatrix} 1 \\ b \end{bmatrix} = 0$$

which is the desired result, (4.4.4).

The HOYW system of equations derived above can be used for frequency estimation in the following way. By replacing the unavailable theoretical covariances

$\{r(k)\}$ in (4.4.5) by the sample covariances $\{\hat{r}(k)\}$, we obtain

$$\begin{bmatrix} \hat{r}(L) & \dots & \hat{r}(1) \\ \vdots & & \vdots \\ \hat{r}(L+M-1) & \dots & \hat{r}(M) \end{bmatrix} \hat{b} \simeq - \begin{bmatrix} \hat{r}(L+1) \\ \vdots \\ \hat{r}(L+M) \end{bmatrix} \tag{4.4.8}$$

Owing to the estimation errors in $\{\hat{r}(k)\}$ the matrix equation (4.4.8) cannot hold exactly in the general case, for any vector $\hat{b}$, which was indicated above by the use of the "approximate equality" symbol $\simeq$. We can solve (4.4.8) for $\hat{b}$ in a sense that is discussed in detail below, then form the polynomial

$$1 + \sum_{k=1}^{L} \hat{b}_k z^{-k} \tag{4.4.9}$$

and finally (in view of (4.2.3) and (4.4.2)) obtain frequency estimates $\{\hat{\omega}_k\}$ as the angular positions of the n roots of (4.4.9) that are located nearest the unit circle.

It may be expected that increasing the values of M and L results in improved frequency estimates. Indeed, by increasing M and L we use higher–lag covariances in (4.4.8), which may bear "additional information" on the data at hand. Increasing M and L also has a second, more subtle, effect that is explained next.

Let Ω denote the $M \times L$ covariance matrix in (4.4.5) and, similarly, let $\hat{\Omega}$ denote the sample covariance matrix in (4.4.8). It can be seen from (4.2.8) that

$$\text{rank}(\Omega) = n \qquad \text{for} \quad M, L \geq n \tag{4.4.10}$$

On the other hand, the matrix $\hat{\Omega}$ has full rank (almost surely)

$$\text{rank}(\hat{\Omega}) = \min(M, L) \tag{4.4.11}$$

owing to the random errors in $\{\hat{r}(k)\}$. However, for reasonably large values of N the matrix $\hat{\Omega}$ is close to the rank–n matrix Ω since the sample covariances $\{\hat{r}(k)\}$ converge to $\{r(k)\}$ as N increases (this is shown in the Complement in Section 4.9.1). Hence, we may expect the linear system (4.4.8) to be *ill–conditioned from a numerical standpoint* (see the discussion in Section A.8.1 in Appendix A). In fact, there is compelling empirical evidence that any LS procedure which determines $\hat{b}$ directly from (4.4.8) has very poor accuracy. In order to overcome the previously described difficulty we can make use of the *a priori rank information* (4.4.10). However, some preparations are required before we shall be able to do so. Let

$$\hat{\Omega} = U \Sigma V^* \triangleq [\underbrace{U_1}_{n} \quad \underbrace{U_2}_{M-n}] \begin{bmatrix} \Sigma_1 & 0 \\ 0 & \Sigma_2 \end{bmatrix} \begin{bmatrix} V_1^* \\ V_2^* \end{bmatrix} \begin{matrix} \}\, n \\ \}\, L-n \end{matrix} \tag{4.4.12}$$

denote the singular value decomposition (SVD) of the matrix $\hat{\Omega}$ (see Section A.4 in Appendix A, and [SÖDERSTRÖM AND STOICA 1989; VAN HUFFEL AND VANDEWALLE 1991] for general discussions on the SVD). In (4.4.12), U is an $M \times M$

unitary matrix, V is an $L \times L$ unitary matrix and Σ is an $M \times L$ diagonal matrix. As $\hat{\Omega}$ is close to a rank–n matrix, Σ_2 in (4.4.12) should be close to zero, which implies that

$$\hat{\Omega}_n \triangleq U_1 \Sigma_1 V_1^* \tag{4.4.13}$$

should be a good approximation for $\hat{\Omega}$. In fact, it can be proven that $\hat{\Omega}_n$ above is the **best** (in the Frobenius–norm sense) *rank–n approximation* of $\hat{\Omega}$ (see Result R18 in Appendix A). Hence, in accordance with the rank information (4.4.10), we can use $\hat{\Omega}_n$ in (4.4.8) in lieu of $\hat{\Omega}$. The so–obtained *rank-truncated HOYW system of equations*:

$$\hat{\Omega}_n \hat{b} \simeq - \begin{bmatrix} \hat{r}(L+1) \\ \vdots \\ \hat{r}(L+M) \end{bmatrix} \tag{4.4.14}$$

can be solved in a numerically sound way by using a simple LS procedure. It is readily verified that

$$\hat{\Omega}_n^{\dagger} = V_1 \Sigma_1^{-1} U_1^* \tag{4.4.15}$$

is the pseudoinverse of $\hat{\Omega}_n$ (see Definition D15 and Result R32). Hence, the LS solution to (4.4.14) is given by

$$\boxed{\hat{b} = -V_1 \Sigma_1^{-1} U_1^* \begin{bmatrix} \hat{r}(L+1) \\ \vdots \\ \hat{r}(L+M) \end{bmatrix}} \tag{4.4.16}$$

The additional bonus for using $\hat{\Omega}_n$ instead of $\hat{\Omega}$ in (4.4.8) is an improvement in the statistical accuracy of the frequency estimates obtained from (4.4.16). This improved accuracy is explained by the fact that $\hat{\Omega}_n$ should be closer to Ω than $\hat{\Omega}$ is; the improved covariance matrix estimate $\hat{\Omega}_n$ obtained by exploitation of the rank information (4.4.10), when used in the HOYW system of equations, should lead to refined frequency estimates.

We remark that a *total least squares* (TLS) solution for $\hat{b}$ can also be obtained from (4.4.8) (see Definition D17 and Result R33 in Appendix A). A TLS solution makes sense because we have errors in both $\hat{\Omega}$ and the right–hand–side vector in equation (4.4.8). In fact the TLS–based estimate of b is often slightly better than the estimate discussed above, which was obtained as the LS solution to the *rank–truncated* system of linear equations in (4.4.14).

We next return to the selection of L and M. As M and L increase, the information brought into the estimation problem under study by the rank condition (4.4.10) is more and more important, and hence the corresponding increase of accuracy is more and more pronounced. (For instance, the information that a 10×10 noisy matrix has rank one in the noise–free case leads to more relations between the matrix elements, and hence to more "noise cleaning", than if the matrix were 2×2.) In fact, for $M = n$ or $L = n$ the rank condition is inactive as $\hat{\Omega}_n = \hat{\Omega}$ in such

Table 4.1. The HOYW Frequency Estimation Method

Step 1. Compute the sample covariances $\{\hat{r}(k)\}_{k=1}^{L+M}$. We may set $L \simeq M$ and select the values of these integers so that $L+M$ is a fraction of the sample length (such as $N/3$). Note that if $L+M$ is set to a value which is too close to N, then the higher–lag covariances required in (4.4.8) cannot be estimated in a reliable way.

Step 2. Compute the SVD of $\hat{\Omega}$, (4.4.12), and determine $\hat{b}$ with (4.4.16).

Step 3. Isolate the n roots of the polynomial (4.4.9) that are closest to the unit circle, and obtain the frequency estimates as the angular positions of these roots.

a case. The previous discussion gives another explanation as to why the accuracy of the frequency estimates obtained from (4.4.16) may be expected to increase with increasing M and L.

Table 4.1 presents the *HOYW frequency estimation method*, in a summarized form. It should be noted that the operation in Step 3 of the HOYW method is implicitly based on the assumption that the estimated "*signal roots*" (*i.e.*, the roots of $A(q)$ in (4.4.2)) are always closer to the unit circle than the estimated "*noise roots*" (*i.e.*, the roots of $\bar{A}(q)$ in (4.4.2)). It can be shown that as $N \to \infty$, all roots of $\bar{A}(q)$ are strictly inside the unit circle (see, *e.g.*, the Complement in Section 6.5.1 and [KUMARESAN AND TUFTS 1983]). While this property cannot be guaranteed in finite samples, there is empirical evidence that it holds most often. In those rare cases where it fails to hold, the HOYW method produces *spurious* (or *false*) *frequency estimates.* The risk of producing spurious estimates is the price paid for the improved accuracy obtained by increasing L (note that for $L = n$ there is no "noise root", and hence no spurious estimate can occur in such a case). The risk for false frequency estimation is a problem that is common to all methods which estimate the frequencies from the roots of a polynomial of degree larger than n, such as the MUSIC and Min–Norm methods to be discussed in the next two sections.

4.5 Pisarenko and MUSIC Methods

The *MUltiple SIgnal Classification* (or *MUltiple SIgnal Characterization*) (MUSIC) method [SCHMIDT 1979; BIENVENU 1979] and Pisarenko's method [PISARENKO 1973] (which is a special case of MUSIC, as explained below) are derived from the covariance model (4.2.7) with $m > n$. Let $\lambda_1 \geq \lambda_2 \geq \ldots \geq \lambda_m$ denote the *eigenvalues* of R in (4.2.7), arranged in nonincreasing order, and let $\{s_1, \ldots, s_n\}$ be the *orthonormal eigenvectors* associated with $\{\lambda_1, \ldots, \lambda_n\}$, and $\{g_1, \ldots, g_{m-n}\}$ a set of *orthonormal eigenvectors* corresponding to $\{\lambda_{n+1}, \ldots, \lambda_m\}$ (see Appendix A).

Since

$$\operatorname{rank}(APA^*) = n \tag{4.5.1}$$

it follows that APA^* has n strictly positive eigenvalues, the remaining $(m - n)$ eigenvalues all being equal to zero. Combining this observation with the fact that (see Result R5 in Appendix A)

$$\lambda_k = \tilde{\lambda}_k + \sigma^2 \qquad (k = 1, \ldots, m) \tag{4.5.2}$$

where $\{\tilde{\lambda}_k\}_{k=1}^m$ are the eigenvalues of APA^* (arranged in nonincreasing order), leads to the following result:

$$\begin{cases} \lambda_k > \sigma^2 & \text{for } k = 1, \ldots, n \\ \lambda_k = \sigma^2 & \text{for } k = n+1, \ldots, m \end{cases} \tag{4.5.3}$$

The set of eigenvalues of R can hence be split into two subsets. Next, we show that the eigenvectors associated with each of these subsets, as introduced above, possess some interesting properties that can be used for frequency estimation.

Let

$$S = [s_1, \ldots, s_n] \quad (m \times n), \qquad G = [g_1, \ldots, g_{m-n}] \quad (m \times (m-n)) \tag{4.5.4}$$

From (4.2.7) and (4.5.3), we get at once:

$$RG = G \begin{bmatrix} \lambda_{n+1} & & 0 \\ & \ddots & \\ 0 & & \lambda_m \end{bmatrix} = \sigma^2 G = APA^*G + \sigma^2 G. \tag{4.5.5}$$

The first equality in (4.5.5) follows from the definition of G and $\{\lambda_k\}_{k=n+1}^m$, the second equality follows from (4.5.3), and the third from (4.2.7). The last equality in equation (4.5.5) implies that $APA^*G = 0$, or (as the matrix AP has full column rank)

$$\boxed{A^*G = 0} \tag{4.5.6}$$

In other words, the columns $\{g_k\}$ of G belong to the *null space* of A^*, a fact which is denoted by $g_k \in \mathcal{N}(A^*)$. Since $\operatorname{rank}(A) = n$, the dimension of $\mathcal{N}(A^*)$ is equal to $m - n$ which is also the dimension of the *range space* of G, $\mathcal{R}(G)$. It follows from this observation and (4.5.6) that

$$\boxed{\mathcal{R}(G) = \mathcal{N}(A^*)} \tag{4.5.7}$$

In words (4.5.7) says that the vectors $\{g_k\}$ span both $\mathcal{R}(G)$ and $\mathcal{N}(A^*)$. Now, since by definition

$$S^*G = 0 \tag{4.5.8}$$

we also have $\mathcal{R}(G) = \mathcal{N}(S^*)$; hence, $\mathcal{N}(S^*) = \mathcal{N}(A^*)$. Since $\mathcal{R}(S)$ and $\mathcal{R}(A)$ are the orthogonal complements to $\mathcal{N}(S^*)$ and $\mathcal{N}(A^*)$, it follows that

$$\boxed{\mathcal{R}(S) = \mathcal{R}(A)} \tag{4.5.9}$$

We can also derive the equality (4.5.9) directly from (4.2.7). Set

$$\mathring{\Lambda} = \begin{bmatrix} \lambda_1 - \sigma^2 & & 0 \\ & \ddots & \\ 0 & & \lambda_n - \sigma^2 \end{bmatrix} \tag{4.5.10}$$

From

$$RS = S \begin{bmatrix} \lambda_1 & & 0 \\ & \ddots & \\ 0 & & \lambda_n \end{bmatrix} = APA^*S + \sigma^2 S \tag{4.5.11}$$

we obtain

$$S = A\left(PA^*S\mathring{\Lambda}^{-1}\right) \tag{4.5.12}$$

which shows that $\mathcal{R}(S) \in \mathcal{R}(A)$. However, $\mathcal{R}(S)$ and $\mathcal{R}(A)$ have the same dimension (equal to n); hence, (4.5.9) follows. Owing to (4.5.9) and (4.5.8), the subspaces $\mathcal{R}(S)$ and $\mathcal{R}(G)$ are sometimes called the ***signal subspace*** and ***noise subspace***, respectively.

The following key result is obtained from (4.5.6).

> The true frequency values $\{\omega_k\}_{k=1}^n$ are the only solutions of the equation
>
> $$a^*(\omega)GG^*a(\omega) = 0 \text{ for any } m > n. \tag{4.5.13}$$

The fact that $\{\omega_k\}$ satisfy the above equation follows from (4.5.6). It only remains to prove that $\{\omega_k\}_{k=1}^n$ are the only solutions to (4.5.13). Let $\tilde{\omega}$ denote another possible solution, with $\tilde{\omega} \neq \omega_k$ $(k = 1, \ldots, n)$. In (4.5.13), GG^* is the *orthogonal projector* onto $\mathcal{R}(G)$ (see Section A.4). Hence, (4.5.13) implies that $a(\tilde{\omega})$ is orthogonal to $\mathcal{R}(G)$, which means that $a(\tilde{\omega}) \in \mathcal{N}(G^*)$. However, the Vandermonde vector $a(\tilde{\omega})$ is linearly independent of $\{a(\omega_k)\}_{k=1}^n$. Since $n+1$ linearly independent vectors cannot belong to an n–dimensional subspace, which is $\mathcal{N}(G^*)$ in the present case, we conclude that no other solution $\tilde{\omega}$ to (4.5.13) can exist; with this, the proof is finished.

The *MUSIC algorithm* uses the previous result to derive frequency estimates in the following steps.

Step 1. Compute the sample covariance matrix

$$\hat{R} = \frac{1}{N}\sum_{t=m}^{N} \tilde{y}(t)\tilde{y}^*(t) \tag{4.5.14}$$

and its eigendecomposition. Let $\hat{S}$ and $\hat{G}$ denote the matrices defined similarly to S and G, but made from the eigenvectors $\{\hat{s}_1, \ldots, \hat{s}_n\}$ and $\{\hat{g}_1, \ldots, \hat{g}_{m-n}\}$ of $\hat{R}$.

Step 2a. *(Spectral MUSIC)* [SCHMIDT 1979; BIENVENU 1979]. Determine frequency estimates as the locations of the n highest peaks of the function

$$\frac{1}{a^*(\omega)\hat{G}\hat{G}^*a(\omega)}, \qquad \omega \in [-\pi, \pi] \tag{4.5.15}$$

(Sometimes (4.5.15) is called a "*pseudospectrum*" since it indicates the presence of sinusoidal components in the studied signal, but it is not a true PSD. This fact may explain the attribute "spectral" attached to this variant of MUSIC.)

OR:

Step 2b. *(Root MUSIC)* [BARABELL 1983]. Determine frequency estimates as the angular positions of the n (pairs of reciprocal) roots of the equation

$$a^T(z^{-1})\hat{G}\hat{G}^*a(z) = 0 \tag{4.5.16}$$

which are located nearest the unit circle. In (4.5.16), $a(z)$ stands for the vector $a(\omega)$, (4.2.4), with $e^{i\omega}$ replaced by z, so

$$a(z) = [1, z^{-1}, \ldots, z^{-(m-1)}]^T$$

For $m = n+1$ (which is the *minimum* possible value) the MUSIC algorithm reduces to the Pisarenko method, which was the earliest proposal for an eigenanalysis–based (or subspace–based) method of frequency estimation [PISARENKO 1973].

$$\boxed{\text{The Pisarenko method is MUSIC with } m = n+1} \tag{4.5.17}$$

In the Pisarenko method, the estimated frequencies are determined from (4.5.16). For $m = n + 1$ this $2(m-1)$–degree equation can be reduced to the following equation of degree $m - 1 = n$:

$$a^T(z^{-1})\hat{g}_1 = 0 \tag{4.5.18}$$

The Pisarenko frequency estimates are obtained as the angular positions of the roots of (4.5.18). The Pisarenko method is the simplest version of MUSIC from a computational standpoint. In addition, unlike MUSIC with $m > n+1$, the Pisarenko procedure does not have the problem of separating the "signal roots" from the "noise roots" (see the discussion on this point at the end of Section 4.4). However, it can be shown that *the accuracy of the MUSIC frequency estimates increases significantly with increasing m*. Hence, the price paid for the computational simplicity of the Pisarenko method may be a relatively poor statistical accuracy.

Regarding the *selection of a value for m*, this parameter may be chosen as large as possible, but not too close to N, in order to still allow a reliable estimation of the covariance matrix (for example, as in (4.5.14)). In some applications, the largest possible value that may be selected for m may also be limited by computational complexity considerations.

Whenever the *tradeoff between statistical accuracy and computational complexity* is an important issue, the following simple ideas may be valuable.

The *finite–sample statistical accuracy* of MUSIC frequency estimates may be improved by modifying the covariance estimator (4.5.14). For instance, $\hat{R}$ is not Toeplitz whereas the true covariance matrix R is. We may correct this situation by replacing the elements in each diagonal of $\hat{R}$ with their average. The so–corrected sample covariance matrix can be shown to be the best (in the Frobenius–norm sense) Toeplitz approximation of $\hat{R}$. Another modification of $\hat{R}$, with the same purpose of improving the finite–sample statistical accuracy, is described in Section 4.8.

The *computational complexity* of MUSIC, for a given m, may be reduced in various ways. Quite often, m is such that $m - n > n$. Then, the computational burdens associated with both Spectral and Root MUSIC may be reduced by using $I - \hat{S}\hat{S}^*$ in (4.5.15) or (4.5.16) in lieu of $\hat{G}\hat{G}^*$. (Note that $\hat{S}\hat{S}^* + \hat{G}\hat{G}^* = I$ by the very definition of the eigenvector matrices.) The computational burden of Root MUSIC may be further reduced as explained in the following. The polynomial in (4.5.16) is a self–reciprocal (or symmetric) one: its roots appear in reciprocal pairs $(\rho e^{i\varphi}, \frac{1}{\rho}e^{i\varphi})$. On the unit circle $z = e^{i\omega}$, (4.5.16) is nonnegative and hence may be interpreted as a PSD. Owing to the properties mentioned above, (4.5.16) can be factored as

$$a^T(z^{-1})\hat{G}\hat{G}^*a(z) = \alpha(z)\alpha^*(1/z^*) \tag{4.5.19}$$

where $\alpha(z)$ is a polynomial of degree $(m-1)$ with all its zeroes located within or on the unit circle. We may then determine the frequency estimates from the n roots of $\alpha(z)$ that are closest to the unit circle. Since there are efficient numerical procedures for spectral factorization, determining $\alpha(z)$ as in (4.5.19) and then computing its zeroes is usually computationally more efficient than finding the (reciprocal) roots of the $2(m-1)$–degree polynomial (4.5.16).

Finally, we address the issue of *spurious frequency estimates*. As implied by the result (4.5.13), for $N \to \infty$ there is no risk of obtaining false frequency estimates. However, in finite samples such a risk always exists. Usually, this risk is quite small but it may become a real problem if m takes on large values. The key result on which the standard MUSIC algorithm, (4.5.15), is based can be used to derive a *modified MUSIC* which does not suffer from the spurious estimate problem. In the following, we only explain the basic ideas leading to the modified MUSIC method without going into details of its implementation (for such details, the interested reader may consult [STOICA AND SHARMAN 1990]). Let $\{c_k\}_{k=1}^n$ denote the coefficients of the polynomial $A(q)$ defined in (4.2.3):

$$A(q) = 1 + c_1 q^{-1} + \ldots + c_n q^{-n} = \prod_{k=1}^{n} (1 - e^{i\omega_k} q^{-1}) \tag{4.5.20}$$

Introduce the following matrix made from $\{c_k\}$:

$$C^* = \begin{bmatrix} 1 & c_1 & \ldots & c_n & & 0 \\ & \ddots & \ddots & & \ddots & \\ 0 & & 1 & c_1 & \ldots & c_n \end{bmatrix}, \qquad (m-n) \times m \tag{4.5.21}$$

It is readily verified that

$$C^* A = 0, \qquad (m-n) \times n \tag{4.5.22}$$

where A is defined in (4.2.4). Combining (4.5.9) and (4.5.22) gives

$$C^* S = 0, \qquad (m-n) \times n \tag{4.5.23}$$

which is the key property here. The matrix equation (4.5.23) can be rewritten in the following form

$$\phi c = \mu \tag{4.5.24}$$

where the $(m-n)n \times n$ matrix ϕ and the $(m-n)n \times 1$ vector μ are entirely determined from the elements of S, and where

$$c = [c_1 \ldots c_n]^T \tag{4.5.25}$$

By replacing the elements of S in ϕ and μ by the corresponding entries of $\hat{S}$, we obtain the sample version of (4.5.24)

$$\hat{\phi}\hat{c} \simeq \hat{\mu} \tag{4.5.26}$$

from which an estimate $\hat{c}$ of c may be obtained by an LS or TLS algorithm; see Section A.8 for details. The frequency estimates can then be derived from the roots of the estimated polynomial (4.5.20) corresponding to $\hat{c}$. Since this polynomial has a (minimal) degree equal to n, there is *no risk* for false frequency estimation.

4.6 Min–Norm Method

MUSIC uses $(m-n)$ linearly independent vectors in $\mathcal{R}(\hat{G})$ to obtain the frequency estimates. Since any vector in $\mathcal{R}(\hat{G})$ is (asymptotically) orthogonal to $\{a(\omega_k)\}_{k=1}^{n}$ (*cf.* (4.5.7)), we may think of using *only one* such vector for frequency estimation. By doing so, we may achieve some computational saving, hopefully without sacrificing too much accuracy.

The Min–Norm method proceeds to estimate the frequencies along these lines [KUMARESAN AND TUFTS 1983]. Let

$$\begin{bmatrix} 1 \\ \hat{g} \end{bmatrix} = \begin{array}{l} \text{the vector in } \mathcal{R}(\hat{G}) \text{, with first element equal to one,} \\ \text{that has minimum Euclidean norm.} \end{array} \tag{4.6.1}$$

Then, the *Min–Norm frequency estimates* are determined as

(*Spectral Min–Norm*). The locations of the n highest peaks in the pseudospectrum

$$\frac{1}{\left| a^*(\omega) \begin{bmatrix} 1 \\ \hat{g} \end{bmatrix} \right|^2} \tag{4.6.2}$$

or, alternatively,

(*Root Min–Norm*). The angular positions of the n roots of the polynomial

$$a^T(z^{-1}) \begin{bmatrix} 1 \\ \hat{g} \end{bmatrix} \tag{4.6.3}$$

that are located nearest the unit circle.

It remains to determine the vector in (4.6.1) and, in particular, to show that its first element can always be normalized to one. We will later comment on the reason behind the specific selection (4.6.1) of a vector in $\mathcal{R}(\hat{G})$. In the following, the Euclidean norm of a vector is denoted by $\|\cdot\|$.

Partition the matrix $\hat{S}$ as

$$\hat{S} = \begin{bmatrix} \alpha^* \\ \bar{S} \end{bmatrix} \begin{array}{l} \}\, 1 \\ \}\, m-1 \end{array} \tag{4.6.4}$$

As $\begin{bmatrix} 1 \\ \hat{g} \end{bmatrix} \in \mathcal{R}(\hat{G})$, it must satisfy the equation

$$\hat{S}^* \begin{bmatrix} 1 \\ \hat{g} \end{bmatrix} = 0 \tag{4.6.5}$$

which, using (4.6.4), can be rewritten as

$$\bar{S}^* \hat{g} = -\alpha \tag{4.6.6}$$

The minimum–norm solution to (4.6.6) is given by (see Result R31 in Appendix A):

$$\hat{g} = -\bar{S}(\bar{S}^* \bar{S})^{-1} \alpha \tag{4.6.7}$$

assuming that the inverse exists. Noting that

$$I = \hat{S}^* \hat{S} = \alpha\alpha^* + \bar{S}^* \bar{S} \tag{4.6.8}$$

and also that one eigenvalue of $I - \alpha\alpha^*$ is equal to $1 - \|\alpha\|^2$ and the remaining $(n-1)$ eigenvalues of $I - \alpha\alpha^*$ are equal to 1, it follows that the inverse in (4.6.7) exists if and only if

$$\|\alpha\|^2 \neq 1 \tag{4.6.9}$$

If the above condition is not satisfied, there will be no vector of the form of (4.6.1) in $\mathcal{R}(\hat{G})$. We postpone the study of (4.6.9) until we obtain a final–form expression for $\hat{g}$.

Under the condition (4.6.9), a simple calculation shows that

$$(\bar{S}^* \bar{S})^{-1}\alpha = (I - \alpha\alpha^*)^{-1}\alpha = \alpha/(1 - \|\alpha\|^2) \tag{4.6.10}$$

Inserting (4.6.10) in (4.6.7) gives

$$\boxed{\hat{g} = -\bar{S}\alpha/(1 - \|\alpha\|^2)} \tag{4.6.11}$$

which expresses $\hat{g}$ as a function of the elements of $\hat{S}$.

We can also obtain $\hat{g}$ as a function of the entries in $\hat{G}$. To do so, partition $\hat{G}$ as

$$\hat{G} = \begin{bmatrix} \beta^* \\ \bar{G} \end{bmatrix} \tag{4.6.12}$$

Since $\hat{S}\hat{S}^* = I - \hat{G}\hat{G}^*$ by the definition of the matrices $\hat{S}$ and $\hat{G}$, it follows that

$$\begin{bmatrix} \|\alpha\|^2 & (\bar{S}\alpha)^* \\ \bar{S}\alpha & \bar{S}\bar{S}^* \end{bmatrix} = \begin{bmatrix} 1 - \|\beta\|^2 & -(\bar{G}\beta)^* \\ -\bar{G}\beta & I - \bar{G}\bar{G}^* \end{bmatrix} \tag{4.6.13}$$

Comparing the blocks in (4.6.13) makes it possible to express $\|\alpha\|^2$ and $\bar{S}\alpha$ as functions of $\bar{G}$ and β, which leads to the following equivalent expression for $\hat{g}$:

$$\boxed{\hat{g} = \bar{G}\beta/\|\beta\|^2} \tag{4.6.14}$$

If $m - n > n$, then it is computationally more advantageous to obtain $\hat{g}$ from (4.6.11); *otherwise*, (4.6.14) should be used.

Next, we return to the condition (4.6.9) that was implicitly assumed to hold in the previous derivations. As already mentioned, this condition is equivalent to $\text{rank}(\bar{S}^*\bar{S}) = n$ which, in turn, holds if and only if

$$\text{rank}(\bar{S}) = n \tag{4.6.15}$$

Now, it follows from (4.5.9) that any block of S made from more than n consecutive rows should have rank equal to n. Hence, (4.6.15) must hold at least for N

sufficiently large. With this observation, the derivation of the Min–Norm frequency estimator is complete.

The statistical accuracy of the Min–Norm method is similar to that corresponding to MUSIC. Hence, Min–Norm achieves MUSIC's performance at a reduced computational cost. It should be noted that the selection (4.6.1) of the vector in $\mathcal{R}(\hat{G})$, used in the Min–Norm algorithm, is critical in obtaining frequency estimates with satisfactory statistical accuracy. Other choices of vectors in $\mathcal{R}(\hat{G})$ may give rather poor accuracy. In addition, there is empirical evidence that the use of the minimum–norm vector in $\mathcal{R}(\hat{G})$, as in (4.6.1), may decrease the risk of spurious frequency estimates compared with other vectors in $\mathcal{R}(\hat{G})$ or even with MUSIC (see the Complement in Section 6.5.1 for details on this aspect).

4.7 ESPRIT Method

Let

$$A_1 = [I_{m-1} \;\; 0]A \qquad (m-1) \times n \tag{4.7.1}$$

and

$$A_2 = [0 \;\; I_{m-1}]A \qquad (m-1) \times n \tag{4.7.2}$$

where I_{m-1} is the identity matrix of dimension $(m-1) \times (m-1)$ and $[I_{m-1} \;\; 0]$ and $[0 \;\; I_{m-1}]$ are $(m-1) \times m$. It is readily verified that

$$A_2 = A_1 D \tag{4.7.3}$$

where

$$D = \begin{bmatrix} e^{-i\omega_1} & & 0 \\ & \ddots & \\ 0 & & e^{-i\omega_n} \end{bmatrix} \tag{4.7.4}$$

Since D is a unitary matrix, the transformation in (4.7.3) is a *rotation*. ESPRIT, *i.e.*, *Estimation of Signal Parameters by Rotational Invariance Techniques* ([Paulraj, Roy, and Kailath 1986; Roy and Kailath 1989]; see also [Kung, Arun, and Rao 1983]), relies on the rotational transformation (4.7.3) as we detail below.

Similarly to (4.7.1) and (4.7.2), define

$$S_1 = [I_{m-1} \;\; 0]S \tag{4.7.5}$$

$$S_2 = [0 \;\; I_{m-1}]S \tag{4.7.6}$$

From (4.5.12), we have that

$$S = AC \tag{4.7.7}$$

where C is the $n \times n$ nonsingular matrix given by

$$C = PA^* S \tilde{\Lambda}^{-1} \tag{4.7.8}$$

(Observe that both S and A in (4.7.7) have full column rank, and hence C must be nonsingular; see Result R2 in Appendix A). The above explicit expression for C actually has no relevance to the present discussion. It is only (4.7.7), and the fact that C is nonsingular, that counts.

By using (4.7.1)–(4.7.3) and (4.7.7), we can write

$$S_2 = A_2 C = A_1 D C = S_1 C^{-1} D C = S_1 \phi \tag{4.7.9}$$

where

$$\phi \triangleq C^{-1} D C \tag{4.7.10}$$

Owing to the Vandermonde structure of A, the matrices A_1 and A_2 have full column rank (equal to n). In view of (4.7.7), S_1 and S_2 must also have full column rank. It then follows from (4.7.9) that the matrix ϕ is uniquely given by

$$\phi = (S_1^* S_1)^{-1} S_1^* S_2 \tag{4.7.11}$$

This formula expresses ϕ as a function of some quantities which can be estimated from the available sample. The importance of being able to estimate ϕ stems from the fact that ϕ and D have the same eigenvalues. (This can be seen from the equation (4.7.10), which is a *similarity transformation* relating ϕ and D, along with Result R6 in Appendix A.)

ESPRIT uses the previous observations to determine frequency estimates as described next.

ESPRIT estimates the frequencies $\{\omega_k\}_{k=1}^n$ as $-\arg(\hat{\nu}_k)$, where $\{\hat{\nu}_k\}_{k=1}^n$ are the eigenvalues of the following (consistent) estimate of the matrix ϕ: $$\hat{\phi} = (\hat{S}_1^* \hat{S}_1)^{-1} \hat{S}_1^* \hat{S}_2$$	(4.7.12)

It should be noted that the above estimate of ϕ is implicitly obtained by solving the following linear system of equations:

$$\hat{S}_1 \hat{\phi} \simeq \hat{S}_2 \tag{4.7.13}$$

by an *LS method.* It was empirically observed that better finite–sample accuracy may be achieved if (4.7.13) is solved for $\hat{\phi}$ by a *Total LS method* (see Section A.8 and [VAN HUFFEL AND VANDEWALLE 1991] for discussions on the TLS approach).

The *statistical accuracy* of ESPRIT is similar to that of the previously described methods: HOYW, MUSIC and Min–Norm. In fact, in most cases, ESPRIT may provide slightly more accurate frequency estimates than the other methods mentioned above; and this at similar computational cost. In addition, unlike these other methods, ESPRIT has *no problem* with separating the "signal roots" from the "noise roots", as can be seen from (4.7.12). Note that this property is shared by the modified MUSIC (discussed in Section 4.5); however, it appears that ESPRIT beats modified MUSIC in terms of statistical performance. All these considerations recommend ESPRIT as the first choice in a frequency estimation application.

4.8 Forward–Backward Approach

The previously described eigenanalysis–based methods (MUSIC, Min–Norm and ESPRIT) derive their frequency estimates from the eigenvectors of the sample covariance matrix $\hat{R}$, (4.5.14), which is restated here for easy reference:

$$\hat{R} = \frac{1}{N}\sum_{t=m}^{N} \begin{bmatrix} y(t) \\ \vdots \\ y(t-m+1) \end{bmatrix} [y^*(t) \dots y^*(t-m+1)] \tag{4.8.1}$$

The $\hat{R}$ above is recognized to be the matrix that appears in the least squares (LS) estimation of the coefficients $\{\alpha_k\}$ of an mth–order *forward* linear predictor of $y^*(t+1)$:

$$\hat{y}^*(t+1) = \alpha_1 y^*(t) + \dots + \alpha_m y^*(t-m+1) \tag{4.8.2}$$

For this reason, the methods which obtain frequency estimates from $\hat{R}$ are named *forward (F) approaches.*

Extensive numerical experience with the aforementioned methods has shown that the corresponding frequency estimation accuracy can be enhanced by using the following modified sample covariance matrix, in lieu of $\hat{R}$,

$$\boxed{\tilde{R} = \frac{1}{2}(\hat{R} + J\hat{R}^T J)} \tag{4.8.3}$$

where

$$J = \begin{bmatrix} 0 & & 1 \\ & \cdot^{\cdot^{\cdot}} & \\ 1 & & 0 \end{bmatrix} \tag{4.8.4}$$

is the so–called "*exchange*" (or "*reversal*") *matrix.* The second term in (4.8.3) has the following detailed form:

$$J\hat{R}^T J = \frac{1}{N}\sum_{t=m}^{N} \begin{bmatrix} y^*(t-m+1) \\ \vdots \\ y^*(t) \end{bmatrix} [y(t-m+1) \dots y(t)] \tag{4.8.5}$$

The matrix (4.8.5) is the one that appears in the LS estimate of the coefficients of an mth–order *backward* linear predictor of $y(t-m)$:

$$\hat{y}(t-m) = \mu_1 y(t-m+1) + \dots + \mu_m y(t) \tag{4.8.6}$$

This observation, along with the previous remark made about $\hat{R}$, suggests the name of *forward–backward (FB) approaches* for methods that determine frequency estimates from $\tilde{R}$ in (4.8.3).

The (i,j) element of $\tilde{R}$ is given by:

$$\begin{aligned}\tilde{R}_{i,j} &= \frac{1}{2N}\sum_{t=m}^{N}[y(t-i)y^*(t-j)+y^*(t-m+1+i)y(t-m+1+j)] \\ &\triangleq T_1+T_2 \qquad (i,j=0,\ldots,m-1)\end{aligned} \tag{4.8.7}$$

Assume that $i \leq j$ (the other case $i \geq j$ can be similarly treated). Let $\hat{r}(j-i)$ denote the usual sample covariance:

$$\hat{r}(j-i) = \frac{1}{N}\sum_{t=(j-i)+1}^{N} y(t)y^*(t-(j-i)) \tag{4.8.8}$$

A straightforward calculation shows that the two terms T_1 and T_2 in (4.8.7) can be written as

$$T_1 = \frac{1}{2N}\sum_{p=m-i}^{N-i} y(p)y^*(p-(j-i)) = \frac{1}{2}\hat{r}(j-i)+O(1/N) \tag{4.8.9}$$

and

$$T_2 = \frac{1}{2N}\sum_{p=j+1}^{N-m+j+1} y(p)y^*(p-(j-i)) = \frac{1}{2}\hat{r}(j-i)+O(1/N) \tag{4.8.10}$$

where $O(1/N)$ denotes a term that tends to zero as $1/N$ when N increases (it is here assumed that $m \ll N$). It follows from (4.8.7)–(4.8.10) that, for large N, the difference between $\tilde{R}_{i,j}$ or $\hat{R}_{i,j}$ and the sample covariance lag $\hat{r}(j-i)$ is "small". Hence, the frequency estimation methods based on $\hat{R}$ or $\tilde{R}$ (or on $[\hat{r}(j-i)]$) may be expected to have similar performances in large samples.

In summary, it follows from the previous discussion that the empirically observed performance superiority of the forward–backward approach over the forward–only approach should only be manifest in samples with relatively small lengths. As such, this superiority cannot be easily established by theoretical means. Let us then argue heuristically.

First, note that the transformation $J(\cdot)^T J$ is such that the following equalities hold:

$$(\hat{R})_{i,j} = (J\hat{R}J)_{m-i,m-j} = (J\hat{R}^T J)_{m-j,m-i} \tag{4.8.11}$$

and

$$(\hat{R})_{m-j,m-i} = (J\hat{R}^T J)_{i,j} \tag{4.8.12}$$

This implies that the (i,j) and $(m-j,m-i)$ elements of $\tilde{R}$ are both given by

$$\tilde{R}_{i,j} = \tilde{R}_{m-j,m-i} = \frac{1}{2}(\hat{R}_{i,j}+\hat{R}_{m-j,m-i}) \tag{4.8.13}$$

Equations (4.8.11)–(4.8.12) imply that $\tilde{R}$ is invariant to the transformation $J(\cdot)^T J$:

$$J\tilde{R}^T J = \tilde{R} \tag{4.8.14}$$

Such a matrix is said to be *persymmetric* (also called *centrosymmetric*). In order to see the reason for this name, note that $\tilde{R}$ is Hermitian (symmetric in the real–valued case) with respect to its main diagonal; *in addition*, $\tilde{R}$ is symmetric about its main antidiagonal. Indeed, the equal elements $\tilde{R}_{i,j}$ and $\tilde{R}_{m-j,m-i}$ of $\tilde{R}$ belong to the same diagonal as $i - j = (m - j) - (m - i)$. They are also symmetrically placed with respect to the main antidiagonal; $\tilde{R}_{i,j}$ lies on antidiagonal $(i + j)$, $\tilde{R}_{m-j,m-i}$ on the $[2m - (j + i)]$th one, and the main antidiagonal is the mth one (and $m = [(i + j) + 2m - (i + j)]/2$).

The theoretical (and unknown) covariance matrix R is Toeplitz and hence persymmetric. Since $\tilde{R}$ is persymmetric like R, whereas $\hat{R}$ is not, we may expect $\tilde{R}$ to be a better estimate of R than $\hat{R}$. In turn, this means that the frequency estimates derived from $\tilde{R}$ are likely to be more accurate than those obtained from $\hat{R}$.

The impact of enforcing the persymmetric property can be seen by examining, say, the $(1, 1)$ and (m, m) elements of $\hat{R}$ and $\tilde{R}$. Both the (1,1) and (m, m) elements of $\hat{R}$ are estimates of $r(0)$; however, the (1,1) element does not use the first $(m - 1)$ lag products $|y(1)|^2, \ldots, |y(m - 1)|^2$, and the (m, m) element does not use the last $(m - 1)$ lag products $|y(N - m + 2)|^2, \ldots, |y(N)|^2$. If $N \gg m$, the omission of these lag products is negligible; for small N, however, this omission may be significant. On the other hand, all lag products of $y(t)$ are used to form the $(1, 1)$ and (m, m) elements of $\tilde{R}$, and in general the (i, j) element of $\tilde{R}$ uses more lag products of $y(t)$ than the corresponding element of $\hat{R}$. For more details on the FB approach, we refer the reader to, *e.g.*, [Rao and Hari 1993; Pillai 1989].

Finally, the reader might wonder why we do not replace $\hat{R}$ by a Toeplitz estimate, obtained for example by averaging the elements along each diagonal of $\hat{R}$. This Toeplitz estimate would at first seem to be a better approximation of R than either $\hat{R}$ or $\tilde{R}$. The reason why we do not "Toeplitz–ize" $\hat{R}$ or $\tilde{R}$ is that for finite N, and infinite signal–to–noise ratio ($\sigma^2 \to 0$), the use of either $\hat{R}$ or $\tilde{R}$ gives exact frequency estimates, whereas the Toeplitz–averaged approximation of R does not. As $\sigma^2 \to 0$, both $\hat{R}$ and $\tilde{R}$ have rank n, but the Toeplitz–averaged approximation of R has full rank in general.

4.9 Complements

4.9.1 Mean Square Convergence of Sample Covariances for Line Spectral Processes

In this complement we prove that

$$\boxed{\lim_{N\to\infty} \hat{r}(k) = r(k) \quad \text{(in a mean square sense)}} \tag{4.9.1}$$

(that is, $\lim_{N\to\infty} E\{|\hat{r}(k) - r(k)|^2\} = 0$). The above result has already been referred to in Section 4.4, in the discussion on the rank properties of $\hat{\Omega}$ and Ω. It is also the basic result from which the consistency of all covariance–based frequency estimators discussed in this chapter can be readily concluded. Note that a signal $\{y(t)\}$ satisfying (4.9.1) is said to be *second-order ergodic* (see [SÖDERSTRÖM AND STOICA 1989; BROCKWELL AND DAVIS 1991] for a more detailed discussion of the ergodicity property).

Solution: A straightforward calculation gives

$$\begin{aligned}\hat{r}(k) &= \frac{1}{N}\sum_{t=k+1}^{N}[x(t)+e(t)][x^*(t-k)+e^*(t-k)] \\ &= \frac{1}{N}\sum_{t=k+1}^{N}[x(t)x^*(t-k)+x(t)e^*(t-k)+e(t)x^*(t-k) \\ &\quad +e(t)e^*(t-k)] \triangleq T_1+T_2+T_3+T_4 \end{aligned} \tag{4.9.2}$$

The limit of T_1 is found as follows. First note that:

$$\begin{aligned}\lim_{N\to\infty} E\{|T_1 - r_x(k)|^2\} &= \lim_{N\to\infty}\left\{\frac{1}{N^2}\sum_{t=k+1}^{N}\sum_{s=k+1}^{N} E\{x(t)x^*(t-k)x^*(s)x(s-k)\}\right. \\ &\quad \left. -\left(\frac{2}{N}\sum_{t=k+1}^{N}|r_x(k)|^2\right)+|r_x(k)|^2\right\} \\ &= \lim_{N\to\infty}\left\{\frac{1}{N^2}\sum_{t=k+1}^{N}\sum_{s=k+1}^{N} E\{x(t)x^*(t-k)x^*(s)x(s-k)\}\right\} \\ &\quad -|r_x(k)|^2\end{aligned}$$

Now,

$$\begin{aligned}E\{x(t)x^*(t-k)x^*(s)x(s-k)\} &= \sum_{p=1}^{n}\sum_{j=1}^{n}\sum_{l=1}^{n}\sum_{m=1}^{n} a_p a_j a_l a_m e^{i(\omega_p-\omega_j)t}e^{i(\omega_m-\omega_l)s} \\ &\quad \cdot e^{i(\omega_j-\omega_m)k}E\{e^{i\varphi_p}e^{-i\varphi_j}e^{i\varphi_m}e^{-i\varphi_l}\} \\ &= \sum_{p=1}^{n}\sum_{j=1}^{n}\sum_{l=1}^{n}\sum_{m=1}^{n} a_p a_j a_l a_m e^{i(\omega_p-\omega_j)t}e^{i(\omega_m-\omega_l)s} \\ &\quad \cdot e^{i(\omega_j-\omega_m)k}(\delta_{p,j}\delta_{m,l}+\delta_{p,l}\delta_{m,j}-\delta_{p,j}\delta_{m,l}\delta_{p,m})\end{aligned}$$

where the last equality follows from the assumed independence of the initial phases $\{\varphi_k\}$. Combining the results of the above two calculations yields:

$$\lim_{N\to\infty} E\{|T_1 - r_x(k)|^2\} = \lim_{N\to\infty}\frac{1}{N^2}\sum_{t=k+1}^{N}\sum_{s=k+1}^{N}\left\{\sum_{p=1}^{n}\sum_{m=1}^{n} a_p^2 a_m^2 e^{i(\omega_p-\omega_m)k}\right.$$

$$
\begin{aligned}
&+\sum_{p=1}^{n}\sum_{m=1}^{n} a_p^2 a_m^2 e^{i(\omega_p-\omega_m)(t-s)} - \left.\sum_{p=1}^{n} a_p^4\right\} - |r_x(k)|^2 \\
&= \sum_{p=1}^{n}\sum_{\substack{m=1\\ m\neq p}}^{n} a_p^2 a_m^2 \lim_{N\to\infty}\frac{1}{N^2}\sum_{\tau=-N}^{N}(N-|\tau|)e^{i(\omega_p-\omega_m)\tau} \\
&= 0 \qquad (4.9.3)
\end{aligned}
$$

It follows that T_1 converges to $r(k)$ (in the mean square sense) as N tends to infinity.

The limits of T_2 and T_3 are equal to zero, as shown below for T_2; the proof for T_3 is similar. Using the fact that $\{x(t)\}$ and $\{e(t)\}$ are by assumption independent random signals, we get

$$
\begin{aligned}
E\left\{|T_2|^2\right\} &= \frac{1}{N^2}\sum_{t=k+1}^{N}\sum_{s=k+1}^{N} E\left\{x(t)e^*(t-k)x^*(s)e(s-k)\right\} \\
&= \frac{\sigma^2}{N^2}\sum_{t=k+1}^{N}\sum_{s=k+1}^{N} E\left\{x(t)x^*(s)\right\}\delta_{t,s} \\
&= \frac{\sigma^2}{N^2}\sum_{t=k+1}^{N} E\left\{|x(t)|^2\right\} = \frac{(N-k)\sigma^2}{N^2}E\left\{|x(t)|^2\right\} \qquad (4.9.4)
\end{aligned}
$$

which tends to zero, as $N\to\infty$. Hence, T_2 (and, similarly, T_3) converges to zero in the mean square sense.

The last term, T_4, in (4.9.2) converges to $\sigma^2\delta_{k,0}$ by the "law of large numbers" (see [Söderström and Stoica 1989; Brockwell and Davis 1991]). In fact, it is readily verified, at least under the Gaussian hypothesis, that

$$
\begin{aligned}
E\left\{|T_4-\sigma^2\delta_{k,0}|^2\right\} &= \frac{1}{N^2}\sum_{t=k+1}^{N}\sum_{s=k+1}^{N} E\left\{e(t)e^*(t-k)e^*(s)e(s-k)\right\} \\
&\quad -\sigma^2\delta_{k,0}\left\{\frac{1}{N}\sum_{t=k+1}^{N} E\left\{e(t)e^*(t-k)+e^*(t)e(t-k)\right\}\right\} \\
&\quad +\sigma^4\delta_{k,0} \\
&= \frac{1}{N^2}\sum_{t=k+1}^{N}\sum_{s=k+1}^{N}\left[\sigma^4\delta_{k,0}+\sigma^4\delta_{t,s}\right] \\
&\quad -2\sigma^4\delta_{k,0}\frac{1}{N}\sum_{t=k+1}^{N}(\delta_{k,0}) + \sigma^4\delta_{k,0} \\
&\to \sigma^4\delta_{k,0} - 2\sigma^4\delta_{k,0} + \sigma^4\delta_{k,0} = 0 \qquad (4.9.5)
\end{aligned}
$$

Hence, T_4 converges to $\sigma^2\delta_{k,0}$ in the mean square sense if $e(t)$ is Gaussian. It can be shown using the law of large numbers that $T_4 \to \sigma^2\delta_{k,0}$ in the mean square sense even if $e(t)$ is non–Gaussian, as long as the fourth–order moment of $e(t)$ is finite.

Next, observe that since, for example, $E\{|T_2|^2\}$ and $E\{|T_3|^2\}$ converge to zero, then $E\{T_2T_3^*\}$ also converges to zero (as $N \to \infty$); this is so because

$$|E\{T_2T_3^*\}| \le \left[E\{|T_2|^2\}\,E\{|T_3|^2\}\right]^{1/2}$$

With this observation, the proof of (4.9.1) is complete.

4.9.2 The Carathéodory Parameterization of a Covariance Matrix

The covariance matrix model in (4.2.7) is more general than it might appear at first sight. Show that for *any* given covariance matrix $R = \{r(i-j)\}_{i,j=1}^{m}$, there exist $n \le m$, σ^2 and $\{\omega_k,\ \alpha_k\}_{k=1}^{n}$ such that R can be written as in (4.2.7). Equation (4.2.7), associated with an arbitrary given covariance matrix R, is named the *Carathéodory parameterization* of R.

Solution: Let σ^2 denote the minimum eigenvalue of R. As σ^2 is not necessarily unique, let $\bar{n}$ denote its multiplicity and set $n = m - \bar{n}$. Define

$$\Gamma = R - \sigma^2 I$$

The matrix Γ is positive semidefinite and Toeplitz and, hence, must be the covariance matrix associated with a stationary signal, say $y(t)$:

$$\Gamma = E\left\{\begin{bmatrix} y(t) \\ \vdots \\ y(t-m+1)\end{bmatrix} [y^*(t)\dots y^*(t-m+1)]\right\}$$

By definition,

$$\text{rank}(\Gamma) = n \tag{4.9.6}$$

which implies that there must exist a linear combination between $\{y(t),\dots,y(t-n)\}$ for all t. Moreover, both $y(t)$ and $y(t-n)$ must appear with nonzero coefficients in that linear combination (otherwise either $\{y(t)\dots y(t-n+1)\}$ or $\{y(t-1)\dots y(t-n)\}$ would be linearly related, and rank(Γ) would be less than n, which would contradict (4.9.6)). Hence $y(t)$ obeys the following homogeneous AR equation:

$$B(q)y(t) = 0 \tag{4.9.7}$$

where q^{-1} is the unit delay operator, and

$$B(q) = 1 + b_1q^{-1} + \cdots + b_nq^{-n}$$

with $b_n \neq 0$. Let $\phi(\omega)$ denote the PSD of $y(t)$. Then we have the following equivalences:

$$\begin{aligned} B(q)y(t) = 0 &\iff \int_{-\pi}^{\pi} |B(\omega)|^2 \ \phi(\omega) \ d\omega = 0 \\ &\iff |B(\omega)|^2 \ \phi(\omega) = 0 \\ &\iff \{\text{If } \phi(\omega) > 0 \ \text{ then } \ B(\omega) = 0\} \\ &\iff \{\phi(\omega) > 0 \text{ for at most } n \text{ values of } \omega\} \end{aligned}$$

Furthermore,

$$\begin{aligned} &\{y(t), \ldots y(t-n+1) \text{ are linearly independent}\} \\ &\iff \left\{E\left\{|g_0 y(t) + \ldots + g_{n-1} y(t-n+1)|^2\right\} > 0 \text{ for every } [g_0 \ldots g_{n-1}]^T \neq 0\right\} \\ &\iff \left\{\int_{-\pi}^{\pi} |G(\omega)|^2 \ \phi(\omega) \ d\omega > 0 \text{ for every } G(z) = \sum_{k=0}^{n-1} g_k z^{-k} \neq 0\right\} \\ &\iff \{\phi(\omega) > 0 \text{ for at least } n \text{ distinct values of } \omega\} \end{aligned}$$

It follows from the two results above that $\phi(\omega) > 0$ for exactly n distinct values of ω. Furthermore, the values of ω for which $\phi(\omega) > 0$ are given by the n roots of the equation $B(\omega) = 0$. A signal $y(t)$ with such a PSD consists of a sum of n sinusoidal components with an $m \times m$ covariance matrix given by

$$\Gamma = APA^* \tag{4.9.8}$$

(*cf.* (4.2.7)). In (4.9.8), the frequencies $\{\omega_k\}_{k=1}^n$ are defined as indicated above, and can be found from Γ using any of the subspace–based frequency estimation methods in this chapter. Once $\{\omega_k\}$ are available, $\{\alpha_i^2\}$ can be determined from Γ. (Show that.) By combining the additive decomposition $R = \Gamma + \sigma^2 I$ and (4.9.8) we obtain (4.2.7). With this observation, the derivation of the Carathéodory parameterization is complete.

It is interesting to note that the sinusoids–in–noise signal which "realizes" a given covariance sequence $\{r(0), \ldots, r(m)\}$ (as described above) also provides a *positive definite extension* of that sequence. More precisely, the covariance lags $\{r(m+1),\ r(m+2), \ldots\}$ derived from the sinusoidal signal equation, when appended to $\{r(0), \ldots, r(m)\}$, provide a positive definite covariance sequence of infinite length. The AR covariance realization is the other well–known method for obtaining a positive definite extension of a given covariance sequence of finite length (see the Complement in Section 3.9.2).

4.9.3 Using the Unwindowed Periodogram for Sine Wave Detection in White Noise

As shown in Section 4.3, the unwindowed periodogram is an accurate frequency estimation method whenever the minimum frequency separation is larger than $1/N$. A simple intuitive explanation as to why the unwindowed periodogram is a better frequency estimator than the windowed periodogram(s) is as follows. The principal effect of a window is to remove the tails of the sample covariance sequence from the periodogram formula; while this is appropriate for signals whose covariance sequence "rapidly" goes to zero, it is inappropriate for sinusoidal signals whose covariance sequence never dies out (for sinusoidal signals, the use of a window is expected to introduce a significant bias in the estimated spectrum). Note, however, that if the data contains sinusoidal components with significantly different amplitudes, then it may be advisable to use a (mildly) windowed periodogram. This will induce bias in the frequency estimates, but, on the other hand, will reduce the leakage and hence make it possible to detect the low–amplitude components.

When using the (unwindowed) periodogram for frequency estimation, an important problem is to infer whether any of the many peaks of the erratic periodogram plot can really be associated with the existence of a sinusoidal component in the data. In order to be more precise, consider the following two hypotheses.

H_0: the data consists of (complex circular Gaussian) white noise only (with unknown variance σ^2).

H_1: the data consists of a sum of sinusoidal components and noise.

Deciding between H_0 and H_1 constitutes the so–called ***(signal) detection problem.*** A solution to the detection problem can be obtained as follows. First, paralleling the calculations in Section 2.4, which led to the variance result (2.4.21), shows that ***under*** H_0

The random variables

$$\{2\hat{\phi}_p(\omega_k)/\sigma^2\}_{k=1}^{N}, \tag{4.9.9}$$

with $\min_{k\neq j}\ |\omega_k - \omega_j| \geq 2\pi/N$, are asymptotically independent and χ^2 distributed with 2 degrees of freedom.

(See, *e.g.*, [PRIESTLEY 1989] and [SÖDERSTRÖM AND STOICA 1989] for the definition and properties of the χ^2 distribution.) Next, make use of the result (4.9.9) to devise a ***significance test*** for

$$\max_k\ \hat{\phi}_p(\omega_k)$$

Use that test to decide whether H_0 should be accepted (with an unknown risk) or rejected (and hence H_1 accepted), with a prespecified risk α.

Solution: From the calculations leading to the result (2.4.21) one can see that the normalized periodogram values in (4.9.9) are independent random variables (under H_0). It remains to derive their distribution. Let

$$\epsilon_r(\omega) = \frac{\sqrt{2}}{\sigma\sqrt{N}} \sum_{t=1}^{N} \ \mathrm{Re}[e(t)e^{-i\omega t}]$$

$$\epsilon_i(\omega) = \frac{\sqrt{2}}{\sigma\sqrt{N}} \sum_{t=1}^{N} \text{Im}[e(t)e^{-i\omega t}]$$

With this notation and under the null hypothesis H_0,

$$2\hat{\phi}_p(\omega)/\sigma^2 = \epsilon_r^2(\omega) + \epsilon_i^2(\omega) \tag{4.9.10}$$

For any two complex scalars z_1 and z_2 we have

$$\text{Re}(z_1)\,\text{Im}(z_2) = \frac{z_1 + z_1^*}{2}\,\frac{z_2 - z_2^*}{2i} = \frac{1}{2}\text{Im}\;(z_1 z_2 + z_1^* z_2) \tag{4.9.11}$$

and, similarly,

$$\text{Re}(z_1)\,\text{Re}(z_2) = \frac{1}{2}\text{Re}(z_1 z_2 + z_1^* z_2) \tag{4.9.12}$$

$$\text{Im}(z_1)\,\text{Im}(z_2) = \frac{1}{2}\text{Re}(-z_1 z_2 + z_1^* z_2) \tag{4.9.13}$$

By making use of (4.9.11)–(4.9.13), we can write

$$\begin{aligned} E\left\{\epsilon_r(\omega)\epsilon_i(\omega)\right\} &= \frac{1}{\sigma^2 N}\,\text{Im}\left\{\sum_{t=1}^{N}\sum_{s=1}^{N} E\left\{e(t)e(s)e^{-i\omega(t+s)} + e^*(t)e(s)e^{i\omega(t-s)}\right\}\right\} \\ &= \text{Im}\{1\} = 0 \end{aligned}$$

$$\begin{aligned} E\left\{\epsilon_r^2(\omega)\right\} &= \frac{1}{\sigma^2 N}\,\text{Re}\left\{\sum_{t=1}^{N}\sum_{s=1}^{N} E\left\{e(t)e(s)e^{-i\omega(t+s)} + e^*(t)e(s)e^{i\omega(t-s)}\right\}\right\} \\ &= \text{Re}\{1\} = 1 \end{aligned} \tag{4.9.14}$$

$$\begin{aligned} E\left\{\epsilon_i^2(\omega)\right\} &= \frac{1}{\sigma^2 N}\,\text{Re}\left\{\sum_{t=1}^{N}\sum_{s=1}^{N} E\left\{-e(t)e(s)e^{-i\omega(t+s)} + e^*(t)e(s)e^{i\omega(t-s)}\right\}\right\} \\ &= \text{Re}\{1\} = 1 \end{aligned} \tag{4.9.15}$$

In addition, note that the random variables $\epsilon_r(\omega)$ and $\epsilon_i(\omega)$ are zero–mean Gaussian distributed because they are linear transformations of the Gaussian white noise sequence. Then, it follows that (4.9.9) holds:

Under H_0*:* $\{2\hat{\phi}_p(\omega_k)/\sigma^2\}_{k=1,2,\ldots}$ are asymptotically independent and χ^2 distributed random variables, each with 2 degrees of freedom. (4.9.16)

It is worth noting that if $\{\omega_k\}$ are equal to the Fourier frequencies $\{2\pi k/N\}_{k=0}^{N-1}$, then the previous distributional result is *exactly valid* (*i.e.*, it holds in samples of finite length; see, for example, equation (2.4.26)). However, this observation is not as important as it might seem at first sight, since σ^2 in (4.9.16) is unknown. When the noise power in (4.9.16) is replaced by a consistent estimate $\hat{\sigma}^2$, the so–obtained normalized periodogram values

$$\{2\hat{\phi}_p(\omega_k)/\hat{\sigma}^2\} \tag{4.9.17}$$

are $\chi^2(2)$ distributed only asymptotically (for $N \gg 1$). A consistent estimate of σ^2 can be obtained as follows. From (4.9.10), (4.9.14), and (4.9.15) we have that under H_0

$$E\left\{\hat{\phi}_p(\omega_k)\right\} = \sigma^2 \qquad \text{for } k = 1, 2, \ldots, N$$

Since $\{\hat{\phi}_p(\omega_k)\}_{k=1}^N$ are independent random variables, a consistent estimate of σ^2 is given by

$$\boxed{\hat{\sigma}^2 = \frac{1}{N} \sum_{k=1}^{N} \hat{\phi}_p\,(\omega_k)}$$

Inserting this expression for $\hat{\sigma}^2$ into (4.9.17) leads to the following "test statistic":

$$\boxed{\mu_k = \frac{2N\hat{\phi}_p(\omega_k)}{\sum_{k=1}^{N} \hat{\phi}_p(\omega_k)}}$$

In accordance with the (asymptotic) χ^2 distribution of $\{\mu_k\}$, we have (for any given $c \geq 0$; see, *e.g.*, [PRIESTLEY 1989]):

$$\Pr(\mu_k \leq c) = \int_0^c \frac{1}{2}\, e^{-x/2}\, dx = 1 - e^{-c/2}. \tag{4.9.18}$$

Let

$$\boxed{\mu = \max_k\,[\mu_k]}$$

Using (4.9.18) and the fact that $\{\mu_k\}$ are independent random variables, gives (for any $c \geq 0$):

$$\begin{aligned} \Pr(\mu > c) &= 1 - \Pr(\mu \leq c) \\ &= 1 - \Pr(\mu_k \leq c \text{ for all } k) \\ &= 1 - (1 - e^{-c/2})^N \qquad (\text{under } H_0) \end{aligned}$$

This result can be used to set a bound on μ that, under H_0, holds with a (high) preassigned probability $1 - \alpha$ (say). More precisely, *let α be given* (*e.g.*, $\alpha = 0.05$) and solve for c from the equation

$$\boxed{(1 - e^{-c/2})^N = 1 - \alpha}$$

Then

- If $\mu \leq c$, accept H_0 with an unknown risk. (That risk depends on the signal–to–noise ratio (SNR). The lower the SNR, the larger the risk of accepting H_0 when it does not hold.)
- If $\mu > c$, reject H_0 with a risk equal to α.

It should be noted that whenever H_0 is rejected by the above test, what we can really infer is that the periodogram peak in question is significant enough to make the existence of a sinusoidal component in the studied data highly probable. However, the previous test does not tell us the number of sinusoidal components in the data. In order to determine that number, the test should be continued by looking at the second highest peak in the periodogram. For a test of the significance of the second highest value of the periodogram, and so on, we refer to [PRIESTLEY 1989].

4.10 Exercises

Exercise 4.1: Speed Measurement by a Doppler Radar as a Frequency Determination Problem

Assume that a radar system transmits a sinusoidal signal towards an object. For the sake of simplicity, further assume that the object moves along a trajectory parallel to the wave propagation direction, at a constant velocity v. Let $\alpha e^{i\omega t}$ denote the signal emitted by the radar. Show that the backscattered signal, measured by the radar system after reflection off the object, is given by:

$$s(t) = \beta e^{i(\omega - \omega^D)t} + e(t) \tag{4.10.1}$$

where $e(t)$ is measurement noise, ω^D is the so–called *Doppler frequency,*

$$\omega^D \triangleq 2\omega v / c$$

and

$$\beta = \mu \alpha e^{-2i\omega r/c}$$

Here c denotes the speed of wave propagation, r is the object range, and μ is an attenuation coefficient.

Conclude from (4.10.1) that the problem of speed measurement can be reduced to one of frequency determination. The latter problem can be solved by using the methods of this chapter.

Exercise 4.2: ACS of Sinusoids with Random Amplitudes or Nonuniform Phases

In some applications, it is not reasonable to assume that the amplitudes of the sinusoidal terms are fixed or that their phases are uniformly distributed. Examples are fast fading in mobile telecommunications (where the amplitudes vary) or sinusoids that have been tracked, so that their phase is random, near zero, but not uniformly distributed. We derive the ACS for such cases.

Let $x(t) = \alpha e^{i(\omega_0 t + \varphi)}$, where α and φ are statistically independent random variables and ω_0 is a constant. Assume α has mean $\bar{\alpha}$ and variance σ_α^2, and that φ is zero mean.

(a) If φ is uniformly distributed on $[-\pi, \pi]$, find $E\{x(t)\}$ and $r_x(k)$. Show also that if α is constant, the expression for $r_x(k)$ reduces to equation (4.1.5).

(b) If φ is not uniformly distributed on $[-\pi, \pi]$, express $E\{x(t)\}$ in terms of the probability density function $p(\varphi)$, and find $r_x(k)$. Find sufficient conditions on $p(\varphi)$ such that $x(t)$ is zero mean, and give an example of such a $p(\varphi)$.

Exercise 4.3: A Nonergodic Sinusoidal Signal

As shown in the Complement in Section 4.9.1, the signal

$$x(t) = \alpha e^{i(\omega t + \varphi)}$$

with α and ω being nonrandom constants and φ being uniformly distributed on $[0,\ 2\pi]$, is second–order ergodic in the sense that the mean and covariances determined from an (infinitely long) temporal realization of the signal coincide with the mean and covariances obtained from an ensemble of (infinitely many) realizations. In the present exercise, assume that α and ω are independent random variables, with ω being uniformly distributed on $[0,\ 2\pi]$; the initial–phase variable φ may be arbitrarily distributed (in particular it can be nonrandom). Show that in such a case,

$$E\{x(t)x^*(t-k)\} = \begin{cases} E\{\alpha^2\} & \text{for } k = 0 \\ 0 & \text{for } k \neq 0 \end{cases} \tag{4.10.2}$$

Also, show that the covariances obtained by "temporal averaging" differ from those given, and hence deduce that the signal is not ergodic. Comment on the behavior of such a signal over the ensemble of realizations and in each realization, respectively.

Exercise 4.4: AR Model–Based Frequency Estimation

Consider the following noisy sinusoidal signal:

$$y(t) = x(t) + e(t)$$

where $x(t) = \alpha e^{i(\omega_0 t + \varphi)}$ (with $\alpha > 0$ and φ uniformly distributed on $[0,\ 2\pi]$), and where $e(t)$ is white noise with zero mean and unit variance. An AR model of order $n \geq 1$ is fitted to $\{y(t)\}$ using the Yule–Walker or LS method. Assuming the limiting case of an infinitely long data sample, the AR coefficients are given by the solution to (3.4.4). Show that the PSD, corresponding to the AR model determined from (3.4.4), has a global peak at $\omega = \omega_0$. Conclude that AR modeling can be used in this case to determine the sinusoidal frequency, in spite of the fact that $\{y(t)\}$ does not satisfy an AR equation of finite order (in the case of multiple sinusoids, the AR frequency estimates are biased). Regarding the estimation of the signal power, however, show that the height of the global peak of the AR spectrum does not directly provide an "estimate" of α^2.

Exercise 4.5: An ARMA Model–Based Derivation of the Pisarenko Method

Let R denote the covariance matrix (4.2.7) with $m = n + 1$, and let g be the eigenvector of R associated with its minimum eigenvalue. The Pisarenko method determines the signal frequencies by exploiting the fact that

$$a^*(\omega)g = 0 \qquad \text{for } \omega = \omega_k,\ k = 1, \ldots, n \tag{4.10.3}$$

(*cf.* (4.5.13) and (4.5.17)). Derive the property (4.10.3) directly from the ARMA model equation (4.2.3).

Exercise 4.6: Frequency Estimation when Some Frequencies are Known

Assume that $y(t)$ is known to have p sinusoidal components at known frequencies $\{\tilde{\omega}_k\}_{k=1}^{p}$ (but with unknown amplitudes and phases), and $n - p$ other sinusoidal components whose frequencies are unknown. Develop a modification of the HOYW method to estimate the unknown frequencies from measurements $\{y(t)\}_{t=1}^{N}$, without estimating the known frequencies.

Exercise 4.7: Chebyshev Inequality and the Convergence of Sample Covariances

Let x be a random variable with finite mean μ and variance σ^2. Show that, for any positive constant c, the so-called Chebyshev inequality holds:

$$\boxed{\Pr(|x - \mu| \geq c\sigma) \leq 1/c^2} \tag{4.10.4}$$

Use (4.10.4) to show that if a sample covariance lag $\hat{r}_N$ (estimated from N data samples) converges to the true value r *in the mean square sense, i.e.*,

$$\lim_{N \to \infty} E\left\{|\hat{r}_N - r|^2\right\} = 0 \tag{4.10.5}$$

then $\hat{r}_N$ also converges to r *in probability*:

$$\lim_{N \to \infty} \Pr(|\hat{r}_N - r| \neq 0) = 0 \tag{4.10.6}$$

For sinusoidal signals, the mean square convergence of $\{\hat{r}_N(k)\}$ to $\{r(k)\}$, as $N \to \infty$, has been proven in the Complement in Section 4.9.1. (In this exercise, we omit the argument k in $\hat{r}_N(k)$ and $r(k)$, for notational simplicity.) Additionally, discuss the use of (4.10.4) to set *bounds* (which hold with a specified probability) on an arbitrary random variable with given mean and variance. Comment on the conservatism of the bounds obtained from (4.10.4) by comparing them with the bounds corresponding to a Gaussian random variable.

Exercise 4.8: More about the Forward–Backward Approach

The sample covariance matrix in (4.8.3), used by the forward–backward approach, is often a better estimate of the theoretical covariance matrix than $\hat{R}$ is (as argued in Section 4.8). Another advantage of (4.8.3) is that the forward–backward sample covariance is always numerically better conditioned than the usual (forward-only) sample covariance matrix $\hat{R}$. To explain this statement, let R be a Hermitian matrix (not necessarily a Toeplitz one, as the R in (4.2.7)). The "condition number" of R is defined as

$$\text{cond}(R) = \lambda_{\max}(R)/\lambda_{\min}(R)$$

where $\lambda_{\max}(R)$ and $\lambda_{\min}(R)$ are the maximum and minimum eigenvalues of R, respectively. The numerical errors that affect many algebraic operations on R, such as inversion, eigendecomposition and so on, are essentially proportional to cond(R). Hence, the smaller cond(R) the better. (See Appendix A for details on this aspect.)

Next, let U be a unitary matrix (the J in (4.8.3) is a special case of such a matrix). Observe that the forward–backward covariance in equation (4.8.3) is of the form $R + U^* R^T U$. Prove that

$$\boxed{\text{cond}(R) \geq \text{cond}(R + U^* R^T U)} \tag{4.10.7}$$

for any unitary matrix U. We note that the result (4.10.7) applies to any Hermitian matrix R and unitary matrix U, and thus is valid in more general cases than the forward–backward approach in Section 4.8, in which R is Toeplitz and $U = J$.

Computer Exercises

Tools for Frequency Estimation:

The text web site `www.prenhall.com/~stoica` contains the following MATLAB functions for use in computing frequency estimates. In each case, `y` is the input data vector and `n` is the desired number of frequency estimates.

- `w=hoyw(y,n,L,M)`
  The HOYW estimator given in Table 4.1; `L` and `M` are the matrix dimensions as in (4.4.8).

- `w=music(y,n,m)`
  The Root MUSIC estimator given by (4.5.12); `m` is the dimension of $a(\omega)$. This function also implements the Pisarenko method by setting $m = n + 1$.

- `w=minnorm(y,n,m)`
  The Root Min–Norm estimator given by (4.6.3); `m` is the dimension of $a(\omega)$.

- `w=esprit(y,n,m)`
  The ESPRIT estimator given by (4.7.12); `m` is the size of the square matrix $\hat{R}$ there, and S_1 and S_2 are chosen as in equations (4.7.5) and (4.7.6).

Exercise C4.9: Resolution Properties of Subspace Methods for Estimation of Line Spectra

In this exercise we test and compare the resolution properties of four subspace methods, Min–Norm, MUSIC, ESPRIT, and HOYW.

Generate realizations of the sinusoidal signal

$$y(t) = 10\sin(0.24\pi t + \varphi_1) + 5\sin(0.26\pi t + \varphi_2) + e(t), \qquad t = 1, \ldots, N$$

where $N = 64$, $e(t)$ is Gaussian white noise with variance σ^2, and where φ_1, φ_2 are independent random variables each uniformly distributed on $[-\pi, \pi]$.

Generate 50 Monte–Carlo realizations of $y(t)$, and present the results from these experiments. The results of frequency estimation can be presented comparing the sample means and variances of the frequency estimates from the various estimators.

(a) Find the exact ACS for $y(t)$. Compute the "true" frequency estimates from the four methods, for $n = 4$ and various choices of the order $m \geq 5$ (and corresponding choices of M and L for HOYW). Which method(s) are able to resolve the two sinusoids, and for what values of m (or M and L)?

(b) Consider now $N = 64$, and set $\sigma^2 = 0$; this corresponds to the finite data length but infinite SNR case. Compute frequency estimates for the four techniques again using $n = 4$ and various choices of m, M and L. Which method(s) are reliably able to resolve the sinusoids? Explain why.

(c) Obtain frequency estimates from the four methods when $N = 64$ and $\sigma^2 = 1$. Use $n = 4$, and experiment with different choices of m, M and L to see the effect on estimation accuracy (*e.g.*, try $m = 5$, 8, and 12 for MUSIC, Min–Norm and ESPRIT, and try $L = M = 4$, 8, and 12 for HOYW). Which method(s) give reliable "super–resolution" estimation of the sinusoids? Is it possible to resolve the two sinusoids in the signal? Discuss how the choices of m, M and L influence the resolution properties. Which method appears to have the best resolution?

    You may want to experiment further by changing the SNR and the relative amplitudes of the sinusoids to gain a better understanding of the differences between the methods.

(d) Compare the estimation results with the AR and ARMA results obtained in Exercise C3.17 in Chapter 3. What are the major differences between the techniques? Which method(s) do you prefer for this problem?

Exercise C4.10: Line Spectral Methods applied to Measured Data

Apply the Min–Norm, MUSIC, ESPRIT, and HOYW frequency estimators to the data in the files `sunspotdata.mat` and `lynxdata.mat` (use both the original `lynx` data and the logarithmically transformed data as in Exercise C2.21). These files can be obtained from the text web site `www.prenhall.com/~stoica`. Try to answer the following questions:

(a) Is the sinusoidal model appropriate for the data sets under study?

(b) Suggest how to choose the number of sinusoids in the model.

(c) What periodicities can you find in the two data sets?

Compare the results you obtain here to the AR(MA) and nonparametric spectral estimation results you obtained in Exercises C2.21 and C3.18.

Chapter 5

FILTER BANK METHODS

5.1 Introduction

The problem of estimating the PSD function $\phi(\omega)$ of a signal from a finite number of observations N is ill posed from a statistical standpoint, *unless* we make some appropriate assumptions on $\phi(\omega)$. More precisely, without any assumption on the PSD we are required to estimate an *infinite* number of independent values $\{\phi(\omega)\}_{\omega=-\pi}^{\pi}$ from a *finite* number of samples. Evidently, we cannot do that in a consistent manner. In order to overcome this problem, we can either

$$\boxed{\text{Parameterize } \{\phi(\omega)\} \text{ by means of a finite–dimensional model}} \tag{5.1.1}$$

or

$$\boxed{\begin{array}{l}\text{Smooth the set } \{\phi(\omega)\}_{\omega=-\pi}^{\pi} \text{ by assuming that } \phi(\omega) \text{ is constant (or} \\ \text{nearly constant) over the band } [\omega-\beta\pi, \omega+\beta\pi], \text{ for some given } \beta \ll 1.\end{array}} \tag{5.1.2}$$

The approach based on (5.1.1) leads to the parametric spectral methods of Chapters 3 and 4, for which the estimation of $\{\phi(\omega)\}$ is reduced to the problem of estimating a number of parameters that is usually much smaller than the data length N.

The other approach to PSD estimation, (5.1.2), leads to the methods to be described in this chapter. The nonparametric methods of Chapter 2 are also (implicitly) based on (5.1.2), as shown in Section 5.2. The approach (5.1.2) should, of course, be used for PSD estimation when we do not have enough information about the studied signal to be able to describe it (and its PSD) by a simple model (such as the ARMA equation in Chapter 3 or the equation of superimposed sinusoidal signals in Chapter 4). On one hand, this implies that the methods derived from (5.1.2) can be used in cases where those based on (5.1.1) cannot.[1] On the other

[1]This statement should be interpreted with some care. One can certainly use, for instance, an ARMA spectral model even if one does not know that the studied signal is really an ARMA signal. However, in such a case one does not only have to estimate the model parameters but must also face the rather difficult task of determining the structure of the parametric model used (for example, the orders of the ARMA model). The nonparametric approach to PSD estimation does not require any structure determination step.

hand, we should expect to pay some price in using (5.1.2) over (5.1.1). Under the assumption in (5.1.2), $\phi(\omega)$ is described by $2\pi/2\pi\beta = 1/\beta$ values. In order to estimate these values from the available data in a consistent manner, we must require that $1/\beta < N$ or

$$\boxed{N\beta > 1} \tag{5.1.3}$$

As β increases, the achievable statistical accuracy of the estimates of $\{\phi(\omega)\}$ should increase (because the number of PSD values estimated from the given N data samples decreases) but the resolution decreases (because $\phi(\omega)$ is assumed to be constant on a larger interval). This *tradeoff between statistical variability and resolution* is the price paid for the generality of the methods derived from (5.1.2). We already met this tradeoff in our discussion of the periodogram–based methods in Chapter 2. Note from (5.1.3) that the *resolution threshold* β of the methods based on (5.1.2) can be lowered down to $1/N$ only if we are going to accept a significant statistical variability for our spectral estimates (because for $\beta = 1/N$ we will have to estimate N spectral values from the available N data samples). The parametric (or model–based) approach embodied in (5.1.1) describes the PSD by a number of parameters that is often much smaller than N, and yet it may achieve better resolution (*i.e.*, a resolution threshold less than $1/N$) compared to the approach derived from (5.1.2).

When taking the approach (5.1.2) to PSD estimation, we are basically following the "definition" (1.1.1) of the spectral estimation problem, which we restate here (in abbreviated form) for easy reference:

From a finite–length data sequence, estimate how the power is distributed over narrow spectral bands.	(5.1.4)

There is an implicit assumption in (5.1.4) that the power is (nearly) constant over "narrow spectral bands", which is a restatement of (5.1.2).

The most natural implementation of the approach to spectral estimation resulting from (5.1.2) and (5.1.4) is depicted in Figure 5.1. The bandpass filter in this figure, which sweeps through the frequency interval of interest, can be viewed as a bank of (bandpass) filters. This observation motivates the name of *filter bank approach* given to the PSD estimation scheme sketched in Figure 5.1. Depending on the bandpass filter chosen, we may obtain various filter bank methods of spectral estimation. Even for a given bandpass filter, we may implement the scheme of Figure 5.1 in different ways, which leads to an even richer class of methods. Examples of bandpass filters that can be used in the scheme of Figure 5.1, as well as specific ways in which they may be implemented, are given in the remainder of this chapter. First, however, we discuss a few more aspects regarding the scheme in Figure 5.1.

As a mathematical motivation of the filter bank approach (FBA) to spectral

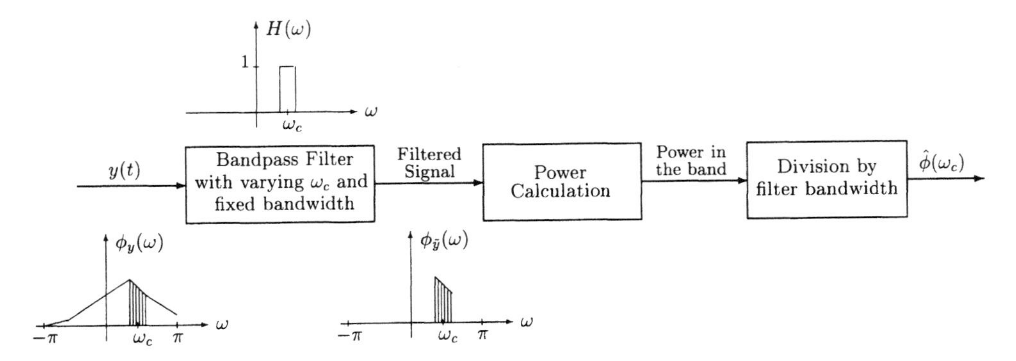

Figure 5.1. The filter bank approach to PSD estimation.

estimation, we prove the following result.

Assume that:

(i) $\phi(\omega)$ is (nearly) constant over the filter passband;

(ii) The filter gain is (nearly) one over the passband and (nearly) zero outside the passband; *and*

(iii) The power of the filtered signal is consistently estimated.

Then:

The PSD estimate, $\hat{\phi}_{\text{FB}}(\omega)$, obtained with the filter bank approach, is a good approximation of $\phi(\omega)$. (5.1.5)

Let $H(\omega)$ denote the transfer function of the bandpass filter, and let $2\pi\beta$ denote its bandwidth. Then by using the formula (1.4.9) and the assumptions (iii), (ii) and (i) (in that order), we can write

$$\begin{aligned}\hat{\phi}_{\text{FB}}(\omega) &\simeq \frac{1}{2\pi\beta}\int_{-\pi}^{\pi}|H(\psi)|^2\phi(\psi)\,d\psi \\ &\simeq \frac{1}{2\pi\beta}\int_{\omega-\beta\pi}^{\omega+\beta\pi}\phi(\psi)\,d\psi \simeq \frac{1}{2\pi\beta}2\pi\beta\phi(\omega) = \phi(\omega)\end{aligned} \tag{5.1.6}$$

where ω denotes the center frequency of the bandpass filter. This was the result which we set out to prove.

If all three assumptions in (5.1.5) could be satisfied, then the FBA methods would produce spectral estimates with high resolution and low statistical variability.

Unfortunately, these assumptions contain conflicting requirements that cannot be met simultaneously. In high–resolution applications, assumption (i) can be satisfied if we use a filter with a *very sharp passband.* According to the time–bandwidth product result (2.6.5), such a filter has a very long impulse response. This implies that we may be able to get only a few samples of the filtered signal (sometimes only one sample, see Section 5.2!). Hence, assumption (iii) cannot be met. In order to satisfy (iii), we need to average many samples of the filtered signal and, therefore, should consider a bandpass filter with a relatively short impulse response and hence a not too narrow passband. Assumption (i) may then be violated or, in other words, the resolution may be sacrificed.

The above discussion has brought once more to light *the compromise between resolution and statistical variability* and the fact that *the resolution is limited by the sample length.* These are the critical issues for any PSD estimation method based on the approach (5.1.2), such as those of Chapter 2 and the ones discussed in the following sections. The previous two issues will always surface within the nonparametric approach to spectral estimation, in many different ways depending on the specific method at hand.

5.2 Filter Bank Interpretation of the Periodogram

The value of the basic periodogram estimator (2.2.1) at a given frequency, say $\tilde{\omega}$, can be expressed as

$$\begin{aligned}\hat{\phi}_p(\tilde{\omega}) &= \frac{1}{N}\left|\sum_{t=1}^{N} y(t)e^{-i\tilde{\omega}t}\right|^2 = \frac{1}{N}\left|\sum_{t=1}^{N} y(t)e^{i\tilde{\omega}(N-t)}\right|^2 \\ &= \frac{1}{\beta}\left|\sum_{k=0}^{N-1} h_k y(N-k)\right|^2 \end{aligned}\tag{5.2.1}$$

where $\beta = 1/N$ and

$$h_k = \frac{1}{N}e^{i\tilde{\omega}k} \qquad k = 0, \dots, N-1 \tag{5.2.2}$$

The *truncated* convolution sum that appears in (5.2.1) can be written as the usual convolution sum associated with a linear causal system, if the weighting sequence in (5.2.2) is padded with zeroes:

$$y_F(N) = \sum_{k=0}^{\infty} h_k y(N-k) \tag{5.2.3}$$

with

$$h_k = \begin{cases} e^{i\tilde{\omega}k}/N & \text{for } k = 0, \dots, N-1 \\ 0 & \text{otherwise} \end{cases} \tag{5.2.4}$$

The transfer function (or the frequency response) of the linear filter corresponding to $\{h_k\}$ in (5.2.4) is readily evaluated:

$$H(\omega) = \sum_{k=0}^{\infty} h_k e^{-i\omega k} = \frac{1}{N} \sum_{k=0}^{N-1} e^{i(\tilde{\omega}-\omega)k} = \frac{1}{N} \frac{e^{iN(\tilde{\omega}-\omega)} - 1}{e^{i(\tilde{\omega}-\omega)} - 1}$$

which gives

$$H(\omega) = \frac{1}{N} \frac{\sin[N(\tilde{\omega}-\omega)/2]}{\sin[(\tilde{\omega}-\omega)/2]} e^{i(N-1)(\tilde{\omega}-\omega)/2} \tag{5.2.5}$$

Figure 5.2 shows $|H(\omega)|$ as a function of $\Delta\omega = \tilde{\omega} - \omega$, for $N = 50$. It can be seen that $H(\omega)$ in (5.2.5) is the transfer function of a bandpass filter with center frequency equal to $\tilde{\omega}$. The *3dB bandwidth* of this filter can be shown to be approximately $2\pi/N$ radians per sampling interval, or $1/N$ cycles per sampling interval. In fact, by comparing (5.2.5) to (2.4.17) we see that $H(\omega)$ resembles the DTFT of the rectangular window, the only differences being the phase term (due to the time offset) and the window lengths ($(2N-1)$ in (2.4.17) versus N in (5.2.5)).

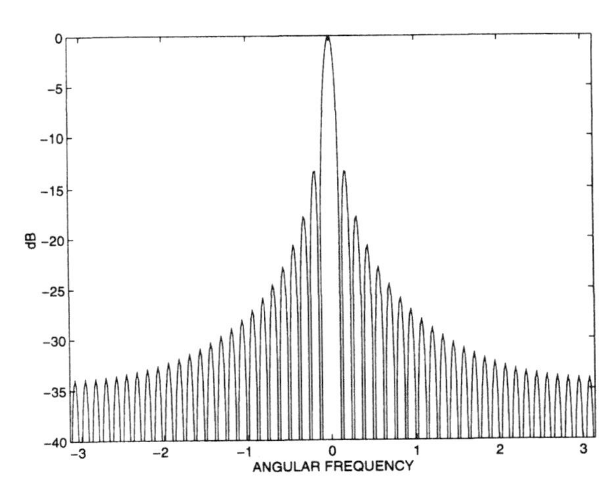

Figure 5.2. The magnitude of the frequency response of the bandpass filter $H(\omega)$ in (5.2.5), associated with the periodogram ($N = 50$), plotted as a function of $(\tilde{\omega} - \omega)$.

Thus, we have proven the following *filter bank interpretation of the basic periodogram.*

> The periodogram $\hat{\phi}_p(\omega)$ can be exactly obtained by the FBA in Figure 5.1, where the bandpass filter's frequency response is given by (5.2.5), its bandwidth is $1/N$ cycles per sampling interval, and the power calculation is done from a *single sample* of the filtered signal. (5.2.6)

This interpretation of $\hat{\phi}_p(\omega)$ highlights a conclusion that was reached, in a different way, in Chapter 2: *the unmodified periodogram sacrifices statistical accuracy for resolution.* Indeed, $\hat{\phi}_p(\omega)$ uses a bandpass filter with the smallest bandwidth afforded by a time aperture of length N. In this way, it achieves a good resolution (see assumption (i) in (5.1.5)). The consequence of doing so is that only one (filtered) data sample is obtained for the power calculation stage, which explains the erratic fluctuations of $\hat{\phi}_p(\omega)$ (owing to violation of assumption (iii) in (5.1.5)).

As explained in Chapter 2, the *modified periodogram methods* (Bartlett, Welch and Daniell) reduce the variance of the periodogram at the expense of increasing the bias (or, equivalently, worsening the resolution). The FBA interpretation of these modified methods provides an interesting explanation of their behavior. In the filter bank context, the basic idea behind all of these modified periodograms is *to improve the power calculation stage* which is done so poorly within the unmodified periodogram.

The Bartlett and Welch methods split the available sample in several stretches which are separately (bandpass) filtered. In principle, the larger the number of stretches, the more samples are averaged in the power calculation stage and the smaller the variance of the estimated PSD, but the worse the resolution (owing to the inability to design an appropriately narrow bandpass filter for a small–aperture stretch).

The Daniell method, on the other hand, does not split the sample of observations but processes it as a whole. This method improves the "power calculation" in a different way. For each value of $\phi(\omega)$ to be estimated, a number of different bandpass filters are employed, each with center frequency near ω. Each bandpass filter yields only one sample of the filtered signal, but as there are several bandpass filters we may get enough information for the power calculation stage. As the number of filters used increases, the variance of the estimated PSD decreases but the resolution becomes worse (since $\phi(\omega)$ is implicitly assumed to be constant over a wider and wider frequency interval centered on the current ω and approximately equal to the union of the filters' passbands).

5.3 Refined Filter Bank Method

The bandpass filter used in the periodogram is nothing but one of many possible choices. Since the periodogram was *not* designed as a filter bank method, we may wonder whether we could not find other better choices of the bandpass filter. In this section, we present a refined filter bank (RFB) approach to spectral estimation. Such an approach was introduced in [THOMSON 1982] and was further developed in [MULLIS AND SCHARF 1991] (more recent references on this approach include [BRONEZ 1992; ONN AND STEINHARDT 1993; RIEDEL AND SIDORENKO 1995]).

For the discussion that follows, it is convenient to use a *baseband filter* in the filter bank approach of Figure 5.1, in lieu of the bandpass filter. Let $H_{\text{BF}}(\omega)$ denote the frequency response of the bandpass filter with center frequency $\tilde{\omega}$ (say), and let

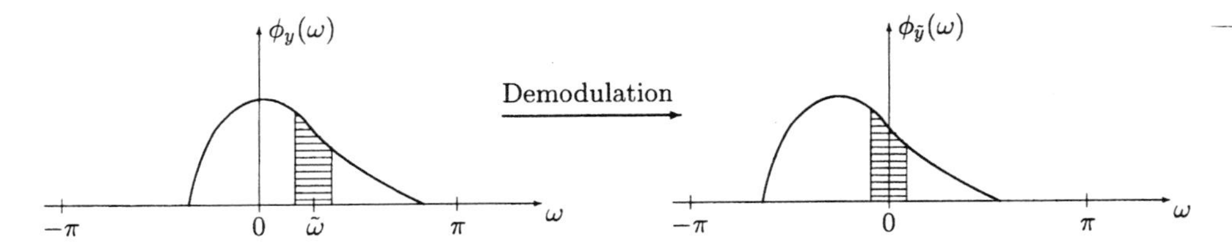

Figure 5.3. The relationship between the PSDs of the original signal $y(t)$ and the demodulated signal $\tilde{y}(t)$.

the baseband filter be defined by:

$$H(\omega) = H_{\mathrm{BF}}(\omega + \tilde{\omega}) \tag{5.3.1}$$

(the center frequency of $H(\omega)$ is equal to zero). If the input to the FBA scheme is also modified in the following way,

$$\boxed{y(t) \longrightarrow \tilde{y}(t) = e^{-i\tilde{\omega}t} y(t)} \tag{5.3.2}$$

then, according to the complex (de)modulation formula (1.4.11), the output of the scheme is left unchanged by the translation in (5.3.1) of the passband down to baseband. In order to help interpret the transformations above, we depict in Figure 5.3 the type of PSD translation implied by the *demodulation process* in (5.3.2). It is clearly seen from this figure that the problem of isolating the band around $\tilde{\omega}$ by bandpass filtering becomes one of baseband filtering. The modified FBA scheme is shown in Figure 5.4. The baseband filter design problem is the subject of the next subsection.

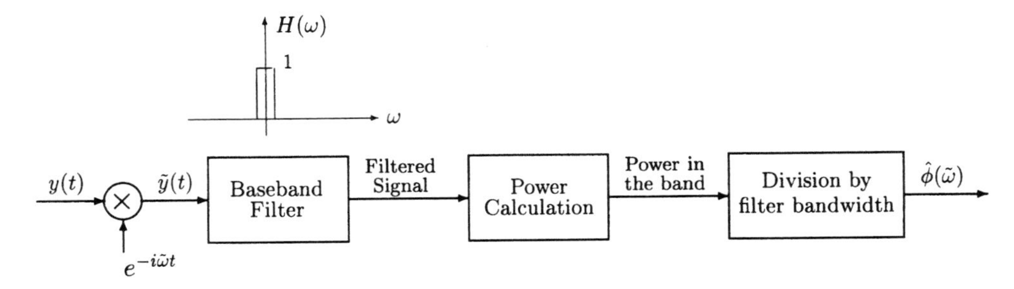

Figure 5.4. The modified filter bank approach to PSD estimation.

5.3.1 Slepian Baseband Filters

In the following, we address the problem of designing a finite impulse response (FIR) baseband filter which passes the *baseband*

$$[-\beta\pi, \beta\pi] \tag{5.3.3}$$

as undistorted as possible, and which attenuates the frequencies outside baseband as much as possible. Let

$$h = [h_0 \dots h_{N-1}]^* \tag{5.3.4}$$

denote the impulse response of such a filter, and let

$$H(\omega) = \sum_{k=0}^{N-1} h_k e^{-i\omega k} = h^* a(\omega)$$

(where $a(\omega) = [1 \; e^{-i\omega} \dots \; e^{-i(N-1)\omega}]^T$) be the corresponding frequency response. The two design objectives can be turned into mathematical specifications in the following way. Let the input to the filter be *white noise* of unit variance. Then the power of the output is:

$$\begin{aligned} \frac{1}{2\pi}\int_{-\pi}^{\pi} |H(\omega)|^2 d\omega &= \sum_{k=0}^{N-1}\sum_{p=0}^{N-1} h_k h_p^* \left[\frac{1}{2\pi}\int_{-\pi}^{\pi} e^{i\omega(p-k)} d\omega\right] \\ &= \sum_{k=0}^{N-1}\sum_{p=0}^{N-1} h_k h_p^* \delta_{k,p} = h^* h \end{aligned} \tag{5.3.5}$$

We note in passing that equation (5.3.5) above can be recognized as the Parseval's theorem (1.2.6). The part of the total power, (5.3.5), that lies in the baseband is given by

$$\frac{1}{2\pi}\int_{-\beta\pi}^{\beta\pi} |H(\omega)|^2 d\omega = h^* \left\{\frac{1}{2\pi}\int_{-\beta\pi}^{\beta\pi} a(\omega)a^*(\omega) d\omega\right\} h \triangleq h^*\Gamma h \tag{5.3.6}$$

The k, p element of the $N \times N$ matrix Γ defined in (5.3.6) is given by

$$\Gamma_{k,p} = \frac{1}{2\pi}\int_{-\beta\pi}^{\beta\pi} e^{-i(k-p)\omega} d\omega = \frac{\sin[(k-p)\beta\pi]}{(k-p)\pi} \tag{5.3.7}$$

which, using the sinc function, can be written as

$$\boxed{\Gamma_{k,p} = \beta \mathrm{sinc}[(k-p)\beta\pi] \triangleq \gamma_{|k-p|}} \tag{5.3.8}$$

Note that the matrix Γ is symmetric and Toeplitz. Also, note that this matrix has already been encountered in the window design example in Section 2.6.3. In fact, as we will shortly see, the window design strategy in that example is quite similar to the baseband filter design method employed here.

Since the filter h must be such that the power of the filtered signal in the baseband is as large as possible relative to the total power, we are led to the following optimization problem:

$$\max_{h} h^* \Gamma h \qquad \text{subject to} \quad h^* h = 1 \tag{5.3.9}$$

The solution to the problem above is given in Result R13 in Appendix A: the maximizing h is equal to the eigenvector of Γ corresponding to its maximum eigenvalue. Hence, we proved the following result.

> The impulse response h of the "most selective" baseband filter (according to the design objectives in (5.3.9)) is given by the dominant eigenvector of Γ, and is called the *first Slepian sequence*. (5.3.10)

The matrix Γ played a key role in the foregoing derivation. In what follows, we look in more detail at the *eigenstructure of* Γ. In particular, we provide an intuitive explanation as to why the first dominant eigenvector of Γ behaves like a baseband filter. We also show that, depending on the relation between β and N, the next dominant eigenvectors of Γ might also be used as baseband filters. Our discussion of these aspects will be partly heuristic. Note that the eigenvectors of Γ are called the *Slepian sequences* [SLEPIAN 1964] (as already indicated in (5.3.10)). We denote these eigenvectors by $\{s_k\}_{k=1}^{N}$.

The theoretical eigenanalysis of Γ is a difficult problem in the case of finite N. (Of course, the eigenvectors and eigenvalues of Γ may always be *computed*, for given β and N; here we are interested in establishing *theoretical expressions* for Γ's eigenelements.) For N sufficiently large, however, "reasonable approximations" to the eigenelements of Γ can be derived. Let $a(\omega)$ be defined as before:

$$a(\omega) = [1 \;\; e^{-i\omega} \ldots \; e^{-i(N-1)\omega}]^T \tag{5.3.11}$$

Assume that β is *chosen larger than* $1/N$, and define

$$K = N\beta \geq 1 \tag{5.3.12}$$

(To simplify the discussion, K and N are assumed to be even integers in what follows.) With these preparations and assuming that N is large, we can approximate the integral in (5.3.6) and write Γ as

$$\begin{aligned} \Gamma &\simeq \frac{1}{2\pi} \sum_{p=-K/2}^{K/2-1} a\left(\frac{2\pi}{N}p\right) a^*\left(\frac{2\pi}{N}p\right) \frac{2\pi}{N} \\ &= \frac{1}{N} \sum_{p=-K/2}^{K/2-1} a\left(\frac{2\pi}{N}p\right) a^*\left(\frac{2\pi}{N}p\right) \triangleq \Gamma_0 \end{aligned} \tag{5.3.13}$$

The vectors $\{a(\frac{2\pi}{N}p)/\sqrt{N}\}_{p=-\frac{N}{2}+1}^{\frac{N}{2}}$, part of which appears in (5.3.13), can be readily shown to form an *orthonormal set*:

$$\frac{1}{N}a^*\left(\frac{2\pi}{N}p\right)a\left(\frac{2\pi}{N}s\right) = \frac{1}{N}\sum_{k=0}^{N-1} e^{i\frac{2\pi}{N}(p-s)k}$$
$$= \begin{cases} \frac{1}{N}\frac{e^{i2\pi(p-s)}-1}{e^{i\frac{2\pi}{N}(p-s)}-1} = 0, & s \neq p \\ 1, & s = p \end{cases} \tag{5.3.14}$$

The eigenvectors of the matrix on the right hand side of equation (5.3.13), Γ_0, are therefore given by $\{a\left(\frac{2\pi}{N}p\right)/\sqrt{N}\}_{p=-N/2+1}^{N/2}$, with eigenvalues of 1 (with multiplicity K) and 0 (with multiplicity $N-K$). The eigenvectors corresponding to the eigenvalues equal to one are $\{a\left(\frac{2\pi}{N}p\right)/\sqrt{N}\}_{p=-K/2+1}^{K/2}$. By paralleling the calculations in (5.2.3)–(5.2.5), it is not hard to show that each of these dominant eigenvectors of Γ_0 is the impulse response of a narrow bandpass filter with bandwidth equal to about $1/N$ and center frequency $\frac{2\pi}{N}p$; the set of these filters therefore covers the interval $[-\beta\pi, \beta\pi]$.

Now, the elements of Γ approach those of Γ_0 as N increases; more precisely, $|[\Gamma]_{i,j} - [\Gamma_0]_{i,j}| = O(1/N)$ for sufficiently large N. However, this does *not* mean that $\|\Gamma - \Gamma_0\| \to 0$, as $N \to \infty$, for any reasonable matrix norm, because Γ and Γ_0 are $(N \times N)$ matrices. Consequently, the eigenelements of Γ do *not* necessarily converge to the eigenelements of Γ_0 as $N \to \infty$. However, based on the previous analysis, we can at least expect that the eigenelements of Γ are not "too different" from those of Γ_0. This observation of the theoretical analysis, backed up with empirical evidence from the computation of the eigenelements of Γ in specific cases, leads us to conclude the following.

> The matrix Γ has K eigenvalues close to one and $(N-K)$ eigenvalues close to zero, provided N is large enough, where K is given by the "time–bandwidth" product (5.3.12). The dominant eigenvectors corresponding to the K largest eigenvalues form a set of orthogonal impulse responses of K bandpass filters that approximately cover the baseband $[-\beta\pi, \beta\pi]$. (5.3.15)

As we argue in the next subsections, in some situations (specified there) we may want to use the whole set of K *Slepian baseband filters*, not only the dominant Slepian filter in this set.

5.3.2 RFB Method for High–Resolution Spectral Analysis

Assume that the spectral analysis problem dealt with is one in which it is important to achieve the maximum resolution afforded by the approach at hand (such a problem appears, for instance, in the case of PSD's with closely spaced peaks). Then we set

$$\boxed{\beta = 1/N \Longleftrightarrow K = 1} \tag{5.3.16}$$

(Note that we cannot set β to a value less than $1/N$ since that choice would lead to $K < 1$, which is meaningless; the fact that we must choose $\beta \geq 1/N$ is one of the many facets of the $1/N$–resolution limit of the nonparametric spectral estimation.)

Since $K = 1$, we can only use the first Slepian sequence as a bandpass filter

$$h = s_1 \tag{5.3.17}$$

The way in which the RFB scheme based on (5.3.17) works is described in the following.

First, note from (5.3.5), (5.3.9) and (5.3.16) that

$$\begin{aligned} 1 = h^* h = \frac{1}{2\pi}\int_{-\pi}^{\pi} |H(\omega)|^2 d\omega &\simeq \frac{1}{2\pi}\int_{-\beta\pi}^{\beta\pi} |H(\omega)|^2 d\omega \\ \simeq \beta |H(0)|^2 = \frac{1}{N}|H(0)|^2 & \end{aligned} \tag{5.3.18}$$

Hence, under the (idealizing) assumption that $H(\omega)$ is different from zero only in the baseband where it takes a constant value, we have

$$|H(0)|^2 \simeq N \tag{5.3.19}$$

Next, consider the sample at the filter's output obtained by the convolution of the whole input sequence $\{\tilde{y}(t)\}_{t=1}^{N}$ with the filter impulse response $\{h_k\}$:

$$x \triangleq \sum_{k=0}^{N-1} h_k \tilde{y}(N-k) = \sum_{t=1}^{N} h_{N-t}\tilde{y}(t) \tag{5.3.20}$$

The power of x should be approximately equal to the PSD value $\phi(\tilde{\omega})$, which is confirmed by the following calculation:

$$\begin{aligned} E\left\{|x|^2\right\} &= \frac{1}{2\pi}\int_{-\pi}^{\pi} |H(\omega)|^2 \phi_{\tilde{y}}(\omega) d\omega \\ &\simeq \frac{N}{2\pi}\int_{-\beta\pi}^{\beta\pi} \phi_{\tilde{y}}(\omega) d\omega = \frac{N}{2\pi}\int_{-\beta\pi}^{\beta\pi} \phi_y(\omega + \tilde{\omega}) d\omega \\ &\simeq \frac{N}{2\pi}\phi_y(\tilde{\omega}) \times 2\pi\beta = N\beta\phi_y(\tilde{\omega}) = \phi_y(\tilde{\omega}) \end{aligned} \tag{5.3.21}$$

The second "equality" above follows from the properties of $H(\omega)$ (see, also, (5.3.19)), the third from the complex demodulation formula (1.4.11), and the fourth from the assumption that $\phi_y(\omega)$ is nearly constant over the passband considered.

In view of (5.3.21), the PSD estimation problem reduces to estimating the power of the filtered signal. Since only one sample, x, of that signal is available, the obvious estimate for the signal power is $|x|^2$. This leads to the following estimate of $\phi(\omega)$:

$$\hat{\phi}(\omega) = \left| \sum_{t=1}^{N} h_{N-t} y(t) e^{-i\omega t} \right|^2 \tag{5.3.22}$$

where $\{h_k\}$ is given by the first Slepian sequence (see (5.3.17)). The reason we did not divide (5.3.22) by the filter bandwidth is that $|H(0)|^2 \simeq N$ by (5.3.19), which differs from assumption (ii) in (5.1.5).

The spectral estimate (5.3.22) is recognized to be a *windowed periodogram* with *temporal window* $\{h_{N-k}\}$. For large values of N, it follows from the analysis in the previous section that h can be expected to be reasonably close to the vector $[1 \ldots 1]^T/\sqrt{N}$. When inserting the latter vector in (5.3.22), we get the unwindowed periodogram. Hence, we reach the conclusion that *for N large enough, the RFB estimate* (5.3.22) *will behave not too differently from the unmodified periodogram* (which is quite natural in view of the fact that we wanted a high–resolution spectral estimator, and the basic periodogram is known to be such an estimator).

Remark: We warn the reader, once again, that the above discussion is heuristic. As explained before (see the discussion related to (5.3.15)), as N increases $\{h_k\}$ may be expected to be "reasonably close" but not necessarily converge to $1/\sqrt{N}$. In addition, even if $\{h_k\}$ in (5.3.22) converges to $1/\sqrt{N}$ as $N \to \infty$, the function in (5.3.22) may not converge to $\hat{\phi}_p(\omega)$ if the convergence *rate* of $\{h_k\}$ is too slow (note that the number of $\{h_k\}$ in (5.3.22) is equal to N). Hence $\hat{\phi}(\omega)$ in (5.3.22) and the periodogram $\hat{\phi}_p(\omega)$ may differ from one another even for large values of N. ■

In any case, even though the two estimators $\hat{\phi}(\omega)$ in (5.3.22) and $\hat{\phi}_p(\omega)$ generally give different PSD values, they both base the power calculation stage of the FBA scheme on only a single sample. Hence, similarly to $\hat{\phi}_p(\omega)$, the RFB estimate (5.3.22) is expected to exhibit erratic fluctuations. The next subsection discusses a way in which the variance of the RFB spectral estimate can be reduced, at the expense of reducing the resolution of this estimate.

5.3.3 RFB Method for Statistically Stable Spectral Analysis

The FBA interpretation of the modified periodogram methods, as explained in Section 5.2, highlighted two approaches to reduce the statistical variability of the spectral estimate (5.3.22). The *first approach* consists of splitting the available sample $\{y(t)\}_{t=1}^N$ into a number of subsequences, computing (5.3.22) for each stretch, and then averaging the so–obtained values. The problem with this way of proceeding is that the values taken by (5.3.22) for different subsequences are not guaranteed to be statistically independent. In fact, if the subsequences overlap then those values may be strongly correlated. The consequence of this fact is that one can never be sure of the "exact" reduction of variance that is achieved by averaging, in a given situation.

The *second approach* to reduce the variance consists of using several bandpass filters, in lieu of only one, which operate on the whole data sample [THOMSON 1982]. This approach aims at producing *statistically independent samples for the power calculation stage.* When this is achieved *the variance is reduced K times,* where K is the number of samples averaged (which equals the number of bandpass filters used).

In the following, we focus on this second approach which appears particularly suitable for the RFB method. We set β to some value larger than $1/N$, which gives (*cf.* (5.3.12))

$$\boxed{K = N\beta > 1} \tag{5.3.23}$$

The larger β (*i.e.*, the lower the resolution), the larger K and hence the larger the reduction in variance that can be achieved. By using the result (5.3.15), we define K baseband filters as

$$\boxed{h_p = [h_{p,0} \dots h_{p,N-1}]^* = s_p, \qquad (p = 1, \dots, K)} \tag{5.3.24}$$

Here h_p denotes the impulse response vector of the pth filter, and s_p is the pth dominant Slepian sequence. Note that s_p is real-valued (see Result R12 in Appendix A), and thus so is h_p. According to the discussion leading to (5.3.15), the set of filters (5.3.24) covers the baseband $[-\beta\pi, \beta\pi]$, with each of these filters passing (roughly speaking) $(1/K)$th of this baseband. Let x_p be defined similarly to x in (5.3.20), but now for the pth filter:

$$x_p = \sum_{k=0}^{N-1} h_{p,k}\tilde{y}(N-k) = \sum_{t=1}^{N} h_{p,N-t}\tilde{y}(t) \tag{5.3.25}$$

The calculation (5.3.21) applies to $\{x_p\}$ in exactly the same way, and hence

$$E\left\{|x_p|^2\right\} \simeq \phi_y(\tilde{\omega}), \qquad p = 1, \dots, K \tag{5.3.26}$$

In addition, a straightforward calculation gives

$$\begin{aligned}
E\{x_p x_k^*\} &= E\left\{\left[\sum_{t=0}^{N-1} h_{p,t}\tilde{y}(N-t)\right]\left[\sum_{s=0}^{N-1} h_{k,s}^*\tilde{y}^*(N-s)\right]\right\} \\
&= \sum_{t=0}^{N-1}\sum_{s=0}^{N-1} h_{p,t}h_{k,s}^* r_{\tilde{y}}(s-t) \\
&= \frac{1}{2\pi}\int_{-\pi}^{\pi}\sum_{t=0}^{N-1}\sum_{s=0}^{N-1} h_{p,t}h_{k,s}^*\phi_{\tilde{y}}(\omega)e^{i(s-t)\omega}d\omega
\end{aligned}$$

$$= \frac{1}{2\pi} \int_{-\pi}^{\pi} H_p(\omega) H_k^*(\omega) \phi_{\tilde{y}}(\omega) d\omega$$

$$\simeq \phi_{\tilde{y}}(0) h_p^* \left[\frac{1}{2\pi} \int_{-\beta\pi}^{\beta\pi} a(\omega) a^*(\omega) d\omega \right] h_k$$

$$= \phi_y(\tilde{\omega}) h_p^* \Gamma h_k = 0 \qquad \text{for } k \neq p \tag{5.3.27}$$

Thus, the random variables x_p and x_k (for $p \neq k$) are approximately uncorrelated under the assumptions made. This implies, at least under the assumption that the $\{x_k\}$ are Gaussian, that $|x_p|^2$ and $|x_k|^2$ are *statistically independent* (for $p \neq k$).

According to the calculations above, $\{|x_p|^2\}_{p=1}^{K}$ can approximately be considered to be independent random variables all with the same mean $\phi_y(\tilde{\omega})$. Then, we can estimate $\phi_y(\tilde{\omega})$ by the following average of $\{|x_p|^2\}$: $\frac{1}{K}\sum_{p=1}^{K} |x_p|^2$, or

$$\boxed{\hat{\phi}(\omega) = \frac{1}{K} \sum_{p=1}^{K} \left| \sum_{t=1}^{N} h_{p,N-t}\, y(t) e^{-i\omega t} \right|^2} \tag{5.3.28}$$

We may suspect that the random variables $\{|x_p|^2\}$ have not only the same mean, but also the same variance (this can, in fact, be readily shown under the Gaussian hypothesis). Whenever this is true, the variance of the average in (5.3.28) is K times smaller than the variance of each of the variables averaged.

The above findings are summarized in the following.

> If the resolution threshold β is increased K times from $\beta = 1/N$ (the lowest value) to $\beta = K/N$, then the variance of the RFB estimate in (5.3.22) may be reduced by a factor K by constructing the spectral estimate as in (5.3.28), where the pth baseband filter's impulse response $\{h_{p,t}\}_{t=0}^{N-1}$ is given by the pth dominant Slepian sequence ($p = 1, \ldots, K$). (5.3.29)

The RFB spectral estimator (5.3.28) can be given two interpretations. First, arguments similar with those following equation (5.3.22) suggest that *for large N the RFB estimate (5.3.28) behaves similarly to the Daniell method of periodogram averaging.* For small or medium-sized values of N, the RFB and Daniell methods behave differently. In such a case, we can relate (5.3.28) to the class of *multiwindow spectral estimators* [THOMSON 1982]. Indeed, the RFB estimate (5.3.28) can be interpreted as the average of K windowed periodograms, where the pth periodogram is computed from the raw data sequence $\{y(t)\}$ windowed with the pth dominant Slepian sequence. Note that since the Slepian sequences are given by the eigenvectors of the real *Toeplitz* matrix Γ, they must be either symmetric: $h_{p,N-t} = h_{p,t-1}$; or skew-symmetric: $h_{p,N-t} = -h_{p,t-1}$ (see Result R25 in Appendix A). This means that (5.3.28) can alternatively be written as

$$\hat{\phi}(\omega) = \frac{1}{K}\sum_{p=1}^{K}\left|\sum_{t=1}^{N} h_{p,t-1}y(t)e^{-i\omega t}\right|^2 \tag{5.3.30}$$

This form of the RFB estimate makes its interpretation as a multiwindow spectrum estimator more direct.

For a second interpretation of the RFB estimate (5.3.28), consider the following (Daniell–type) spectrally smoothed periodogram estimator of $\phi(\tilde{\omega})$:

$$\begin{aligned}
\hat{\phi}(\tilde{\omega}) &= \frac{1}{2\pi\beta}\int_{\tilde{\omega}-\beta\pi}^{\tilde{\omega}+\beta\pi} \hat{\phi}_p(\omega)d\omega = \frac{1}{2\pi\beta}\int_{-\beta\pi}^{\beta\pi} \hat{\phi}_p(\omega+\tilde{\omega})d\omega \\
&= \frac{1}{2\pi\beta}\int_{-\beta\pi}^{\beta\pi} \frac{1}{N}\left|\sum_{t=1}^{N} y(t)e^{-i(\omega+\tilde{\omega})t}\right|^2 d\omega \\
&= \frac{1}{2\pi K}\int_{-\beta\pi}^{\beta\pi} \sum_{t=1}^{N}\sum_{s=1}^{N} \tilde{y}(t)\tilde{y}^*(s)e^{-i\omega t}e^{i\omega s}d\omega \\
&= \frac{1}{K}[\tilde{y}^*(1) \;\ldots\; \tilde{y}^*(N)] \\
&\quad \cdot\left\{\frac{1}{2\pi}\int_{-\beta\pi}^{\beta\pi}\begin{bmatrix} 1 \\ e^{i\omega} \\ \vdots \\ e^{i(N-1)\omega}\end{bmatrix}[1 \;\; e^{-i\omega} \ldots e^{-i(N-1)\omega}]d\omega\right\}\begin{bmatrix}\tilde{y}(1) \\ \vdots \\ \tilde{y}(N)\end{bmatrix} \\
&= \frac{1}{K}[\tilde{y}^*(1) \;\ldots\; \tilde{y}^*(N)]\Gamma\begin{bmatrix}\tilde{y}(1) \\ \vdots \\ \tilde{y}(N)\end{bmatrix}
\end{aligned} \tag{5.3.31}$$

where we made use of the fact that Γ is real–valued. It follows from the result (5.3.15) that Γ can be approximated by the *rank–K* matrix:

$$\Gamma \simeq \sum_{p=1}^{K} s_p s_p^T = \sum_{p=1}^{K} h_p h_p^T \tag{5.3.32}$$

Inserting (5.3.32) into (5.3.31) and using the fact that the Slepian sequences $s_p = h_p$ are real–valued leads to the following PSD estimator:

$$\hat{\phi}(\tilde{\omega}) \simeq \frac{1}{K}\sum_{p=1}^{K}\left|\sum_{t=1}^{N} h_{p,t-1}\tilde{y}(t)\right|^2 \tag{5.3.33}$$

which is precisely the RFB estimator (5.3.30). Hence, the RFB estimate of the PSD can also be interpreted as a *reduced–rank smoothed periodogram.*

We might think of using the full–rank smoothed periodogram (5.3.31) as an estimator for PSD, in lieu of the reduced–rank smoothed periodogram (5.3.33) which coincides with the RFB estimate. However, from a theoretical standpoint we have no strong reason to do so. Moreover, from a practical standpoint we have clear reasons against such an idea. We can explain this briefly as follows. The K dominant eigenvectors of Γ can be *precomputed* with satisfactory numerical accuracy. Then, evaluation of (5.3.33) can be done by using an FFT algorithm in approximately $\frac{1}{2}KN\log_2 N = \frac{1}{2}\beta N^2 \log_2 N$ flops. On the other hand, a direct evaluation of (5.3.31) would require N^2 flops for each value of ω, which leads to a prohibitively large total computational burden. A computationally efficient evaluation of (5.3.31) would require some factorization of Γ to be performed, such as the eigendecomposition of Γ. However, Γ is an extremely ill–conditioned matrix (recall that $N - K = N(1-\beta)$ of its eigenvalues are close to zero), which means that such a complete factorization cannot easily be performed with satisfactory numerical accuracy. In any case even if we were able to precompute the eigendecomposition of Γ, evaluation of (5.3.31) would require $\frac{1}{2}N^2 \log_2 N$ flops, which is still larger by a factor of $1/\beta$ than what is required for (5.3.33).

5.4 Capon Method

The periodogram was previously shown to be a filter bank approach which uses a bandpass filter whose impulse response vector is given by the standard Fourier transform vector (*i.e.*, $[1, e^{-i\tilde{\omega}}, \ldots, e^{-i(N-1)\tilde{\omega}}]^T$). In the periodogram approach there was *no attempt to purposely design* the bandpass filter to achieve some desired characteristics (see, however, Section 5.5). The RFB method, on the other hand, uses a bandpass filter specifically designed to be "*as selective as possible*" *for a white noise input* (see (5.3.5) and the discussion preceding it). The RFB's filter is still *data independent* in the sense that it does not adapt to the processed data in any way. Presumably, it might be valuable to take the data properties into consideration when designing the bandpass filter. In other words, the filter should be designed to be "as selective as possible" (according to a criterion to be specified) not for a fictitious white noise input, but for the input consisting of the studied data themselves. This is the basic idea behind the Capon method, which is an FBA procedure based on a *data–dependent bandpass filter* [Capon 1969; Lacoss 1971].

5.4.1 Derivation of the Capon Method

The Capon method (CM), in contrast to the RFB estimator (5.3.28), uses only *one bandpass filter* for computing one estimated spectrum value. This suggests that if the CM is to provide statistically stable spectral estimates, then it should make use of the other approach which affords this: *splitting the raw sample into subsequences* and averaging the results obtained from each subsequence. Indeed, as we shall see the Capon method is essentially based on this second approach.

Consider a filter with a finite impulse response of length m, denoted by

$$h = [h_0 \;\; h_1 \ldots h_m]^* \tag{5.4.1}$$

where m is a positive integer that is unspecified for the moment. The output of the filter at time t, when the input is the raw data sequence $\{y(t)\}$, is given by

$$\begin{aligned} y_F(t) &= \sum_{k=0}^{m} h_k y(t-k) \\ &= h^* \begin{bmatrix} y(t) \\ y(t-1) \\ \vdots \\ y(t-m) \end{bmatrix} \end{aligned} \tag{5.4.2}$$

Let R denote the covariance matrix of the data vector in (5.4.2). Then the power of the filter output can be written as:

$$E\left\{|y_F(t)|^2\right\} = h^* R h \tag{5.4.3}$$

where, according to the definition above,

$$R = E\left\{ \begin{bmatrix} y(t) \\ \vdots \\ y(t-m) \end{bmatrix} [y^*(t) \ldots y^*(t-m)] \right\} \tag{5.4.4}$$

The response of the filter (5.4.2) to a sinusoidal component of frequency ω (say) is determined by the filter's frequency response:

$$H(\omega) = \sum_{k=0}^{m} h_k e^{-i\omega k} = h^* a(\omega) \tag{5.4.5}$$

where

$$a(\omega) = [1 \;\; e^{-i\omega} \ldots e^{-im\omega}]^T \tag{5.4.6}$$

If we want to make the filter as selective as possible for a frequency band around the current value ω, then we may think of minimizing the total power in (5.4.3) subject to the constraint that the filter passes the frequency ω undistorted. This idea leads to the following optimization problem:

$$\boxed{\min_h h^* R h \quad \text{subject to } h^* a(\omega) = 1} \tag{5.4.7}$$

The solution to (5.4.7) is given in Result R35 in Appendix A:

$$\boxed{h = R^{-1} a(\omega) / a^*(\omega) R^{-1} a(\omega)} \tag{5.4.8}$$

Inserting (5.4.8) into (5.4.3) gives

$$E\left\{|y_F(t)|^2\right\} = 1/a^*(\omega)R^{-1}a(\omega) \tag{5.4.9}$$

This is the power of $y(t)$ in a passband centered on ω. Then, assuming that the (idealized) conditions (i) and (ii) in (5.1.5) hold, we can approximately determine the value of the PSD of $y(t)$ at the passband's center frequency as

$$\phi(\omega) \simeq \frac{E\left\{|y_F(t)|^2\right\}}{\beta} = \frac{1}{\beta a^*(\omega)R^{-1}a(\omega)} \tag{5.4.10}$$

where β denotes the frequency bandwidth of the filter given by (5.4.8). The division by β, as above, is sometimes omitted in the literature, but it is required to complete the FBA scheme in Figure 5.1. Note that since the bandpass filter (5.4.8) is data dependent, its bandwidth β is not necessarily data independent, nor is it necessarily frequency independent. Hence, the division by β in (5.4.10) may not represent a simple scaling of $E\left\{|y_F(t)|^2\right\}$, but it may change the shape of this quantity as a function of ω.

There are various possibilities for determining the bandwidth β, depending on the degree of precision we are aiming for. The simplest possibility is to set

$$\beta = 1/(m+1) \tag{5.4.11}$$

This choice is motivated by the time–bandwidth product result (2.6.5), which says that for a filter whose temporal aperture is equal to $(m+1)$, the bandwidth should roughly be given by $1/(m+1)$. By inserting (5.4.11) in (5.4.10), we obtain

$$\phi(\omega) \simeq \frac{(m+1)}{a^*(\omega)R^{-1}a(\omega)} \tag{5.4.12}$$

Note that if $y(t)$ is white noise of variance σ^2, (5.4.12) takes the correct value: $\phi(\omega) = \sigma^2$. In the general case, however, (5.4.11) gives only a rough indication of the filter's bandwidth, as the time–bandwidth product result does not apply exactly to the present situation (see the conditions under which (2.6.5) has been derived).

An often more exact expression for β can be obtained as follows [LAGUNAS, SANTAMARIA, GASULL, AND MORENO 1986]. The (equivalent) bandwidth of a bandpass filter can be defined as the support of the rectangle centered on ω (the filter's center frequency) that concentrates the whole energy in the filter's frequency response. According to this definition, β can be assumed to satisfy:

$$\int_{-\pi}^{\pi} |H(\psi)|^2 d\psi = |H(\omega)|^2 2\pi\beta \tag{5.4.13}$$

Since in the present case $H(\omega) = 1$ (see (5.4.7)), we obtain from (5.4.13):

$$\beta = \frac{1}{2\pi}\int_{-\pi}^{\pi} |h^* a(\psi)|^2 d\psi = h^* \left[\frac{1}{2\pi}\int_{-\pi}^{\pi} a(\psi)a^*(\psi)d\psi\right] h \tag{5.4.14}$$

The (k,p) element of the central matrix in the above quadratic form is given by

$$\frac{1}{2\pi}\int_{-\pi}^{\pi} e^{-i\psi(k-p)}d\psi = \delta_{k,p} \tag{5.4.15}$$

With this observation and (5.4.8), (5.4.14) leads to

$$\beta = h^*h = \frac{a^*(\omega)R^{-2}a(\omega)}{[a^*(\omega)R^{-1}a(\omega)]^2} \tag{5.4.16}$$

Note that this expression of the bandwidth is both data and frequency dependent (as was alluded to previously). Inserting (5.4.16) in (5.4.10) gives

$$\phi(\omega) \simeq \frac{a^*(\omega)R^{-1}a(\omega)}{a^*(\omega)R^{-2}a(\omega)} \tag{5.4.17}$$

Remark: The expression for β in (5.4.16) is based on the assumption that most of the area under the curve of $|H(\psi)|^2 = |h^*a(\psi)|^2$ (for $\psi \in [-\pi,\pi]$) is located around the center frequency ω. This assumption is often true, but not always true. For instance, consider a data sequence $\{y(t)\}$ consisting of a number of sinusoidal components with frequencies $\{\omega_k\}$ in noise with *small* power. Then the Capon filter (5.4.8) with center frequency ω will likely place nulls at $\{\psi = \omega_k\}$ to annihilate the strong sinusoidal components in the data, but will pay little attention to the weak noise component. The consequence is that $|H(\psi)|^2$ will be nearly zero at $\{\psi = \omega_k\}$, and one at $\psi = \omega$ (by (5.4.7)), but may take rather large values at other frequencies (see, for example, the numerical examples in [Li and Stoica 1996a], which demonstrate this behavior of the Capon filter). In such a case, the formula (5.4.16) may significantly overestimate the "true" bandwidth, and hence the spectral formula (5.4.17) may significantly underestimate the PSD $\phi(\omega)$. ■

In the derivations above, the true data covariance matrix R has been assumed available. In order to turn the previous PSD formulas into practical spectral estimation algorithms, we must replace R in these formulas by a sample estimate, for instance by

$$\hat{R} = \frac{1}{N-m}\sum_{t=m+1}^{N}\begin{bmatrix} y(t) \\ \vdots \\ y(t-m)\end{bmatrix}[y^*(t)\dots y^*(t-m)] \tag{5.4.18}$$

Doing so, we obtain the following two spectral estimators corresponding to (5.4.12) and (5.4.17), respectively:

$$\boxed{\text{CM–Version 1:}\quad \hat{\phi}(\omega) = \frac{m+1}{a^*(\omega)\hat{R}^{-1}a(\omega)}} \tag{5.4.19}$$

$$\text{CM–Version 2:} \quad \hat{\phi}(\omega) = \frac{a^*(\omega)\hat{R}^{-1}a(\omega)}{a^*(\omega)\hat{R}^{-2}a(\omega)} \tag{5.4.20}$$

There is an implicit assumption in both (5.4.19) and (5.4.20) that $\hat{R}^{-1}$ exists. This assumption sets a limit on the maximum value that can be chosen for m:

$$m < N/2 \tag{5.4.21}$$

(Observe that $\text{rank}(\hat{R}) \leq N - m$, which is less than $\dim(\hat{R}) = m + 1$ if (5.4.21) is violated.) The inequality (5.4.21) is important since it sets a limit on the resolution achievable by the Capon method. Indeed, since the Capon method is based on a bandpass filter with impulse response's aperture equal to m, we may expect its resolution threshold to be on the order of $1/m > 2/N$ (with the inequality following from (5.4.21)).

As m is decreased, we can expect the resolution of Capon method to become worse (*cf.* the previous discussion). On the other hand, the accuracy with which $\hat{R}$ is determined increases with decreasing m (since more outer products are averaged in (5.4.18)). The main consequence of the increased accuracy of $\hat{R}$ is to statistically stabilize the spectral estimate (5.4.19) or (5.4.20). Hence, the choice of m should be done with the ubiquitous tradeoff between resolution and statistical accuracy in mind. It is interesting to note that for the Capon method both the filter design and power calculation stages are data dependent. The accuracy of both these stages may worsen if m is chosen too large. In applications, the maximum value that can be chosen for m might also be limited from considerations of computational complexity.

Empirical studies have shown that *the ability of the Capon method to resolve fine details of a PSD, such as closely spaced peaks, is superior to the corresponding performance of the periodogram–based methods.* This superiority may be attributed to the higher statistical stability of Capon method, as explained next. For m smaller than $N/2$ (see (5.4.21)), we may expect the Capon method to possess worse resolution but better statistical accuracy compared with the unwindowed or "mildly windowed" periodogram method. It should be stressed that *the notion of "resolution" refers to the ability of the theoretically averaged spectral estimate $E\{\hat{\phi}(\omega)\}$ to resolve fine details in the true PSD $\phi(\omega)$.* This resolution is roughly inversely proportional to the window's length or the bandpass filter impulse response's aperture. *The "resolving power" corresponding to the estimate $\hat{\phi}(\omega)$ is more difficult to quantify,* but — of course — it is what interests the most. It should be clear that the resolving power of $\hat{\phi}(\omega)$ depends not only on the bias of this estimate (*i.e.*, on $E\{\hat{\phi}(\omega)\}$), but also on its variance. *A spectral estimator with low bias–based resolution but high statistical accuracy may be better able to resolve finer details in a studied PSD than can a high resolution/low accuracy estimator.* Since the periodogram may achieve better bias–based resolution than the Capon method, the higher (empirically observed) "resolving power" of the latter should be due to a better statistical accuracy (*i.e.*, a lower variance).

In the context of the previous discussion, it is interesting to note that the Blackman–Tukey periodogram with a Bartlett window of length $2m+1$, which is given by (see (2.5.1)):

$$\hat{\phi}_{\mathrm{BT}}(\omega) = \sum_{k=-m}^{m} \frac{(m+1-|k|)}{m+1} \hat{r}(k) e^{-i\omega k}$$

can be written in a form that bears some resemblance with the form (5.4.19) of the CM–Version 1 estimator. A straightforward calculation gives

$$\hat{\phi}_{\mathrm{BT}}(\omega) = \sum_{t=0}^{m}\sum_{s=0}^{m} \hat{r}(t-s) e^{-i\omega(t-s)}/(m+1) \tag{5.4.22}$$

$$= \frac{1}{m+1} a^*(\omega)\hat{R}a(\omega) \tag{5.4.23}$$

where $a(\omega)$ is as defined in (5.4.6), and $\hat{R}$ is the Hermitian Toeplitz sample covariance matrix

$$\hat{R} = \begin{bmatrix} \hat{r}(0) & \hat{r}(1) & \dots & \hat{r}(m) \\ \hat{r}^*(1) & \hat{r}(0) & \ddots & \vdots \\ \vdots & \ddots & \ddots & \hat{r}(1) \\ \hat{r}^*(m) & \dots & \hat{r}^*(1) & \hat{r}(0) \end{bmatrix}$$

Comparing the above expression for $\hat{\phi}_{\mathrm{BT}}(\omega)$ with (5.4.19), it is seen that the *CM–Version 1 can be obtained from Blackman–Tukey estimator by replacing $\hat{R}$ in the Blackman–Tukey estimator with $\hat{R}^{-1}$, and then inverting the so–obtained quadratic form.* Below we provide a brief explanation as to why this replacement and inversion make sense. That is, if we ignore for a moment the technically sound filter bank derivation of the Capon method, then why should the above way of obtaining CM–Version 1 from the Blackman–Tukey method provide a reasonable spectral estimator? We begin by noting that (*cf.* Section 1.3.2):

$$\lim_{m\to\infty} E\left\{ \frac{1}{m+1} \left| \sum_{t=0}^{m} y(t) e^{-i\omega t} \right|^2 \right\} = \phi(\omega)$$

However, a simple calculation shows that

$$E\left\{ \frac{1}{m+1} \left| \sum_{t=0}^{m} y(t) e^{-i\omega t} \right|^2 \right\} = \frac{1}{m+1} \sum_{t=0}^{m}\sum_{s=0}^{m} r(t-s) e^{-i\omega t} e^{i\omega s} = \frac{1}{m+1} a^*(\omega) R a(\omega)$$

Hence,

$$\lim_{m\to\infty} \frac{1}{m+1} a^*(\omega) R a(\omega) = \phi(\omega) \tag{5.4.24}$$

Similarly, one can show that

$$\lim_{m\to\infty} \frac{1}{m+1} a^*(\omega) R^{-1} a(\omega) = \phi^{-1}(\omega) \tag{5.4.25}$$

(see, *e.g.*, [HANNAN AND WAHLBERG 1989]). Comparing (5.4.24) with (5.4.25) provides the explanation we were looking for. Observe that the CM–Version 1 estimator is a finite–sample approximation to equation (5.4.25), whereas the Blackman–Tukey estimator is a finite–sample approximation to equation (5.4.24).

The Capon method has also been compared with the AR method of spectral estimation (see Section 3.2). It has been empirically observed that *the CM–Version 1 possesses less variance but worse resolution than the AR spectral estimator.* This may be explained by making use of the relationship that exists between the CM–Version 1 and AR spectral estimators; see the next subsection (and also [BURG 1972]). The CM–Version 2 spectral estimator is less well studied and hence its properties are not so well understood. In the following subsection, we also relate the CM–Version 2 to the AR spectral estimator. In the case of CM–Version 2, the relationship is more involved, hence leaving less room for intuitive explanations.

5.4.2 Relationship between Capon and AR Methods

The AR method of spectral estimation has been described in Chapter 3. In the following we consider the covariance matrix estimate in (5.4.18). The AR method corresponding to this sample covariance matrix is the LS method discussed in Section 3.4.2. Let us denote the matrix $\hat{R}$ in (5.4.18) by $\hat{R}_{m+1}$ and its principal lower–right $k \times k$ block by $\hat{R}_k$ ($k = 1, \ldots, m+1$), as shown below:

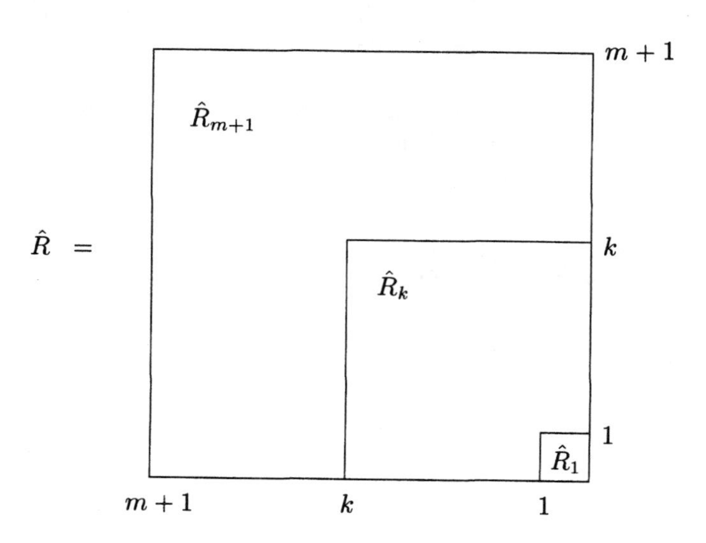

(5.4.26)

With this notation, the coefficient vector θ_k and the residual power σ_k^2 of the *kth-order AR model* fitted to the data $\{y(t)\}$ are obtained as the solutions to the following matrix equation (refer to (3.4.6)):

$$\hat{R}_{k+1} \begin{bmatrix} 1 \\ \hat{\theta}_k^c \end{bmatrix} = \begin{bmatrix} \hat{\sigma}_k^2 \\ 0 \end{bmatrix} \tag{5.4.27}$$

(the complex conjugate in (5.4.27) appears owing to the fact that $\hat{R}_k$ above is equal to the complex conjugate of the sample covariance matrix used in Chapter 3). The nested structure of (5.4.26) along with the defining equation (5.4.27) imply:

$$\hat{R}_{m+1} \begin{bmatrix} 1 & 0 & \cdots & 0 & 0 \\ & 1 & & \vdots & \vdots \\ & & \ddots & 0 & \\ & & & 1 & 0 \\ \hat{\theta}_m^c & \hat{\theta}_{m-1}^c & & \hat{\theta}_1^c & 1 \end{bmatrix} = \begin{bmatrix} \hat{\sigma}_m^2 & x & \cdots & x \\ 0 & \hat{\sigma}_{m-1}^2 & \ddots & \vdots \\ \vdots & \ddots & \ddots & x \\ 0 & \cdots & 0 & \hat{\sigma}_0^2 \end{bmatrix} \tag{5.4.28}$$

where "x" stands for undetermined elements. Let

$$\hat{\mathcal{H}} = \begin{bmatrix} 1 & 0 & \cdots & 0 & 0 \\ & 1 & & \vdots & \vdots \\ & & \ddots & 0 & \\ & & & 1 & 0 \\ \hat{\theta}_m^c & \hat{\theta}_{m-1}^c & & \hat{\theta}_1^c & 1 \end{bmatrix} \tag{5.4.29}$$

It follows from (5.4.28) that

$$\hat{\mathcal{H}}^* \hat{R}_{m+1} \hat{\mathcal{H}} = \begin{bmatrix} \hat{\sigma}_m^2 & x & \cdots & x \\ & \hat{\sigma}_{m-1}^2 & \ddots & \vdots \\ 0 & & \ddots & x \\ & & & \hat{\sigma}_0^2 \end{bmatrix} \tag{5.4.30}$$

(where, once more, x denotes undetermined elements). Since $\hat{\mathcal{H}}^* \hat{R}_{m+1} \hat{\mathcal{H}}$ is a Hermitian matrix, the elements designated by "x" in (5.4.30) must be equal to zero. Hence, we proved the following result which is essential in establishing a relation between the AR and Capon methods of spectral estimation (this result extends the one in Exercise 3.7 to the non–Toeplitz covariance case).

> The parameters $\{\hat{\theta}_k, \hat{\sigma}_k^2\}$ of the AR models of orders $k = 1, 2, \ldots, m$ determine the following factorization of the inverse (sample) covariance matrix:
>
> $$\hat{R}_{m+1}^{-1} = \hat{\mathcal{H}}\hat{\Sigma}^{-1}\hat{\mathcal{H}}^* \quad ; \quad \hat{\Sigma} = \begin{bmatrix} \hat{\sigma}_m^2 & & & 0 \\ & \hat{\sigma}_{m-1}^2 & & \\ & & \ddots & \\ 0 & & & \hat{\sigma}_0^2 \end{bmatrix} \tag{5.4.31}$$

Let

$$\hat{A}_k(\omega) = [1 \;\; e^{-i\omega} \ldots e^{-ik\omega}] \begin{bmatrix} 1 \\ \hat{\theta}_k \end{bmatrix} \tag{5.4.32}$$

denote the polynomial corresponding to the kth–order AR model, and let

$$\hat{\phi}_k^{\text{AR}}(\omega) = \frac{\hat{\sigma}_k^2}{|\hat{A}_k(\omega)|^2} \tag{5.4.33}$$

denote its associated PSD (see Chapter 3). It is readily verified that

$$\begin{aligned} a^*(\omega)\hat{\mathcal{H}} &= [1 \;\; e^{i\omega} \ldots e^{im\omega}] \begin{bmatrix} 1 & 0 & \ldots & 0 & 0 \\ & 1 & & \vdots & \vdots \\ & & \ddots & 0 & \\ & & & 1 & 0 \\ \hat{\theta}_m^c & \hat{\theta}_{m-1}^c & & \hat{\theta}_1^c & 1 \end{bmatrix} \\ &= [\hat{A}_m^*(\omega), \; e^{i\omega}\hat{A}_{m-1}^*(\omega), \; \ldots, \; e^{im\omega}\hat{A}_0^*(\omega)] \end{aligned} \tag{5.4.34}$$

It follows from (5.4.31) and (5.4.34) that the quadratic form in the denominator of the CM–Version 1 spectral estimator can be written as

$$\begin{aligned} a^*(\omega)\hat{R}^{-1}a(\omega) &= a^*(\omega)\hat{\mathcal{H}}\hat{\Sigma}^{-1}\hat{\mathcal{H}}^*a(\omega) \\ &= \sum_{k=0}^{m} |\hat{A}_k(\omega)|^2/\hat{\sigma}_k^2 = \sum_{k=0}^{m} 1/\hat{\phi}_k^{\text{AR}}(\omega) \end{aligned} \tag{5.4.35}$$

which leads at once to the following result:

> $$\hat{\phi}_{\text{CM-1}}(\omega) = \frac{1}{\dfrac{1}{m+1}\displaystyle\sum_{k=0}^{m} 1/\hat{\phi}_k^{\text{AR}}(\omega)} \tag{5.4.36}$$

This is the desired relation between the CM–Version 1 and the AR spectral estimates. This relation says that the inverse of the CM–Version 1 spectral estimator can be obtained by averaging the inverse estimated AR spectra of orders from 0 to m. In view of the averaging operation in (5.4.36), it is not difficult to understand why the CM–Version 1 possesses less statistical variability than the AR estimator. Moreover, the fact that the CM–Version 1 has also been found to have worse resolution and bias properties than the AR spectral estimate should be due to the presence of low–order AR models in (5.4.36).

Next, consider the CM–Version 2. The previous analysis of CM–Version 1 already provides a relation between the numerator in the spectral estimate corresponding to CM–Version 2, (5.4.20), and the AR spectra. In order to obtain a similar expression for the denominator in (5.4.20), some preparations are required. The (sample) covariance matrix $\hat{R}$ can be used to define $m+1$ *AR models of order* m, depending on which coefficient of the AR equation

$$\hat{a}_0 y(t) + \hat{a}_1 y(t-1) + \ldots + \hat{a}_m y(t-m) = \text{residuals} \tag{5.4.37}$$

we choose to set to one. The AR model $\{\hat{\theta}_m, \hat{\sigma}_m^2\}$ used in the previous analysis corresponds to setting $\hat{a}_0 = 1$ in (5.4.37). However, in principle, any other AR coefficient in (5.4.37) may be normalized to one. The mth–order LS AR model obtained by setting $\hat{a}_k = 1$ in (5.4.37) is denoted by $\{\hat{\mu}_k = \text{coefficient vector and } \hat{\gamma}_k = \text{residual variance}\}$, and is given by the solution to the following linear system of equations (compare with (5.4.27)):

$$\hat{R}_{m+1}\hat{\mu}_k^c = \hat{\gamma}_k u_k \tag{5.4.38}$$

where the $(k+1)$st component of $\hat{\mu}_k$ is equal to one $(k = 0, \ldots, m)$, and where u_k stands for the $(k+1)$st column of the $(m+1) \times (m+1)$ identity matrix:

$$u_k = [\underbrace{0 \ldots 0}_{k} \;\; 1 \;\; \underbrace{0 \ldots 0}_{m-k}]^T \tag{5.4.39}$$

Evidently, $[1 \; \hat{\theta}_m^T]^T = \hat{\mu}_0$ and $\hat{\sigma}_m^2 = \hat{\gamma}_0$.

Similarly to (5.4.32) and (5.4.33), the (estimated) PSD corresponding to the kth mth–order AR model given by (5.4.38) is obtained as

$$\hat{\phi}_k^{\text{AR}(m)}(\omega) = \frac{\hat{\gamma}_k}{|a^*(\omega)\hat{\mu}_k^c|^2} \tag{5.4.40}$$

It is shown in the following calculation that the denominator in (5.4.20) can be expressed as a (weighted) average of the AR spectra in (5.4.40):

$$\begin{aligned} \sum_{k=0}^{m} \frac{1}{\hat{\gamma}_k \hat{\phi}_k^{\text{AR}(m)}(\omega)} &= \sum_{k=0}^{m} \frac{|a^*(\omega)\hat{\mu}_k^c|^2}{\hat{\gamma}_k^2} = a^*(\omega)\left[\sum_{k=0}^{m} \frac{\hat{\mu}_k^c \hat{\mu}_k^T}{\hat{\gamma}_k^2}\right] a(\omega) \\ &= a^*(\omega)\hat{R}^{-1}\left[\sum_{k=0}^{m} u_k u_k^*\right]\hat{R}^{-1}a(\omega) = a^*(\omega)\hat{R}^{-2}a(\omega) \end{aligned} \tag{5.4.41}$$

Combining (5.4.35) and (5.4.41) gives

$$\hat{\phi}_{\text{CM-2}}(\omega) = \frac{\sum_{k=0}^{m} 1/\hat{\phi}_k^{\text{AR}}(\omega)}{\sum_{k=0}^{m} 1/\hat{\gamma}_k \hat{\phi}_k^{\text{AR}(m)}(\omega)} \tag{5.4.42}$$

The above relation appears to be more involved, and hence more difficult to interpret, than the similar relation (5.4.36) corresponding to CM–Version 1. Nevertheless, since (5.4.42) is still obtained by averaging various AR spectra, we may expect that *the CM–Version 2 estimator, like the CM–Version 1 estimator, is more statistically stable but has poorer resolution than the AR spectral estimator.*

5.5 Filter Bank Reinterpretation of the Periodogram

As we saw in Section 5.2, the basic periodogram spectral estimator can be interpreted as an FBA method with a *preimposed* bandpass filter (whose impulse response is equal to the Fourier transform vector). In contrast, RFB and Capon are FBA methods based on *designed* bandpass filters. The filter used in the RFB method is data independent, whereas it is a function of the data covariances in the Capon method. The use of a data–dependent bandpass filter, such as in the Capon method, is intuitively appealing but it also leads to the following drawback: since we need to consistently estimate the filter impulse response, the temporal aperture of the filter should be chosen (much) smaller than the sample length, which sets a rather hard limit on the achievable spectral resolution. In addition, it appears that any other filter design methodology, except the one originally suggested by Capon, will most likely lead to a problem (such as an eigenanalysis) that should be solved for each value of the center frequency; which — of course — would be a rather prohibitive computational task. With these difficulties of the data–dependent design in mind, we may content ourselves with a "well–designed" data–independent filter. The purpose of this section is to show that *the basic periodogram and the Daniell method can be interpreted as FBA methods based on well–designed data–independent filters*, similar to the RFB method. As we will see, the bandpass filters used by the aforementioned periodogram methods are obtained *by combining the design procedures employed in the RFB and Capon methods.*

The following result is required (see R35 in Appendix A for a proof). Let R, H, A and C be matrices of dimensions $(m \times m)$, $(m \times K)$, $(m \times n)$ and $(K \times n)$, respectively. Assume that R is positive definite and A has full column rank equal to n (hence, $m \geq n$). Then the solution to the following quadratic optimization problem with linear constraints:

$$\min_H (H^* R H) \quad \text{subject to} \quad H^* A = C$$

is given by

$$H = R^{-1} A (A^* R^{-1} A)^{-1} C^* \tag{5.5.1}$$

We can now proceed to derive our "new" FBA–based spectral estimation method (as we will see below, it turns out that this method is not really new!). We would like this method to possess a facility for compromising between the bias and variance of the estimated PSD. As explained in the previous sections of this chapter, there are two main ways of doing this within the FBA: we either (i) use a bandpass filter with temporal aperture less than N, obtain the allowed number of samples of the filtered signal and then calculate the power from these samples; or (ii) use a set of K bandpass filters with length–N impulse responses, that cover a band centered on the current frequency value, obtain one sample of the filtered signals for each filter in the set and calculate the power by averaging these K samples. As argued in Section 5.3, approach (ii) may be more effective than (i) in reducing the variance of the estimated PSD, while keeping the bias low. In the sequel, we follow approach (ii).

Let $\beta \geq 1/N$ be the *prespecified (desired) resolution* and let K be defined by equation (5.3.12): $K = \beta N$. According to the time–bandwidth product result, a bandpass filter with a length–N impulse response may be expected to have a bandwidth on the order of $1/N$ (but not less). Hence, we can cover the preimposed passband

$$[\tilde{\omega} - \beta\pi, \tilde{\omega} + \beta\pi] \tag{5.5.2}$$

(here $\tilde{\omega}$ stands for the current frequency value) by using $2\pi\beta/(2\pi/N) = K$ filters, which pass essentially nonoverlapping $1/N$–length frequency bands in the interval (5.5.2). *The requirement that the filters' passbands are (nearly) nonoverlapping is a key condition for variance reduction.* In order to see this, let x_p denote the sample obtained at the output of the pth filter:

$$x_p = \sum_{k=0}^{N-1} h_{p,k} y(N-k) = \sum_{t=1}^{N} h_{p,N-t} y(t) \tag{5.5.3}$$

Here $\{h_{p,k}\}_{k=0}^{N-1}$ is the pth filter's impulse response. The associated frequency response is denoted by $H_p(\omega)$. Note that in the present case we consider bandpass filters operating on the raw data, in lieu of baseband filters operating on demodulated data (as in RFB). Assume that the *center–frequency gain* of each filter is normalized so that

$$H_p(\tilde{\omega}) = 1, \qquad p = 1, \ldots, K \tag{5.5.4}$$

Then, we can write

$$\begin{aligned} E\left\{|x_p|^2\right\} &= \frac{1}{2\pi}\int_{-\pi}^{\pi} |H_p(\omega)|^2 \phi(\omega) d\omega \\ &\simeq \frac{1}{2\pi}\int_{\tilde{\omega}-\pi/N}^{\tilde{\omega}+\pi/N} \phi(\omega) d\omega \simeq \frac{2\pi/N}{2\pi}\phi(\tilde{\omega}) = \frac{1}{N}\phi(\tilde{\omega}) \end{aligned} \tag{5.5.5}$$

The second "equality" in (5.5.5) follows from (5.5.4) and the *assumed bandpass characteristics of $H_p(\omega)$*, and the third equality results from the assumption that

$\phi(\omega)$ is approximately constant over the passband. (Note that the angular frequency passband of $H_p(\omega)$ is $2\pi/N$, as explained before.) In view of (5.5.5), we can estimate $\phi(\tilde{\omega})$ by averaging over the squared magnitudes of the filtered samples $\{x_p\}_{p=1}^{K}$. By doing so, we may achieve a reduction in variance by a factor K, *provided* $\{x_p\}$ are statistically independent (see Section 5.3 for details). Under the assumption that the filters $\{H_p(\omega)\}$ pass essentially nonoverlapping frequency bands, we readily get (compare (5.3.27)):

$$E\left\{x_p x_k^*\right\} = \frac{1}{2\pi}\int_{-\pi}^{\pi} H_p(\omega)H_k^*(\omega)\phi(\omega)d\omega \simeq 0 \tag{5.5.6}$$

which implies that the random variables $\{|x_p|^2\}$ are independent at least under the Gaussian hypothesis. Without the previous assumption on $\{H_p(\omega)\}$, the filtered samples $\{x_p\}$ may be strongly correlated and, therefore, a reduction in variance by a factor K cannot be guaranteed.

The conclusion from the previous (more or less heuristic) discussion is summarized in the following.

If the passbands of the filters used to cover the prespecified interval (5.5.2) do not overlap, then by using all filters' output samples — as contrasted to using the output sample of only one filter — we achieve a reduction in the variance of the estimated PSD by a factor equal to the number of filters. The maximum number of such filters that can be found is given by $K = \beta N$.	(5.5.7)

By using the insights provided by the above discussion, as summarized in (5.5.7), we can now approach the bandpass filters design problem. We sample the frequency axis as in the FFT (as almost any practical implementation of a spectral estimation method does):

$$\tilde{\omega}_s = \frac{2\pi}{N}s \qquad s = 0,\ldots,N-1 \tag{5.5.8}$$

The frequency samples that fall within the passband (5.5.2) are readily seen to be the following:

$$\frac{2\pi}{N}(s+p) \qquad p = -K/2,\ldots,0,\ldots,K/2-1 \tag{5.5.9}$$

(to simplify the discussion we assume that K is an even integer). Let

$$H = [h_1 \ldots h_K] \qquad (N \times K) \tag{5.5.10}$$

denote the matrix whose pth column is equal to the impulse response vector corresponding to the pth bandpass filter. We assume that *the input to the filters is white noise (as in RFB) and design the filters so as to minimize the output power under the constraint that each filter passes undistorted one (and only one) of the frequencies in (5.5.9) (as in Capon).* These design objectives lead to the following optimization problem:

$$\boxed{\begin{array}{l} \min_{H}(H^*H) \text{ subject to } H^*A = I \\ \text{where } A = \left[a\left(\frac{2\pi}{N}\left(s-\frac{K}{2}\right)\right), \ldots, a\left(\frac{2\pi}{N}\left(s+\frac{K}{2}-1\right)\right)\right] \end{array}} \tag{5.5.11}$$

and where $a(\omega) = [1 \;\; e^{-i\omega} \ldots e^{-i(N-1)\omega}]^T$. Note that the constraint in (5.5.11) guarantees that each frequency in the passband (5.5.9) is passed undistorted by one filter in the set, and it is annihilated by all the other $(K-1)$ filters. In particular, observe that (5.5.11) implies (5.5.4).

The solution to (5.5.11) follows at once from the result (5.5.1): the minimizing H matrix is given by

$$H = A(A^*A)^{-1} \tag{5.5.12}$$

However, the columns in A are orthogonal

$$A^*A = NI$$

(see (4.3.15)); therefore, (5.5.12) simplifies to

$$\boxed{H = \frac{1}{N}A} \tag{5.5.13}$$

which is the solution of the filter design problem previously formulated.

By using (5.5.13) in (5.5.3), we get

$$\begin{aligned} |x_p|^2 &= \frac{1}{N^2}\left|\sum_{t=1}^{N} e^{i(N-t)\frac{2\pi}{N}(s+p)}y(t)\right|^2 \\ &= \frac{1}{N^2}\left|\sum_{t=1}^{N} y(t)e^{-i\frac{2\pi}{N}(s+p)t}\right|^2 \\ &= \frac{1}{N}\hat{\phi}_p\left(\frac{2\pi}{N}(s+p)\right) \qquad p = -K/2, \ldots, K/2-1 \end{aligned} \tag{5.5.14}$$

where the dependence of $|x_p|^2$ on s (and hence on $\tilde{\omega}_s$) is omitted to simplify the notation, and where $\hat{\phi}_p(\omega)$ is the standard periodogram. Finally, (5.5.14) along with (5.5.5) lead to the following *FBA spectral estimator*:

$$\boxed{\hat{\phi}\left(\frac{2\pi}{N}s\right) = \frac{1}{K}\sum_{p=-K/2}^{K/2-1} N|x_p|^2 = \frac{1}{K}\sum_{l=s-K/2}^{s+K/2-1} \hat{\phi}_p\left(\frac{2\pi}{N}l\right)} \tag{5.5.15}$$

which coincides with the Daniell periodogram estimator (2.7.16). Furthermore, *for* $K = 1$ (*i.e.*, $\beta = 1/N$, which is the choice suitable for "high–resolution" applications), *(5.5.15) reduces to the unmodified periodogram.* Recall also that the RFB method in Section 5.3, for large data lengths, is expected to have similar performance to the Daniell method for $K > 1$ and to the basic periodogram for $K = 1$. Hence, in the family of nonparametric spectral estimation methods the periodograms "are doing well".

5.6 Complements

5.6.1 Another Relationship between the Capon and AR Methods

The relationship between the AR and Capon spectra established in Section 5.4.2 involves all AR spectral models of orders 0 through m. Another interesting relationship, which involves the AR spectrum of order m alone, is presented in this complement.

Let $\hat{\theta} = [\hat{a}_0\ \hat{a}_1 \ldots \hat{a}_m]^T$ (with $\hat{a}_0 = 1$) denote the vector of the coefficients of the mth–order AR model fitted to the data sample covariances, and let $\hat{\sigma}^2$ denote the corresponding residual variance (see Chapter 3 and (5.4.27)). Then the mth–order AR spectrum is given by:

$$\hat{\phi}_{AR}(\omega) = \frac{\hat{\sigma}^2}{|a^*(\omega)\hat{\theta}^c|^2} = \frac{\hat{\sigma}^2}{|\sum_{k=0}^{m} \hat{a}_k e^{-i\omega k}|^2} \tag{5.6.1}$$

Show, by a simple calculation, that $\hat{\phi}_{AR}(\omega)$ above can be rewritten in the following form:

$$\boxed{\hat{\phi}_{AR}(\omega) = \frac{\hat{\sigma}^2}{\sum_{s=-m}^{m} \hat{\rho}(s) e^{i\omega s}}} \tag{5.6.2}$$

where

$$\boxed{\hat{\rho}(s) = \sum_{k=0}^{m-s} \hat{a}_k \hat{a}_{k+s}^* = \hat{\rho}^*(-s), \qquad s = 0, \ldots, m.} \tag{5.6.3}$$

Next, assume that the (sample) covariance matrix $\hat{R}$ is Toeplitz. (We note in passing that this is a minor restriction for the temporal spectral estimation problem of this chapter, but it may be quite a restrictive assumption for the spatial problem of the next chapter.) Then show that the Capon spectrum in equation (5.4.19) (with the factor $m + 1$ omitted, for convenience) can be written as:

$$\boxed{\hat{\phi}_{CM}(\omega) = \frac{\hat{\sigma}^2}{\sum_{s=-m}^{m} \hat{\mu}(s) e^{i\omega s}}} \tag{5.6.4}$$

where

$$\boxed{\hat{\mu}(s) = \sum_{k=0}^{m-s} (m + 1 - 2k - s)\hat{a}_k \hat{a}_{k+s}^* = \hat{\mu}^*(-s), \qquad s = 0, \ldots, m} \tag{5.6.5}$$

To prove (5.6.4) make use of the Gohberg–Semencul (GS) formula derived in the Complement in Section 3.9.4, which is repeated here for convenience:

$$\hat{\sigma}^2 \hat{R}^{-1} = \begin{bmatrix} 1 & \cdots & \cdots & 0 \\ \hat{a}_1^* & \ddots & & \vdots \\ \vdots & \ddots & \ddots & \vdots \\ \hat{a}_m^* & \cdots & \hat{a}_1^* & 1 \end{bmatrix} \begin{bmatrix} 1 & \hat{a}_1 & \cdots & \hat{a}_m \\ \vdots & \ddots & \ddots & \vdots \\ \vdots & & \ddots & \hat{a}_1 \\ 0 & \cdots & \cdots & 1 \end{bmatrix}$$

$$- \begin{bmatrix} 0 & \cdots & \cdots & 0 \\ \hat{a}_m & \ddots & & \vdots \\ \vdots & \ddots & \ddots & \vdots \\ \hat{a}_1 & \cdots & \hat{a}_m & 0 \end{bmatrix} \begin{bmatrix} 0 & \hat{a}_m^* & \cdots & \hat{a}_1^* \\ \vdots & \ddots & \ddots & \vdots \\ \vdots & & \ddots & \hat{a}_m^* \\ 0 & \cdots & \cdots & 0 \end{bmatrix}$$

(The above formula is in fact the complex conjugate of the GS formula in Section 3.9.4 because the matrix $\hat{R}$ above is the complex conjugate of the one considered in Chapter 3).

Comment on the relation between $\hat{\phi}_{AR}(\omega)$ and $\hat{\phi}_{CM}(\omega)$ afforded by equations (5.6.2) and (5.6.4). Furthermore, outline a computationally efficient way to evaluate the Capon spectrum by using the formula (5.6.4).

Solution: We consider (5.6.4) first. For the sake of convenience, let $\hat{a}_k = 0$ for $k \notin [0, m]$. By making use of this convention, and of the GS formula, we obtain:

$$\begin{aligned} f(\omega) &\triangleq \hat{\sigma}^2 a^*(\omega) \hat{R}^{-1} a(\omega) \\ &= \sum_{p=0}^{m} \left\{ \left| \sum_{k=0}^{m} \hat{a}_{k-p}\, e^{-i\omega k} \right|^2 - \left| \sum_{k=0}^{m} \hat{a}_{m+1-k+p}^* e^{-i\omega k} \right|^2 \right\} \\ &= \sum_{p=0}^{m} \sum_{k=0}^{m} \sum_{\ell=0}^{m} (\hat{a}_{k-p} \hat{a}_{\ell-p}^* - \hat{a}_{m+1+p-k}^* \hat{a}_{m+1-\ell+p}) e^{i\omega(\ell-k)} \\ &= \sum_{\ell=0}^{m} \sum_{p=0}^{m} \sum_{s=\ell-m}^{\ell} (\hat{a}_{\ell-s-p} \hat{a}_{\ell-p}^* - \hat{a}_{m+1-\ell+s+p}^* \hat{a}_{m+1+p-\ell}) e^{i\omega s} \end{aligned} \tag{5.6.6}$$

where the last equality has been obtained by the substitution $s = \ell - k$. Next, make the substitution $j = \ell - p$ in (5.6.6) to obtain:

$$f(\omega) = \sum_{\ell=0}^{m} \sum_{j=\ell-m}^{\ell} \sum_{s=\ell-m}^{\ell} (\hat{a}_{j-s} \hat{a}_j^* - \hat{a}_{m+1-j} \hat{a}_{m+1+s-j}^*) e^{i\omega s} \tag{5.6.7}$$

Since $\hat{a}_{j-s} = 0$ and $\hat{a}_{m+1+s-j}^* = 0$ for $s > j$, we can extend the summation over s in (5.6.7) up to $s = m$. Furthermore, the summand in (5.6.7) is zero for $j < 0$, and hence we can truncate the summation over j to the interval $[0,\ \ell]$. These two observations yield:

$$f(\omega) = \sum_{\ell=0}^{m} \sum_{j=0}^{\ell} \sum_{s=\ell-m}^{m} (\hat{a}_{j-s}\hat{a}_j^* - \hat{a}_{m+1-j}\hat{a}_{m+1+s-j}^*)e^{i\omega s} \tag{5.6.8}$$

Next, decompose $f(\omega)$ additively as follows:

$$f(\omega) = T_1(\omega) + T_2(\omega)$$

where

$$T_1(\omega) = \sum_{\ell=0}^{m} \sum_{j=0}^{\ell} \sum_{s=0}^{m} (\hat{a}_{j-s}\hat{a}_j^* - \hat{a}_{m+1-j}\hat{a}_{m+1+s-j}^*)e^{i\omega s}$$

$$T_2(\omega) = \sum_{\ell=0}^{m} \sum_{j=0}^{\ell} \sum_{s=\ell-m}^{-1} (\hat{a}_{j-s}\hat{a}_j^* - \hat{a}_{m+1-j}\hat{a}_{m+1+s-j}^*)e^{i\omega s}$$

(The term in T_2 corresponding to $\ell = m$ is zero.) Let

$$\hat{\mu}(s) \triangleq \sum_{\ell=0}^{m} \sum_{j=0}^{\ell} (\hat{a}_{j-s}\hat{a}_j^* - \hat{a}_{m+1-j}\hat{a}_{m+1+s-j}^*) \tag{5.6.9}$$

By using this notation, we can write $T_1(\omega)$ as

$$T_1(\omega) = \sum_{s=0}^{m} \hat{\mu}(s)e^{i\omega s}$$

Since $f(\omega)$ is real–valued for any $\omega \in [-\pi,\ \pi]$, we must also have

$$T_2(\omega) = \sum_{s=-1}^{-m} \hat{\mu}^*(-s)e^{i\omega s}$$

As the summand in (5.6.9) does not depend on ℓ, we readily obtain

$$\begin{aligned}
\hat{\mu}(s) &= \sum_{j=0}^{m} (m+1-j)\ (\hat{a}_{j-s}\hat{a}_j^* - \hat{a}_{m+1-j}\hat{a}_{m+1+s-j}^*) \\
&= \sum_{k=0}^{m-s} (m+1-k-s)\ \hat{a}_k\hat{a}_{k+s}^* - \sum_{k=1}^{m} k\hat{a}_k\hat{a}_{k+s}^* \\
&= \sum_{k=0}^{m-s} (m+1-2k-s)\hat{a}_k\hat{a}_{k+s}^*
\end{aligned}$$

which coincides with (5.6.5). Thus, the proof of (5.6.4) is concluded.

Next, note that (5.6.2) can be proven by a straightforward calculation:

$$
\begin{aligned}
\left| \sum_{k=0}^{m} \hat{a}_k e^{-i\omega k} \right|^2 &= \sum_{k=0}^{m} \sum_{p=0}^{m} \hat{a}_k \hat{a}_p^* e^{-i\omega(k-p)} = \sum_{k=0}^{m} \sum_{s=k-m}^{k} \hat{a}_k \hat{a}_{k-s}^* e^{-i\omega s} \\
&= \sum_{k=0}^{m} \sum_{s=-m}^{m} \hat{a}_k \hat{a}_{k-s}^* e^{-i\omega s} = \sum_{s=-m}^{m} \sum_{k=0}^{m} \hat{a}_k \hat{a}_{k+s}^* e^{i\omega s} \\
&= \sum_{s=-m}^{m} \left(\sum_{k=0}^{m-s} \hat{a}_k \hat{a}_{k+s}^* \right) e^{i\omega s}
\end{aligned}
$$

Remark: The reader may wonder what happens with the formulas derived above if the AR model parameters are calculated by using the same sample covariance matrix as in the Capon estimator. In such a case, the parameters $\{\hat{a}_k\}$ in (5.6.1) and in the GS formula above should be replaced by $\{\hat{a}_k^*\}$ (see (5.4.27)). Consequently both (5.6.2)–(5.6.3) and (5.6.4)–(5.6.5) continue to hold but with $\{\hat{a}_k\}$ replaced by $\{\hat{a}_k^*\}$ (and $\{\hat{a}_k^*\}$ replaced by $\{\hat{a}_k\}$, of course). ■

By comparing (5.6.2) and (5.6.4) we see that the reciprocals of both $\hat{\phi}_{AR}(\omega)$ and $\hat{\phi}_{CM}(\omega)$ have the form of a Blackman–Tukey spectral estimate associated with the "covariance sequences" $\{\hat{\rho}(s)\}$ and $\{\hat{\mu}(s)\}$, respectively. The only difference between $\hat{\phi}_{AR}(\omega)$ and $\hat{\phi}_{CM}(\omega)$ is that the sequence $\{\hat{\mu}(s)\}$ corresponding to $\hat{\phi}_{CM}(\omega)$ is a "linearly tapered" version of the sequence $\{\hat{\rho}(s)\}$ corresponding to $\hat{\phi}_{AR}(\omega)$. Similar to the interpretation in Section 5.4.2, the previous observation can be used to intuitively understand why the Capon spectral estimates are smoother and have poorer resolution than the AR estimates of the same order. (For more details on this aspect and other aspects related to the discussion in this complement, see [Musicus 1985].)

We remark in passing that the name "covariance sequence" given, for example, to $\{\hat{\rho}(s)\}$ is not coincidental: $\{\hat{\rho}(s)\}$ are so–called *sample inverse covariances* associated with $\hat{R}$ and they can be shown to possess a number of interesting and useful properties (see, *e.g.*, [Cleveland 1972; Bhansali 1980]).

The formula (5.6.4) can be used for the computation of $\hat{\phi}_{CM}(\omega)$, as we now show. Assuming that $\hat{R}$ is already available, we can use the Levinson–Durbin algorithm to compute $\{\hat{a}_k\}$ and $\hat{\sigma}^2$, and then $\{\hat{\mu}(s)\}$ in $O(m^2)$ flops. Then (5.6.4) can be evaluated at M Fourier frequencies (say) by using the FFT. The resulting total computational burden is on the order of $O(m^2 + M \log_2 M)$ flops. For commonly encountered values of m and M, this is about m times smaller than the burden associated with the eigendecomposition–based computational procedure of Exercise 5.5. Note, however, that the latter algorithm can be applied to a general $\hat{R}$

matrix, whereas the one derived in this complement is limited to Toeplitz covariance matrices.

5.7 Exercises

Exercise 5.1: Multiwindow Interpretation of Bartlett and Welch Methods

Equation (5.3.30) allows us to interpret the RFB method as a *multiwindow* (or *multitaper*) approach. Indeed, according to equation (5.3.30), we can write the RFB spectral estimator as:

$$\hat{\phi}(\omega) = \frac{1}{K} \sum_{p=1}^{K} \left| \sum_{t=1}^{N} w_{p,t} y(t) e^{-i\omega t} \right|^2 \tag{5.7.1}$$

where K is the number of data windows (or tapers), and where in the case of RFB the $w_{p,t}$ are obtained from the pth dominant Slepian sequence ($p = 1, \ldots, K$).

Show that the Bartlett and Welch methods can also be cast into the previous multiwindow framework. Make use of the multiwindow interpretation of these methods to compare them with one another and with the RFB approach.

Exercise 5.2: An Alternative Statistically Stable RFB Estimate

In Section 5.3.3 we developed a statistically stable RFB spectral estimator using a bank of narrow bandpass filters. In Section 5.4 we derived the Capon method, which employs a shorter filter length than the RFB. In this exercise we derive the RFB analog of the Capon approach and show its correspondence with the Welch and Blackman–Tukey estimators.

As an alternative technique to the filter in (5.3.4), consider a passband filter of shorter length:

$$h = [h_0, \ldots, h_m]^* \tag{5.7.2}$$

for some $m < N$. The optimal h will be the first Slepian sequence in (5.3.10) found using a Γ matrix of size $m \times m$. In this case, the filtered output

$$y_F(t) = \sum_{k=0}^{m} h_k \tilde{y}(t-k) \tag{5.7.3}$$

(with $\tilde{y}(t) = y(t)e^{-i\omega t}$) can be computed for $t = m+1, \ldots, N$. The resulting RFB spectral estimate is given by

$$\hat{\phi}(\omega) = \frac{1}{N-m} \sum_{t=m+1}^{N} |y_F(t)|^2 \tag{5.7.4}$$

(a) Show that the estimator in (5.7.4) is an unbiased estimate of $\phi(\omega)$, under the standard assumptions considered in this chapter.

(b) Show that $\hat{\phi}(\omega)$ can be written as

$$\hat{\phi}(\omega) = \frac{1}{m+1} h^*(\omega) \hat{R}\, h(\omega) \tag{5.7.5}$$

where $\hat{R}$ is an $(m+1) \times (m+1)$ Hermitian (but not Toeplitz) estimate of the covariance matrix of $y(t)$. Find the corresponding filter $h(\omega)$.

(c) Compare (5.7.5) with the Blackman–Tukey estimate in equation (5.4.22). Discuss how the two compare when N is large.

(d) Interpret $\hat{\phi}(\omega)$ as a Welch–type estimator. What is the overlap parameter K in the corresponding Welch method?

Exercise 5.3: Another Derivation of the Capon FIR Filter

The Capon FIR filter design problem can be restated as follows:

$$\min_h \; h^* R h / |h^* a(\omega)|^2 \tag{5.7.6}$$

Make use of the Cauchy–Schwartz inequality (Result R22 in Appendix A) to obtain a simple proof of the fact that h given by (5.4.8) is a solution to the optimization problem above.

Exercise 5.4: The Capon Filter is a Matched Filter

Compare the Capon filter design problem (5.4.7) with the following classical *matched filter design.*

- Filter: A causal FIR filter with an $(m+1)$–dimensional impulse response vector denoted by h.

- Signal–in–noise model: $y(t) = \alpha e^{i\omega t} + \varepsilon(t)$, which gives the following expression for the input vector to the filter:

$$z(t) = \alpha a(\omega) e^{i\omega t} + e(t) \tag{5.7.7}$$

where $a(\omega)$ is as defined in (5.4.6), $\alpha e^{i\omega t}$ is a sinusoidal signal,

$$z(t) = [y(t), y(t-1), \ldots, y(t-m)]^T$$

and $e(t)$ is a possibly colored noise vector defined similarly to $z(t)$. The signal and noise terms above are assumed to be uncorrelated.

- Design goal: Maximize the signal–to–noise ratio in the filter's output,

$$\max_h |h^* a(\omega)|^2 / h^* Q h \qquad (5.7.8)$$

where Q is the noise covariance matrix.

Show that the Capon filter is identical to the matched filter which solves the above design problem. The adjective "matched" attached to the above filter is motivated by the fact that the filter impulse response vector h depends on, and hence is "matched to", the signal term in (5.7.7).

Exercise 5.5: Computation of the Capon Spectrum

The Capon spectral estimators are defined in equations (5.4.19) and (5.4.20). The bulk of the computation of either estimator consists in the evaluation of an expression of the form $a^*(\omega)Qa(\omega)$, where Q is a given positive definite matrix, at a number of points on the frequency axis. Let these evaluation points be given by $\{\omega_k = 2\pi k/M\}_{k=0}^{M-1}$ for some sufficiently large M value (which we assume to be a power of two). The direct evaluation of $a^*(\omega_k)Qa(\omega_k)$, for $k = 0, \ldots, M-1$, would require $O(Mm^2)$ flops. Show that an evaluation based on the eigendecomposition of Q and the use of FFT is usually much more efficient computationally.

Exercise 5.6: A Relationship between the Capon Method and MUSIC (Pseudo)Spectra

Assume that the covariance matrix R, entering the Capon spectrum formula, has the expression (4.2.7) in the frequency estimation application. Then, show that

$$\lim_{\sigma^2 \to 0} (\sigma^2 R^{-1}) = I - A(A^*A)^{-1}A^* \qquad (5.7.9)$$

Conclude that the limiting (for $N \gg 1$) Capon and MUSIC (pseudo)spectra, associated with the frequency estimation data, are close to one another, provided that all signal–to–noise ratios are large enough.

Exercise 5.7: A Capon–like Implementation of MUSIC

The Capon and MUSIC (pseudo)spectra, as the data length N increases, are given by the functions in equations (5.4.12) and (4.5.13), respectively. Recall that the columns of the matrix G in (4.5.13) are equal to the $(m-n)$ eigenvectors corresponding to the smallest eigenvalues of the covariance matrix R in (5.4.12).

Consider the following Capon–like pseudospectrum:

$$g_k(\omega) = a^*(\omega)R^{-k}a(\omega)\lambda^k \qquad (5.7.10)$$

where λ is the minimum eigenvalue of R; the covariance matrix R is assumed to have the form (4.2.7) postulated by MUSIC. Show that, under this assumption,

$$\boxed{\lim_{k \to \infty} g_k(\omega) = a^*(\omega)GG^*a(\omega) = (4.5.13)} \qquad (5.7.11)$$

(where the convergence is uniform in ω). Explain why the convergence in (5.7.11) may be slow in difficult scenarios, such as those with closely spaced frequencies, and hence the use of (5.7.10) with a large k to approximate the MUSIC pseudospectrum may be computationally inefficient. However, the use of (5.7.10) for frequency estimation has a potential advantage over MUSIC that may outweigh its computational inefficiency. Find and comment on that advantage.

Exercise 5.8: Capon Estimate of the Parameters of a Single Sine Wave

Assume that the data under study consists of a sinusoidal signal observed in white noise. In such a case, the covariance matrix R is given by (*cf.* (4.2.7)):

$$R = \alpha^2 a(\omega_0)a(\omega_0)^* + \sigma^2 I, \qquad (m \times m)$$

where ω_0 denotes the true frequency value. Show that the limiting (as $N \to \infty$) Capon spectrum (5.4.12) peaks at $\omega = \omega_0$. Derive the height of the peak and show that it is not equal to α^2 (as might have been expected) but is given by a function of α^2, m and σ^2. Conclude that the Capon method can be used to obtain a consistent estimate of the frequency of a *single* sinusoidal signal in white noise (but not of the signal power).

We note that, for two or more sinusoidal signals, the Capon frequency estimates are inconsistent. Hence the Capon frequency estimator behaves somewhat similarly to the AR frequency estimation method in this respect; see Exercise 4.4.

Exercise 5.9: An Alternative Derivation of the Relationship between the Capon and AR Methods

Make use of the equation (3.8.5) relating R_{m+1}^{-1} to R_m^{-1} to obtain a simple proof of the formula (5.4.36) relating the Capon and AR spectral estimators.

Computer Exercises

Tools for Filter Bank Spectral Estimation:

The text web site `www.prenhall.com/~stoica` contains the following MATLAB functions for use in computing filter bank spectral estimates.

- `h=slepian(N,K,J)`
  Returns the first `J` Slepian sequences given `N` and `K` as defined in Section 5.3; `h` is an $N \times J$ matrix whose ith column gives the ith Slepian sequence.

- `phi=rfb(y,K,L)`
  The RFB spectral estimator. The vector `y` is the input data vector, `L` controls the frequency sample spacing of the output, and the output vector `phi`$= \hat{\phi}(\omega_k)$ where $\omega_k = \frac{2\pi k}{L}$. For $K = 1$, this function implements the high resolution RFB method in equation (5.3.22), and for $K > 1$ it implements the statistically stable RFB method.

- `phi=capon(y,m,L)`
  The CM Version–1 spectral estimator in equation (5.4.19); `y`, `L`, and `phi` are as for the RFB spectral estimator, and `m` is the size of the square matrix $\hat{R}$.

Exercise C5.10: Slepian Window Sequences

We consider the Slepian window sequences for both $K = 1$ (high resolution) and $K = 4$ (lower resolution, higher statistical stability) and compare them with classical window sequences.

(a) Evaluate and plot the first 8 Slepian window sequences and their Fourier transforms for $K = 1$ and 4 and for $N = 32$, 64, and 128 (and perhaps other values, too). Qualitatively describe the filter passbands of these first 8 Slepian sequences for $K = 1$ and $K = 4$. Which act as lowpass filters and which act as "other" types of filters?

(b) In this chapter we showed that for "large N" and $K = 1$, the first Slepian sequence is "reasonably close to" the rectangular window; compare the first Slepian sequence and its Fourier transform for $N = 32$, 64, and 128 to the rectangular window and its Fourier transform. How do they compare as a function of N? Based on this comparison, how do you expect the high resolution RFB PSD estimator to perform relative to the periodogram?

Exercise C5.11: Resolution of Refined Filter Bank Methods

We will compare the resolving power of the RFB spectral estimator with $K = 1$ to that of the periodogram. To do so we look at the spectral estimates of sequences which are made up of two sinusoids in noise, and where we vary the frequency difference.

Generate the sequences

$$y_\alpha(t) = 10\sin(0.2 \cdot 2\pi t) + 5\sin((0.2 + \alpha/N)2\pi t)$$

for various values of α near 1. Compare the resolving ability of the RFB power spectral estimate for $K = 1$ and of the periodogram for both $N = 32$ and $N = 128$. Discuss your results in relation to the theoretical comparisons between the two estimators. Do the results echo the theoretical predictions based on the analysis of Slepian sequences?

Exercise C5.12: The Statistically Stable RFB Power Spectral Estimator

In this exercise we will compare the RFB power spectral estimator when $K = 4$ to the Blackman–Tukey and Daniell estimators. We will use the narrowband and broadband processes considered in Exercise C2.20.

Broadband ARMA Process:

(a) Generate 50 realizations of the broadband ARMA process in Exercise C2.20, using $N = 256$. Estimate the spectrum using:

- The RFB method with $K = 4$.
- The Blackman–Tukey method with an appropriate window (such as the Bartlett window) and window length M. Choose M to obtain similar performance to the RFB method (you can select an appropriate value of M off–line and verify it in your experiments).
- The Daniell method with $\tilde{N} = 8N$ and an appropriate choice of J. Choose J to obtain similar performance to the RFB method (you can select J off–line and verify it in your experiments).

(b) Evaluate the relative performance of the three estimators in terms of bias and variance. Are the comparisons in agreement with the theoretical predictions?

Narrowband ARMA Process: Repeat parts (a) and (b) above using 50 realizations (with $N = 256$) of the narrowband ARMA process in Exercise C2.20.

Exercise C5.13: The Capon Method

In this exercise we compare the Capon method to the RFB and AR methods. Consider the sinusoidal data sequence in equation (2.9.19) from Exercise C2.17, with $N = 64$.

(a) We first compare the data filters corresponding to a RFB method (in which the filter is data independent) with the filter corresponding to the CM Version–1 method using both $m = N/4$ and $m = N/2 - 1$; we choose the Slepian RFB method with $K = 1$ and $K = 4$ for this comparison. For two estimation frequencies, $\omega = 0$ and $\omega = 2\pi \cdot 0.1$, plot the frequency response of the five filters (1 for $K = 1$ and 4 for $K = 4$) shown in the first block of Figure 5.1 for the two RFB methods, and also plot the response of the two Capon filters (one for each value of m; see (5.4.5) and (5.4.8)). What are their characteristic features in relation to the data? Based on these plots, discuss how data dependence can improve spectral estimation performance.

(b) Compare the two Capon estimators with the RFB estimator for both $K = 1$ and $K = 4$. Generate 50 Monte–Carlo realizations of the data and overlay plots of the 50 spectral estimates for each estimator. Discuss the similarities and differences between the RFB and Capon estimators.

(c) Compare Capon and Least Squares AR spectral estimates, again by generating 50 Monte–Carlo realizations of the data and overlaying plots of the 50 spectral estimates. Use $m = 8$, 16, and 31 for both the Capon method and the AR model order. How do the two methods compare in terms of resolution and variance? What are your main summarizing conclusions? Explain your results in terms of the data characteristics.

Chapter 6

SPATIAL METHODS

6.1 Introduction

In this chapter, we consider the problem of *locating n radiating sources by using an array of m passive sensors*, as shown in Figure 6.1. The emitted energy from the sources may for example be acoustic, electromagnetic, and so on, and the receiving sensors may be any transducers that convert the received energy to electrical signals. Examples of sensors include electromagnetic antennas, hydrophones, and seismometers. This type of problem finds applications in *radar and sonar systems, communications, astrophysics, biomedical research, seismology, underwater surveillance* (also called passive listening) and many other fields. This problem basically consists of determining how the "energy" is distributed over *space* (which may be air, water or the earth), with the source positions representing points in space with high concentrations of energy. Hence, it can be named a *spatial spectral estimation problem.* This name is also motivated by the fact that there are close ties between the source location problem and the problem of temporal spectral estimation treated in Chapters 1–5. In fact, as we will see, almost any of the methods encountered in the previous chapters may be used to derive a solution for the source location problem.

The emphasis in this chapter will be on *developing a model for the output signal of the receiving sensor array.* When this model is derived, the source location problem is turned into a parameter estimation problem that is quite similar to the temporal–frequency finding application discussed in Chapter 4. Hence, as we shall see, most of the methods developed for frequency estimation can be used to solve the spatial problem of source location.

The sources in Figure 6.1 generate a *wave field* that travels through space and is *sampled, in both space and time, by the sensor array.* By making analogy with temporal sampling, we may expect that the spatial sampling done by the array provides more and more information on the incoming waves as the *array's aperture* increases. The array's aperture is the space occupied by the array, as measured in units of signal wavelength. It is then no surprise that an array of sensors may provide significantly enhanced location performance as compared to the use of a *single antenna* (which was the system used in the early applications of the source

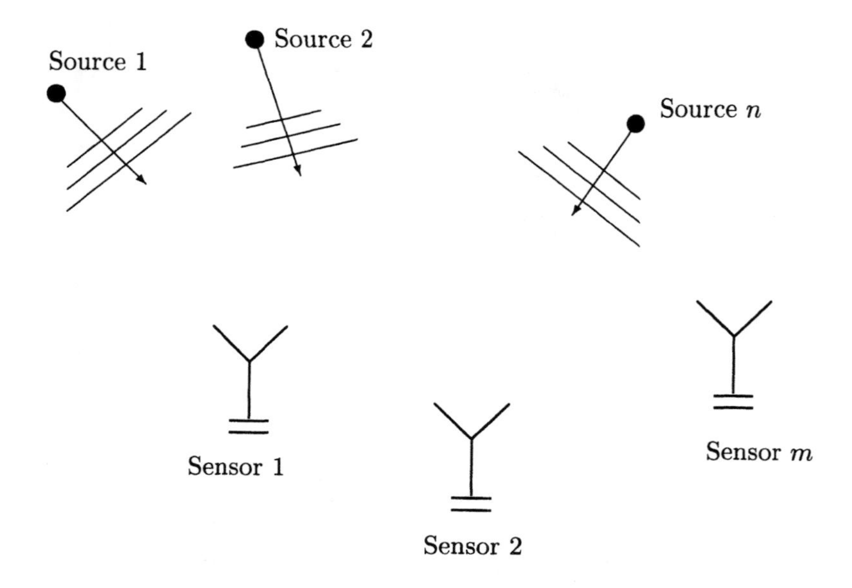

Figure 6.1. The setup of the source location problem.

location problem.)

The development of the array model in the next section is based on a number of simplifying assumptions. Some of these assumptions, which have a more general character, are listed below. The sources are assumed to be situated in the *far field* of the array. Furthermore, we assume that both the sources and the sensors in the array are in the *same plane* and that the sources are *point emitters*. In addition, it is assumed that the *propagation medium is homogeneous (i.e., not dispersive)* so that the waves arriving at the array can be considered to be *planar*. Under these assumptions, the only parameter that characterizes the source locations is the so–called *angle of arrival*, or *direction of arrival* (DOA); the DOA will be formally defined later on.

The above assumptions may be relaxed at the expense of significantly complicating the array model. Note that in the general case of a near–field source and a three–dimensional array, three parameters are required to define the position of one source, for instance the *azimuth*, *elevation* and *range*. Nevertheless, if the assumption of planar waves is maintained then we can treat the case of several unknown parameters per source without complicating the model too much. However, in order to keep the discussion as simple as possible, we will only consider the case of one parameter per source.

In the following discussion, it is also assumed that *the number of sources n is known.* The estimation of n, when it is unknown, is a problem of significant importance for many applications, which is often referred to as the *detection problem.* For solutions to the detection problem (which is analogous to the problem of order estimation in signal and system modeling), the reader is referred to [WAX AND KAILATH 1985; FUCHS 1988; VIBERG, OTTERSTEN, AND KAILATH 1991; FUCHS 1992].

Finally, it is assumed that the sensors in the array can be modeled as linear (time–invariant) systems; and that their transfer characteristics as well as their locations are known. In short, we say that *the array is assumed to be calibrated.*

6.2 Array Model

We begin by considering the case of a *single source.* Once we establish a model of the array for this case, the general model for the multiple source case is simply obtained by the superposition principle.

Suppose that a single waveform impinges upon the array and let $x(t)$ denote the value of the signal waveform as measured at some *reference point*, at time t. The "reference point" may be one of the sensors in the array, or any other point placed near enough to the array so that the previously made assumption of planar wave propagation holds true. The physical signals received by the array are *continuous time waveforms* and hence t is a continuous variable here, unless otherwise stated.

Let τ_k denote the time needed for the wave to travel from the reference point to sensor k ($k = 1, \ldots, m$). Then the output of sensor k can be written as

$$\bar{y}_k(t) = \bar{h}_k(t) * x(t - \tau_k) + \bar{e}_k(t) \tag{6.2.1}$$

where $\bar{h}_k(t)$ is the impulse response of the kth sensor, "$*$" denotes the convolution operation, and $\bar{e}_k(t)$ is an *additive noise.* The noise may enter in equation (6.2.1) either as "thermal noise" generated by the sensor's circuitry, as "random background radiation" impinging on the array, or in other ways. In (6.2.1), $\bar{h}_k(t)$ is assumed known and the "input" signal $x(t)$ as well as the delay τ_k are unknown. The parameters characterizing the source location enter in (6.2.1) through $\{\tau_k\}$. Hence, the source location problem is basically one of *time–delay estimation for the unknown input case.*

The model equation (6.2.1) can be simplified significantly if *the signals are assumed to be narrowband.* In order to show how this can be done, a number of preliminaries are required.

Let $X(\omega)$ denote the Fourier transform of the (continuous–time) signal $x(t)$:

$$X(\omega) = \int_{-\infty}^{\infty} x(t) e^{-i\omega t} dt \tag{6.2.2}$$

(which is assumed to exist and be finite for all $\omega \in (-\infty, \infty)$). The inverse transform, which expresses $x(t)$ as a linear functional of $X(\omega)$, is given by

$$x(t) = \frac{1}{2\pi} \int_{-\infty}^{\infty} X(\omega) e^{i\omega t} d\omega \tag{6.2.3}$$

Similarly, we define the transfer function $\bar{H}_k(\omega)$ of the kth sensor as the Fourier transform of $\bar{h}_k(t)$. In addition, let $\bar{Y}_k(\omega)$ and $\bar{E}_k(\omega)$ denote the Fourier transforms of the signal $\bar{y}_k(t)$ and noise $\bar{e}_k(t)$ in (6.2.1). By using this notation and the properties of the Fourier transform, $\bar{Y}_k(\omega)$ can be written as

$$\bar{Y}_k(\omega) = \bar{H}_k(\omega) X(\omega) e^{-i\omega\tau_k} + \bar{E}_k(\omega) \tag{6.2.4}$$

For a general class of physical signals, such as carrier modulated signals encountered in communications, the energy spectral density of $x(t)$ has the form shown in Figure 6.2. There, ω_c denotes the *center (or carrier) frequency* which is usually the center of the frequency band occupied by the signal (hence its name). A signal having an energy spectrum of the form depicted in Figure 6.2 is called a ***bandpass signal*** (by direct analogy with the notion of bandpass filters).

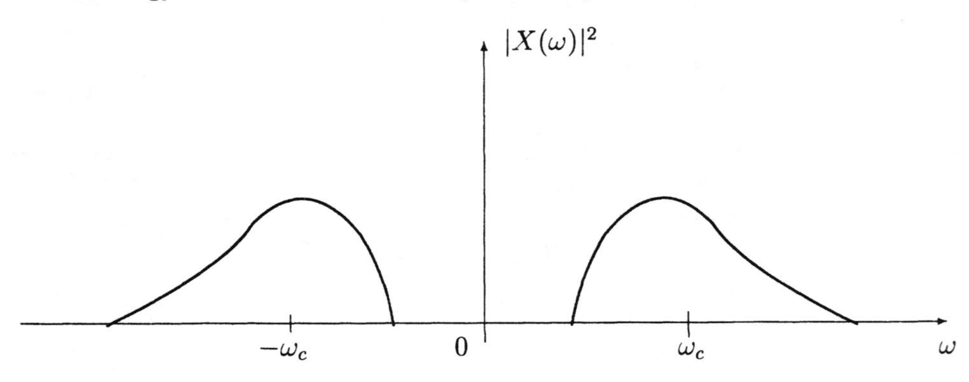

Figure 6.2. The energy spectrum of a bandpass signal.

For now, ***we assume that the received signal $x(t)$ is bandpass.*** It is clear from Figure 6.2 that the spectrum of such a signal is completely defined by the spectrum of a corresponding *baseband (or lowpass) signal.* The baseband spectrum, say $|S(\omega)|^2$, corresponding to the one in Figure 6.2, is displayed in Figure 6.3. Let $s(t)$ denote the baseband signal associated with $x(t)$. The process of obtaining $x(t)$ from $s(t)$ is called *modulation*, whereas the inverse process is named *demodulation*. In the following we make a number of comments on the modulation and demodulation processes, which — while not being strictly relevant to the source location problem — may be helpful in clarifying some claims in the text.

6.2.1 The Modulation–Transmission–Demodulation Process

The physical signal $x(t)$ is real–valued and hence its spectrum $|X(\omega)|^2$ should be even (*i.e.*, symmetric about $\omega = 0$; see, for instance, Figure 6.2). On the other hand,

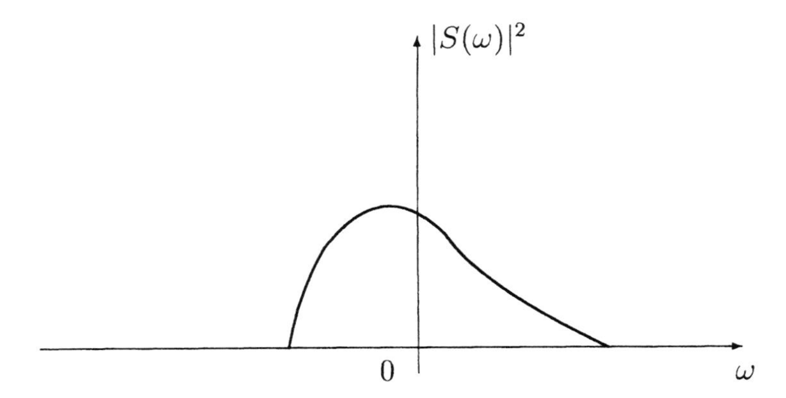

Figure 6.3. The baseband spectrum that gives rise to the bandpass spectrum in Figure 6.2.

the spectrum of the demodulated signal $s(t)$ may not be even (as indicated in Figure 6.3) and hence $s(t)$ may be complex–valued. The way in which this may happen is explained as follows. The *transmitted* signal is, of course, obtained by modulating a real–valued signal. Hence, in the spectrum of the transmitted signal the baseband spectrum is symmetric about $\omega = \omega_c$. The characteristics of the *transmission channel* (or the *propagation medium*), however, most often are asymmetric about $\omega = \omega_c$. This results in a *received* bandpass signal with an associated baseband spectrum that is not even. Hence, the demodulated received signal is complex–valued. This observation supports a claim made in Chapter 1 that complex–valued signals are not uncommon in spectral estimation problems.

The Modulation Process: If $s(t)$ is multiplied by $e^{i\omega_c t}$, then the Fourier transform of $s(t)$ is translated in frequency to the right by ω_c (assumed to be positive), as is verified by

$$\int_{-\infty}^{\infty} s(t)e^{i\omega_c t}e^{-i\omega t}d\omega = \int_{-\infty}^{\infty} s(t)e^{-i(\omega-\omega_c)t}d\omega = S(\omega - \omega_c) \tag{6.2.5}$$

The above formula describes the essence of the so–called *complex modulation process*. (An analogous formula for random discrete–time signals is given by equation (1.4.11) in Chapter 1.) The output of the complex modulation process is always complex–valued (hence the name of this form of modulation). If the modulated signal is real–valued, as $x(t)$ is, then it must have an even spectrum. In such a case the translation of $S(\omega)$ to the right by ω_c, as in (6.2.5), must be accompanied by a translation to the left (also by ω_c) of the folded and complex–conjugated baseband spectrum. This process results in the following expression for $X(\omega)$:

$$\boxed{X(\omega) = S(\omega - \omega_c) + S^*(-(\omega + \omega_c))} \tag{6.2.6}$$

It is readily verified that in the time domain, the *real modulation process* leading to (6.2.6) corresponds to taking the real part of the complex–modulated signal $s(t)e^{i\omega_c t}$:

$$\begin{aligned} x(t) &= \frac{1}{2\pi}\int_{-\infty}^{\infty}[S(\omega-\omega_c)+S^*(-\omega-\omega_c)]e^{i\omega t}d\omega \\ &= \frac{1}{2\pi}\int_{-\infty}^{\infty} S(\omega-\omega_c)e^{i(\omega-\omega_c)t}e^{i\omega_c t}d\omega \\ &\quad + \left[\frac{1}{2\pi}\int_{-\infty}^{\infty} S(-\omega-\omega_c)e^{-i(\omega+\omega_c)t}e^{i\omega_c t}d\omega\right]^* \\ &= s(t)e^{i\omega_c t} + [s(t)e^{i\omega_c t}]^* \end{aligned}$$

which gives

$$\boxed{x(t) = 2\text{Re}[s(t)e^{i\omega_c t}]} \tag{6.2.7}$$

or

$$\boxed{x(t) = 2\alpha(t)\cos(\omega_c t + \varphi(t))} \tag{6.2.8}$$

where $\alpha(t)$ and $\varphi(t)$ are the amplitude and phase of $s(t)$, respectively:

$$s(t) = \alpha(t)e^{i\varphi(t)}$$

If we let $s_I(t)$ and $s_Q(t)$ denote the real and imaginary parts of $s(t)$, then we can also write (6.2.7) as

$$\boxed{x(t) = 2[s_I(t)\cos(\omega_c t) - s_Q(t)\sin(\omega_c t)]} \tag{6.2.9}$$

We note in passing the following terminology associated with the equivalent time–domain representations (6.2.7)–(6.2.9) of a bandpass signal: $s(t)$ is called *the complex envelope* of $x(t)$; and $s_I(t)$ and $s_Q(t)$ are said to be *the in–phase and quadrature components* of $x(t)$.

The Demodulation Process: A calculation similar to (6.2.5) shows that the Fourier transform of $x(t)e^{-i\omega_c t}$ is given by

$$[S(\omega) + S^*(-\omega - 2\omega_c)]$$

which is simply $X(\omega)$ translated in frequency to the left by ω_c. The baseband (or lowpass) signal $s(t)$ can then be obtained by filtering $x(t)e^{-i\omega_c t}$ with a *baseband (or lowpass) filter* whose bandwidth is matched to that of $S(\omega)$. The hardware implementation of the demodulation process is presented later on, in block form, in Figure 6.4.

6.2.2 Derivation of the Model Equation

Given the background of the previous subsection, we return to equation (6.2.4) describing the output of sensor k. Since $x(t)$ is assumed to be a bandpass signal, $X(\omega)$ is given by (6.2.6) which, when inserted in (6.2.4), leads to

$$\bar{Y}_k(\omega) = \bar{H}_k(\omega)[S(\omega - \omega_c) + S^*(-\omega - \omega_c)]e^{-i\omega\tau_k} + \bar{E}_k(\omega) \tag{6.2.10}$$

Let $\tilde{y}_k(t)$ denote the demodulated signal:

$$\tilde{y}_k(t) = \bar{y}_k(t)e^{-i\omega_c t}$$

It follows from (6.2.10) and the previous discussion on the demodulation process that the Fourier transform of $\tilde{y}_k(t)$ is given by

$$\begin{aligned}\tilde{Y}_k(\omega) = {} & \bar{H}_k(\omega + \omega_c)[S(\omega) + S^*(-\omega - 2\omega_c)]e^{-i(\omega+\omega_c)\tau_k} \\ & + \bar{E}_k(\omega + \omega_c)\end{aligned} \tag{6.2.11}$$

When $\tilde{y}_k(t)$ is passed through a lowpass filter with bandwidth matched to $S(\omega)$, in the filter output (say, $y_k(t)$) the component in (6.2.11) centered at $\omega = -2\omega_c$ is eliminated along with all the other frequency components that fall in the stopband of the lowpass filter. Hence, we obtain:

$$Y_k(\omega) = H_k(\omega + \omega_c)S(\omega)e^{-i(\omega+\omega_c)\tau_k} + E_k(\omega + \omega_c) \tag{6.2.12}$$

where $H_k(\omega+\omega_c)$ and $E_k(\omega+\omega_c)$ denote the parts of $\bar{H}_k(\omega+\omega_c)$ and $\bar{E}_k(\omega+\omega_c)$ that fall within the lowpass filter's passband, Ω, and where the frequency ω is restricted to Ω.

We now make the following *key assumption.*

The received signals are narrowband, so that $\lvert S(\omega)\rvert$ decreases rapidly with increasing $\lvert\omega\rvert$.	(6.2.13)

Under the assumption above, (6.2.12) reduces (in an approximate way) to the following equation:

$$Y_k(\omega) = H_k(\omega_c)S(\omega)e^{-i\omega_c\tau_k} + E_k(\omega + \omega_c) \qquad \text{for } \omega \in \Omega \tag{6.2.14}$$

Because $H_k(\omega_c)$ must be different from zero, the sensor transfer function $\bar{H}_k(\omega)$ should pass frequencies near $\omega = \omega_c$ (as expected, since ω_c is the center frequency of the received signal). Also note that we do not replace $E_k(\omega + \omega_c)$ in (6.2.14) by $E_k(\omega_c)$ since this term might not be (nearly) constant over the signal bandwidth (for instance, this would be the case when the noise term in (6.2.12) contains a narrowband interference with the same center frequency as the signal).

Remark: It is sometimes claimed that (6.2.12) can be reduced to (6.2.14) even if the *signals are broadband* but *the sensors in the array are narrowband* with center frequency $\omega = \omega_c$. Under such an assumption, $|H_k(\omega + \omega_c)|$ goes quickly to zero as $|\omega|$ increases and hence (6.2.12) becomes

$$Y_k(\omega) = H_k(\omega + \omega_c)S(0)e^{-i\omega_c\tau_k} + E_k(\omega + \omega_c) \tag{6.2.15}$$

which apparently is different from (6.2.14). In order to obtain (6.2.14) from (6.2.12) under the previous conditions, we need to make some additional assumptions. Hence, if we further assume that *the sensor frequency response is flat over the passband* (so that $H_k(\omega + \omega_c) = H_k(\omega_c)$) and that *the signal spectrum varies over the sensor passband* (so that $S(\omega)$ differs quite a bit from $S(0)$ over the passband in question) then we can still obtain (6.2.14) from (6.2.12). ■

The model of the array is derived in a straightforward manner from equation (6.2.14). The time–domain counterpart of (6.2.14) is the following:

$$y_k(t) = H_k(\omega_c)e^{-i\omega_c\tau_k}s(t) + e_k(t) \tag{6.2.16}$$

where $y_k(t)$ and $e_k(t)$ are the inverse Fourier transforms of the corresponding terms in (6.2.14) (by a slight abuse of notation, $e_k(t)$ is associated with $E_k(\omega + \omega_c)$, not $E_k(\omega)$).

The hardware implementation required to obtain $\{y_k(t)\}$, as defined above, is indicated in Figure 6.4. Note that the scheme in Figure 6.4 generates samples of the real and imaginary components of $y_k(t)$. These samples are paired in the digital machine following the analog scheme of Figure 6.4 to obtain samples of the complex–valued signal $y_k(t)$. (We stress once more that all physical analog signals are real–valued.) Note that the continuous–time signal in (6.2.16) is *bandlimited*: according to (6.2.14) (and the related discussion), $Y_k(\omega)$ is approximately equal to zero for $\omega \notin \Omega$. Here Ω is the support of $S(\omega)$ (recall that the filter bandwidth was matched to the signal bandwidth), and hence it is a narrow interval. Consequently we can sample (6.2.16) with a rather low sampling frequency.

The sampled version of $\{y_k(t)\}$ is used by the "digital processing equipment" for the purpose of DOA estimation. Of course, the *digital form* of $\{y_k(t)\}$ satisfies an equation directly analogous to (6.2.16). In fact, to avoid a complication of notation by the introduction of a new discrete–time variable, *from here on we consider that t in equation (6.2.16) takes discrete values*

$$t = 1, 2, \ldots, N \tag{6.2.17}$$

(as usual, we choose the sampling period as the unit of the time axis). We remark once again that the scheme in Figure 6.4 samples the baseband signal, which may be done using lower sampling rates compared to those needed for the bandpass signal (see also [PROAKIS, RADER, LING, AND NIKIAS 1992]).

Next, we introduce the so–called *array transfer vector* (or *direction vector*):

$$\boxed{a(\theta) = [H_1(\omega_c)e^{-i\omega_c\tau_1} \ldots H_m(\omega_c)e^{-i\omega_c\tau_m}]^T} \tag{6.2.18}$$

Here, θ denotes the *source's direction of arrival* which is the parameter of interest in our problem. Note that since the transfer characteristics and positions of the

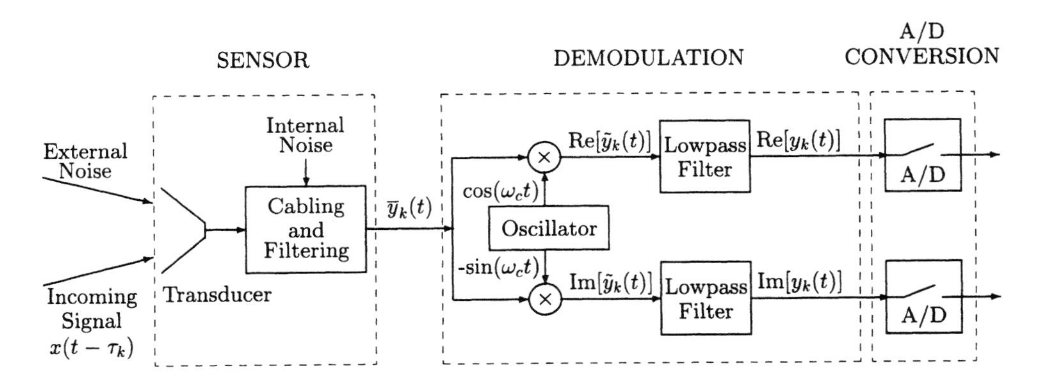

Figure 6.4. A simplified block diagram of the analog processing in a receiving array element.

sensors in the array are assumed to be known, the vector in (6.2.18) is a function of θ only, as indicated by notation (this fact will be illustrated shortly by means of a particular form of array). By making use of (6.2.18), we can write equation (6.2.16) as

$$y(t) = a(\theta)s(t) + e(t) \tag{6.2.19}$$

where

$$\begin{aligned} y(t) &= [y_1(t) \ldots y_m(t)]^T \\ e(t) &= [e_1(t) \ldots e_m(t)]^T \end{aligned}$$

denote the *array's output vector* and the *additive noise vector*, respectively. It should be noted that θ enters in (6.2.18) not only through $\{\tau_k\}$ but also through $\{H_k(\omega_c)\}$. In some cases, the sensors may be considered to be *omnidirectional* over the DOA range of interest, and then $\{H_k(\omega_c)\}_{k=1}^m$ are independent of θ. Sometimes, the sensors may also be assumed to be *identical.* Then by *redefining the signal* ($H(\omega_c)s(t)$ is redefined as $s(t)$) and *selecting the first sensor as the reference point,* the expression (6.2.18) can be simplified to the following form:

$$\boxed{a(\theta) = [1 \;\; e^{-i\omega_c\tau_2} \ldots e^{-i\omega_c\tau_m}]^T} \tag{6.2.20}$$

The extension of equation (6.2.19) to the case of *multiple sources* is straightforward. Since the sensors in the array were assumed to be linear elements, a direct application of the *superposition principle* leads to the following *model of the array.*

$$\begin{aligned} y(t) &= [a(\theta_1)\dots a(\theta_n)] \begin{bmatrix} s_1(t) \\ \vdots \\ s_n(t) \end{bmatrix} + e(t) \triangleq As(t) + e(t) \\ \theta_k &= \text{the DOA of the } k\text{th source} \\ s_k(t) &= \text{the signal corresponding to the } k\text{th source} \end{aligned} \tag{6.2.21}$$

It is interesting to note that the above model equation mainly relies on the *narrowband assumption* (6.2.13). The *planar wave* assumption made in the introductory part of this chapter has *not* been used so far. *This assumption is to be used when deriving the explicit dependence of* $\{\tau_k\}$ *as a function of* θ, as is illustrated in the following for an array with a special geometry.

Uniform Linear Array: Consider the array of m identical sensors uniformly spaced on a line, depicted in Figure 6.5. Such an array is commonly referred to as a uniform linear array (ULA). Let d denote the distance between two consecutive sensors, and let θ denote the DOA of the signal illuminating the array, as measured (counterclockwise) with respect to the normal to the line of sensors. Then, under the planar wave hypothesis and the assumption that the first sensor in the array is chosen as the reference point, we find that

$$\tau_k = (k-1)\frac{d\sin\theta}{c} \qquad \text{for } \theta \in [-90°, 90°] \tag{6.2.22}$$

where c is the propagation velocity of the impinging waveform (for example, the speed of light in the case of electromagnetic waves). Inserting (6.2.22) into (6.2.20) gives

$$a(\theta) = \left[1, e^{-i\omega_c d\sin\theta/c}, \dots, e^{-i(m-1)\omega_c d\sin\theta/c}\right]^T \tag{6.2.23}$$

The restriction of θ to lie in the interval $[-90°, 90°]$ is a limitation of ULAs: two sources at locations symmetric with respect to the array line yield identical sets of delays $\{\tau_k\}$ and hence cannot be distinguished from one another. In practice this ambiguity of ULAs is eliminated by using sensors that only pass signals whose DOAs are in $[-90°, 90°]$.

Let λ denote the *signal wavelength*:

$$\lambda = c/f_c, \qquad f_c = \omega_c/2\pi \tag{6.2.24}$$

(which is the distance traveled by the waveform in one period of the carrier). Define

$$f_s = f_c\frac{d\sin\theta}{c} = \frac{d\sin\theta}{\lambda} \tag{6.2.25}$$

and

$$\boxed{\omega_s = 2\pi f_s = \omega_c\frac{d\sin\theta}{c}} \tag{6.2.26}$$

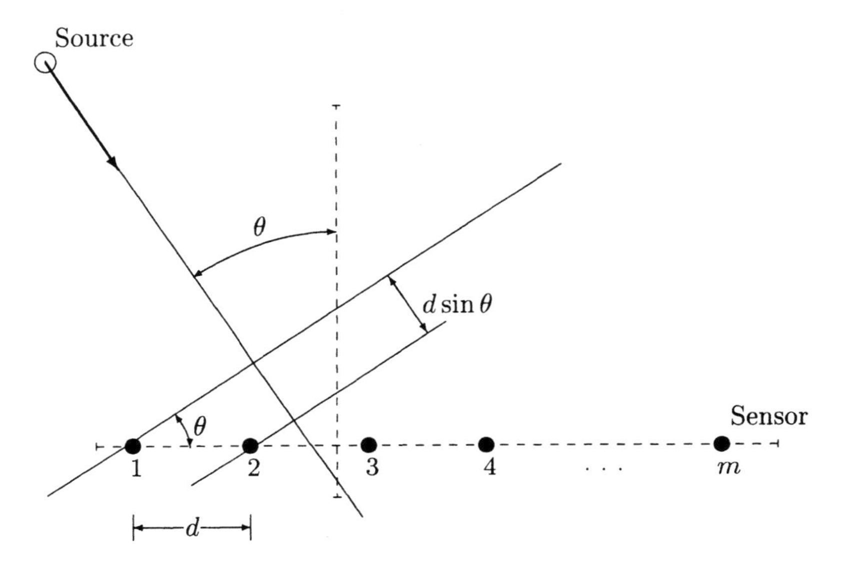

Figure 6.5. The uniform linear array scenario.

With this notation, the transfer vector (6.2.23) can be rewritten as:

$$a(\theta) = [1 \;\; e^{-i\omega_s} \ldots e^{-i(m-1)\omega_s}]^T \tag{6.2.27}$$

This is a Vandermonde vector which is completely analogous with the vector made from the uniform samples of the sinusoidal signal $\{e^{-i\omega_s t}\}$. Let us explore this analogy a bit further.

First, by the above analogy, ω_s is called the *spatial frequency*.

Second, if we were to sample a continuous–time sinusoidal signal with frequency ω_c then, in order to avoid aliasing effects, the sampling frequency f_0 should satisfy (by the Shannon sampling theorem):

$$f_0 \geq 2f_c \tag{6.2.28}$$

Now, in the ULA case considered in this example, we see from (6.2.27) that the vector $a(\theta)$ is uniquely defined (*i.e.*, there is no "spatial aliasing") if and only if ω_s is constrained as follows:

$$|\omega_s| \leq \pi \tag{6.2.29}$$

However, (6.2.29) is equivalent to

$$|f_s| \leq 1/2 \iff d|\sin\theta| \leq \lambda/2 \tag{6.2.30}$$

which is satisfied if

$$d \leq \lambda/2 \tag{6.2.31}$$

Since we may think of the ULA as performing a uniform spatial sampling of the wavefield, equation (6.2.31) simply says that the (spatial) sampling period d should be smaller than half of the signal wavelength. By analogy with (6.2.28), this result may be interpreted as a *spatial Shannon sampling theorem.*

Equipped with the array model (6.2.21) derived previously, we can reduce the problem of DOA finding to that of estimating the parameters $\{\theta_k\}$ in (6.2.21). As there is a *direct analogy between (6.2.21) and the model (4.2.6) for sinusoidal signals in noise*, we may expect that most of the methods developed in Chapter 4 for (temporal) frequency estimation can also be used for DOA estimation. This is shown to be the case in the following sections, which briefly review the most important DOA finding methods.

6.3 Nonparametric Methods

The methods to be described in this section *do not make any assumption on the covariance structure of the data.* As such, they may be considered to be "nonparametric". On the other hand, they assume that *the functional form of the array's transfer vector $a(\theta)$ is known.* Can we then still categorize them as "nonparametric methods"? The array performs a spatial sampling of the incoming wavefront, which is analogous to the temporal sampling done by the tapped–delay line implementation of a (temporal) finite impulse response (FIR) filter, see Figure 6.6. Thus, assuming that the form of $a(\theta)$ is available is no more restrictive than making the same assumption for $a(\omega)$ in Figure 6.6a. In conclusion, the functional form of $a(\theta)$ characterizes the array as a *spatial sampling device*, and assuming it is known should not be considered to be parametric (or model–based) information. As already mentioned, an array for which the functional form of $a(\theta)$ is know is said to be *calibrated.*

Figure 6.6 also makes an analogy between *temporal FIR filtering* and *spatial filtering* using an array of sensors. In what follows, we comment briefly on this analogy since it is of interest for the nonparametric approach to DOA finding. In the time series case, a FIR filter is defined by the relation

$$y_F(t) = \sum_{k=0}^{m-1} h_k u(t-k) \triangleq h^* y(t) \tag{6.3.1}$$

where $\{h_k\}$ are the filter weights, $u(t)$ is the input to the filter and

$$h = [h_0 \dots h_{m-1}]^* \tag{6.3.2}$$

$$y(t) = [u(t) \dots u(t-m+1)]^T \tag{6.3.3}$$

Similarly, we can use the spatial samples $\{y_k(t)\}_{k=1}^{m}$ obtained with a sensor array to define a *spatial filter*:

$$\boxed{y_F(t) = h^* y(t)} \tag{6.3.4}$$

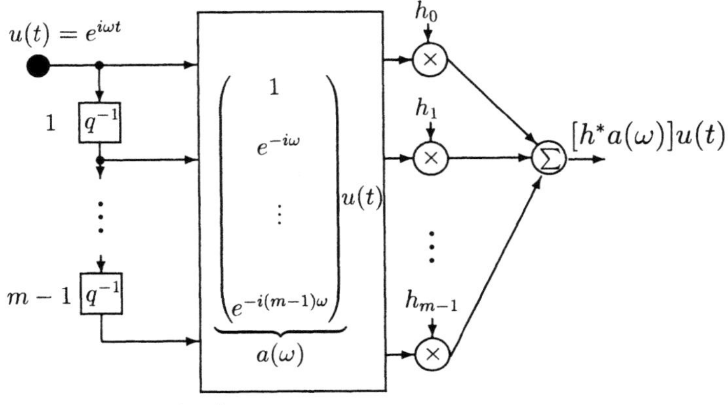

(a) Temporal filter

narrowband source with DOA=θ

1

2

m

reference point

1

$e^{-i\omega\tau_2(\theta)}$

$e^{-i\omega\tau_m(\theta)}$

$s(t)$

$a(\theta)$

h_0

h_1

h_{m-1}

$[h^* a(\theta)]s(t)$

(Spatial sampling)

(b) Spatial filter

Figure 6.6. Analogy between temporal sampling and filtering and the corresponding (spatial) operations performed by an array of sensors.

A temporal filter can be made to enhance or attenuate some selected frequency bands by appropriately choosing the vector h. More precisely, since the filter output for a sinusoidal input $u(t)$ is given by

$$y_F(t) = [h^* a(\omega)]u(t) \tag{6.3.5}$$

(where $a(\omega)$ is as defined, for instance, in Figure 6.6), then by selecting h so that $h^* a(\omega)$ is large (small) we can enhance (attenuate) the power of $y_F(t)$ at frequency ω.

In direct analogy with (6.3.5), the (noise–free) spatially filtered output (as in (6.3.4)) of an array illuminated by a narrowband wavefront with complex envelope $s(t)$ and DOA equal to θ is given by (*cf.* (6.2.19)):

$$\boxed{y_F(t) = [h^* a(\theta)]s(t)} \tag{6.3.6}$$

This equation clearly shows that *the spatial filter can be selected to enhance (attenuate) the signals coming from a given direction* θ, by making $h^* a(\theta)$ in (6.3.6) large (small). This observation lies at the basis of the DOA finding methods to be described in this section. All of these methods can be derived by using the *filter bank approach* of Chapter 5. More specifically, assume that a filter h has been found such that

(6.3.7)

- (i) It passes undistorted the signals with a given DOA θ; and
- (ii) It attenuates all the other DOAs different from θ as much as possible.

Then, the power of the spatially filtered signal in (6.3.4),

$$\boxed{E\left\{|y_F(t)|^2\right\} = h^* R h, \qquad R = E\left\{y(t)y^*(t)\right\}} \tag{6.3.8}$$

should give a good indication of the energy coming from direction θ. (Note that θ enters in (6.3.8) via h.) Hence, *$h^* R h$ should peak at the DOAs of the sources located in the array's viewing field* when evaluated over the DOA range of interest. This fact may be exploited for the purpose of DOA finding. Depending on the specific way in which the (loose) design objectives in (6.3.7) are formulated, the above approach can lead to different DOA estimation methods. In the following, we present *spatial extensions of the periodogram and Capon techniques*. The *RFB method* of Chapter 5 may also be extended to the spatial processing case, provided the array's geometry is such that the transfer vector $a(\theta + \alpha)$ can be factored as

$$a(\theta + \alpha) = D(\theta)a(\alpha) \tag{6.3.9}$$

where D is a unitary (possibly diagonal) matrix. Without such a property, the RFB spatial filter should be computed, *for each* θ, by solving an $m \times m$ eigendecomposition problem, which would be computationally prohibitive in most applications.

Since it is not *a priori* obvious that an arbitrary array satisfies (6.3.9), we do not consider the RFB approach in what follows.[1] Finally, we remark that a spatial filter satisfying the design objectives in (6.3.7) can be viewed as *forming a (reception) beam* in the direction θ, as pictorially indicated in Figure 6.7. Because of this interpretation, the methods resulting from this approach to the DOA finding problem, in particular the method of the next subsection, are called *beamforming methods* [VAN VEEN AND BUCKLEY 1988; JOHNSON AND DUDGEON 1992].

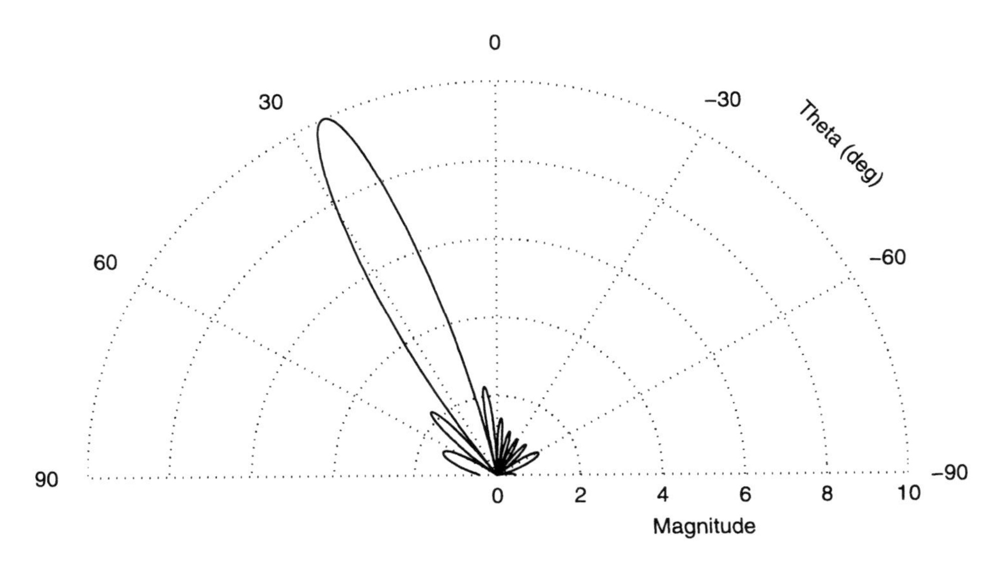

Figure 6.7. The response magnitude $|h^* a(\theta)|$, versus θ, of a spatial filter (or beamformer). Here, $h = a(\theta_0)$, where $\theta_0 = 25°$ is the DOA of interest; the array is a 10–element ULA with $d = \lambda/2$.

6.3.1 Beamforming

In view of (6.3.6), *condition* (i) of the filter design problem (6.3.7) can be formulated as:

$$h^* a(\theta) = 1 \tag{6.3.10}$$

In what follows, we assume that the transfer vector $a(\theta)$ has been normalized so that

$$a^*(\theta) a(\theta) = m \tag{6.3.11}$$

Note that in the case of an array with identical sensors, the condition (6.3.11) is automatically met (*cf.* (6.2.20)).

[1] Referring back to Chapter 5 may prove useful for understanding these comments on RFB and for several other discussions in this section.

Regarding *condition* (ii) in (6.3.7), if $y(t)$ in (6.3.8) were *spatially white with* $R = I$, then we would obtain the following expression for the power of the filtered signal:

$$E\left\{|y_F(t)|^2\right\} = h^*h \tag{6.3.12}$$

which is different from zero for every θ (note that we cannot have $h = 0$, because of condition (6.3.10)). This fact indicates that *a spatially white signal in the array output can be considered as impinging on the array with equal power from all directions* θ (in the same manner as a temporally white signal in the array output contains equal power in all frequency bands). We deduce from this observation that a natural mathematical formulation of condition (ii) would be to require that h minimizes the power in (6.3.12). Hence, we are led to the following design problem:

$$\min_h h^*h \quad \text{subject to} \quad h^*a(\theta) = 1 \tag{6.3.13}$$

As (6.3.13) is a special case of the optimization problem (5.4.7) in Chapter 5, we obtain the solution to (6.3.13) from (5.4.8) as:

$$h = a(\theta)/a^*(\theta)a(\theta) \tag{6.3.14}$$

By making use of (6.3.11), (6.3.14) reduces to

$$h = a(\theta)/m \tag{6.3.15}$$

which, when inserted in (6.3.8), gives

$$E\left\{|y_F(t)|^2\right\} = a^*(\theta)Ra(\theta)/m^2 \tag{6.3.16}$$

The theoretical covariance matrix R in (6.3.16) cannot be (exactly) determined from the available finite sample $\{y(t)\}_{t=1}^N$ and hence it must be replaced by some estimate, such as

$$\hat{R} = \frac{1}{N}\sum_{t=1}^{N} y(t)y^*(t) \tag{6.3.17}$$

By doing so and omitting the factor $1/m^2$ in (6.3.16), which has no influence on the DOA estimates, we obtain the Beamforming method which determines the DOAs as summarized in the next box.

The Beamforming DOA estimates are given by the locations of the n highest peaks of the function (6.3.18)

$$a^*(\theta)\hat{R}a(\theta)$$

When the estimated spatial spectrum in (6.3.18) is compared to the expression derived in Section 5.4 for the Blackman–Tukey periodogram, it is seen that

Beamforming is a direct (spatial) extension of the periodogram. In fact, the function in (6.3.18) may be thought of as being obtained by averaging the "spatial periodograms"

$$|a^*(\theta)y(t)|^2 \tag{6.3.19}$$

over the set of available "snapshots" ($t = 1, \ldots, N$).

The connection established in the previous paragraph, between Beamforming and the (averaged) periodogram, suggests that the *resolution properties* of the Beamforming method are analogous to those of the periodogram method. In fact, by an analysis similar to that in Chapters 2 and 5 it can be shown that the *beamwidth*[2] of the spatial filter used by Beamforming is approximately equal to the inverse of the array's aperture (as measured in signal wavelengths). This sets a limit on the resolution achievable with Beamforming, as indicated below (see Exercise 6.2):

$$\boxed{\text{Beamforming DOA resolution limit} \simeq \text{wavelength / array "length"}} \tag{6.3.20}$$

Next, we note that as N increases, the sample spatial spectrum in (6.3.18) converges (under mild conditions) to (6.3.16), uniformly in θ. Hence the Beamforming estimates of the DOAs converge to the n maximum points of (6.3.16), as N tends to infinity. *If the array model (6.2.21) holds* (it has not been used so far!), *the noise $e(t)$ is spatially white and has the same power σ^2 in all sensors*, and *if there is only one source* (with DOA denoted by θ_0, for convenience), then R in (6.3.16) is given by

$$R = a(\theta_0)a^*(\theta_0)P + \sigma^2 I \tag{6.3.21}$$

where $P = E\left\{|s(t)|^2\right\}$ denotes the signal power. Hence,

$$\begin{aligned} a^*(\theta)Ra(\theta) &= |a^*(\theta)a(\theta_0)|^2 P + a^*(\theta)a(\theta)\sigma^2 \\ &\leq |a^*(\theta)a(\theta)||a^*(\theta_0)a(\theta_0)|P + \sigma^2 a^*(\theta)a(\theta) \\ &= m(mP + \sigma^2) \end{aligned} \tag{6.3.22}$$

where the inequality follows from the Cauchy–Schwartz lemma (see Result R22 in Appendix A) and the last equality from (6.3.11). The upper bound in (6.3.22) is achieved for $a(\theta) = a(\theta_0)$ which, under mild conditions, implies $\theta = \theta_0$. In conclusion, the *Beamforming DOA estimate is consistent under the previous assumptions* ($n = 1$, *etc.*). *In the general case of multiple sources, however, the DOA estimates obtained with Beamforming are inconsistent.* The (asymptotic) bias of these estimates may be significant if the sources are strongly correlated or closely spaced.

As explained above, Beamforming is the spatial analog of the Blackman–Tukey periodogram (with a certain covariance estimate) and the Bartlett periodogram (if we interpret the m–dimensional snapshots in (6.3.19) as "subsamples" of the available "sample" $[y^T(1), \ldots, y^T(N)]^T$). Note, however, that the value of m in the periodogram methods can be chosen by the user, whereas in the Beamforming

[2]The beamwidth is the spatial counterpart of the temporal notion of bandwidth associated with a bandpass filter.

method m is fixed. This difference might seem small at first, but it has a significant impact on the consistency properties of Beamforming. More precisely, it can be shown that, for instance, the Bartlett periodogram estimates of *temporal frequencies* are *consistent* under the model (4.2.7), *provided that* m increases without bound as the number of samples N tends to infinity (*e.g.*, we can set $m = N$, which yields the unmodified periodogram).[3] For Beamforming, on the other hand, the value of m (*i.e.*, the number of array elements) is *limited* by physical considerations. This prevents Beamforming from providing consistent DOA estimates in the multiple signal case. An additional difficulty is that in the spatial scenario the signals can be correlated with one another, whereas they are always uncorrelated in the temporal frequency estimation case. Explaining why this is so and completing a consistency analysis of the Beamforming DOA estimates is left as an exercise for the reader.

Now, if the model (6.2.21) holds, if the minimum DOA separation is larger than the array beamwidth (which implies that m is sufficiently large), if the signals are uncorrelated, and if the noise is spatially white, then it is readily seen that the multiple–source spectrum (6.3.16) decouples (approximately) in n single–source spectra; this means that Beamforming may provide reasonably accurate DOA estimates in such a case. In fact, in this case Beamforming can be shown to provide an approximation to the nonlinear LS DOA estimation method discussed in Section 6.4.1; see the remark in that section.

6.3.2 Capon

The derivation of the Capon method for array signal processing is entirely analogous with the derivation of the Capon method for the time series data case developed in Section 5.4 [Capon 1969; Lacoss 1971]. The Capon spatial filter design problem is the following:

$$\min_{h} h^* R h \quad \text{subject to} \quad h^* a(\theta) = 1 \tag{6.3.23}$$

Hence, objective (i) in the general design problem (6.3.7) is ensured by constraining the filter exactly as in the Beamforming approach (see (6.3.10)). Objective (ii) in (6.3.7), however, is accomplished in a more sound way: by requiring the filter to minimize the output power, when fed with the actual array data $\{y(t)\}$. Hence, in the Capon approach, objective (ii) is formulated in a "data–dependent" way, whereas it was formulated independently of the data in the Beamforming method. As a consequence, the goal of the Capon filter steered to a certain direction θ is to attenuate any other signal that *actually impinges on the array* from a DOA $\neq$

[3]The unmodified periodogram in an *inconsistent estimator* for *continuous* PSDs (as shown in Chapter 2). However, as asserted above, the plain periodogram estimates of *discrete (or line)* PSDs are *consistent*. Showing this is left as an exercise to the reader. (Make use of the covariance matrix model (4.2.7) with $m \to \infty$, and the fact that the Fourier (or Vandermonde) vectors, at different frequencies, become orthogonal to one another as their dimension increases.)

θ, whereas the Beamforming filter pays uniform attention to *all other* DOAs $\neq \theta$, even though there might be no incoming signal for many of those DOAs.

The solution to (6.3.23), as derived in Section 5.4, is given by

$$h = \frac{R^{-1}a(\theta)}{a^*(\theta)R^{-1}a(\theta)} \tag{6.3.24}$$

which, when inserted in the output power formula (6.3.8), leads to

$$E\left\{|y_F(t)|^2\right\} = \frac{1}{a^*(\theta)R^{-1}a(\theta)} \tag{6.3.25}$$

It only remains to replace R in (6.3.25) by a sample estimate, such as $\hat{R}$ in (6.3.17), to obtain the Capon DOA estimator.

The Capon DOA estimates are obtained as the locations of the n largest peaks of the following function:

$$\frac{1}{a^*(\theta)\hat{R}^{-1}a(\theta)} \tag{6.3.26}$$

There is an implicit assumption in (6.3.26) that $\hat{R}^{-1}$ exists, but this can be ensured under weak conditions (in particular, $\hat{R}^{-1}$ exists with probability 1 if $N \geq m$ and if the noise term has a positive definite spatial covariance matrix). Note that the "spatial spectrum" in (6.3.26) corresponds to the "CM–Version 1" PSD in the time series case (see equation (5.4.12) in Section 5.4). A Capon spatial spectrum similar to the "CM–Version 2" PSD formula (see (5.4.17)) might also be derived, but it appears to be more complicated than the time series formula if the array is not a ULA.

Capon DOA estimation has been empirically found to possess superior performance as compared with Beamforming. The common advantage of these two nonparametric methods is that they do not assume anything about the statistical properties of the data and, therefore, they can be used in situations where we lack information about these properties. On the other hand, in the cases where such information is available, for example in the form of a covariance model of the data, a nonparametric approach does not give the performance that one can achieve with a parametric (model–based) approach. The parametric approach to DOA estimation is the subject of the next section.

6.4 Parametric Methods

In this section, *we postulate the array model (6.2.21)*. Furthermore, the noise $e(t)$ is assumed to be spatially white with components having identical variance:

$$E\left\{e(t)e^*(t)\right\} = \sigma^2 I \tag{6.4.1}$$

In addition, the signal covariance matrix

$$P = E\{s(t)s^*(t)\} \tag{6.4.2}$$

is assumed to be *nonsingular* (but not necessarily diagonal; hence the signals may be (partially) correlated). When the signals are fully correlated, so that P is singular, they are said to be *coherent*. Finally, we assume that the signals and the noise are uncorrelated with one another.

Under the previous assumptions, the theoretical covariance matrix of the array output vector is given by

$$\boxed{R = E\{y(t)y^*(t)\} = APA^* + \sigma^2 I} \tag{6.4.3}$$

There is a direct analogy between the array models above, (6.2.21) and (6.4.3), and the corresponding models encountered in our discussion of the sinusoids–in–noise case in Chapter 4. More specifically, the "nonlinear regression" model (6.2.21) of the array is analogous to (4.2.6), and the array covariance model (6.4.3) is much the same as (4.2.7). The consequence of these analogies is that *all methods introduced in Chapter 4 for frequency estimation can also be used for DOA estimation* without any essential modification. In the following, we briefly review these methods with a view of pointing out any differences from the frequency estimation application. When the assumed array model is a good representation of reality, the parametric DOA estimation methods reviewed in the sequel provide highly accurate DOA estimates, even in adverse situations (such as low SNR scenarios). As our main thrust in this text has been the understanding of the basic ideas behind the presented spectral estimation methodologies, we do not dwell on the details of the analysis required to establish the statistical properties of the DOA estimators discussed in the following; see, however, Appendix B for a discussion on the Cramér–Rao bound and the best accuracy achievable in DOA estimation problems. Such analysis details are available in [Stoica and Nehorai 1989a; Stoica and Nehorai 1990; Stoica and Sharman 1990; Stoica and Nehorai 1991; Viberg and Ottersten 1991; Rao and Hari 1993]. For reviews of many of the recent advances in spatial spectral analysis, the reader can consult [Pillai 1989] and [Ottersten, Viberg, Stoica, and Nehorai 1993].

6.4.1 Nonlinear Least Squares Method

This method determines the unknown DOAs as the minimizing elements of the following function

$$f = \frac{1}{N}\sum_{t=1}^{N} \| y(t) - As(t) \|^2 \tag{6.4.4}$$

Minimization with respect to $\{s(t)\}$ gives (see Result R32 in Appendix A)

$$s(t) = (A^*A)^{-1}A^*y(t) \qquad t = 1, \ldots, N \tag{6.4.5}$$

By inserting (6.4.5) into (6.4.4), we get the following concentrated nonlinear least squares (LS) criterion:

$$
\begin{aligned}
f &= \frac{1}{N}\sum_{t=1}^{N} \| \{I - A(A^*A)^{-1}A^*\}y(t) \|^2 \\
&= \frac{1}{N}\sum_{t=1}^{N} y^*(t)[I - A(A^*A)^{-1}A^*]y(t) \\
&= \mathrm{tr}\{[I - A(A^*A)^{-1}A^*]\hat{R}\} \qquad (6.4.6)
\end{aligned}
$$

The second equality in (6.4.6) follows from the fact that the matrix $I - A(A^*A)^{-1}A^*$ is idempotent (it is the orthogonal projector onto $\mathcal{N}(A^*)$), and the third from the properties of the trace operator (see Result R8 in Appendix A). It follows from (6.4.6) that the nonlinear LS DOA estimates are given by

$$
\boxed{\{\hat{\theta}_k\} = \arg\max_{\{\theta_k\}} \mathrm{tr}[A(A^*A)^{-1}A^*\hat{R}]} \qquad (6.4.7)
$$

Remark: Similar to the frequency estimation case, it can be shown that Beamforming provides an approximate solution to the previous nonlinear LS problem whenever the DOAs are known to be well separated. To see this, let us assume that we restrict the search for the maximizers of (6.4.7) to a set of well-separated DOAs (according to the *a priori* information that the true DOAs belong to this set.) In such a set, $A^*A \simeq mI$ under weak conditions, and hence the function in (6.4.7) can approximately be written as:

$$
\mathrm{tr}\left[A(A^*A)^{-1}A^*\hat{R}\right] \simeq \frac{1}{m}\sum_{k=1}^{n} a^*(\theta_k)\hat{R}a(\theta_k)
$$

Paralleling the discussion following equation (4.3.16) in Chapter 4 we can show that the Beamforming DOA estimates maximize the right–hand side of the above equation over the set under consideration. With this observation, the proof of the fact that the computationally efficient Beamforming method provides an approximate solution to (6.4.7) in scenarios with well–separated DOAs is concluded. ■

One difference between (6.4.7) and the corresponding optimization problem in the frequency estimation application (see (4.3.8) in Section 4.3) lies in the fact that in the frequency estimation application only one "snapshot" of data is available, in contrast to the N snapshots available in the DOA estimation application. Another, more important difference is that for non–ULA cases the matrix A in (6.4.7) does not have the Vandermonde structure of the corresponding matrix in (4.3.8). As a consequence, several of the algorithms used to (approximately) solve the frequency estimation problem (such as the one in [Kumaresan, Scharf, and Shaw 1986] and [Bresler and Macovski 1986]) are no longer applicable to solving (6.4.7) unless the array is a ULA.

6.4.2 Yule–Walker Method

The matrix Γ, which lies at the basis of the Yule–Walker method (see Section 4.4), can be constructed from any block of R in (6.4.3) that does not include diagonal elements. To be more precise, partition the array model (6.2.21) into the following two nonoverlapping parts:

$$y(t) = \begin{bmatrix} \bar{y}(t) \\ \tilde{y}(t) \end{bmatrix} = \begin{bmatrix} \bar{A} \\ \tilde{A} \end{bmatrix} s(t) + \begin{bmatrix} \bar{e}(t) \\ \tilde{e}(t) \end{bmatrix} \tag{6.4.8}$$

Since $\bar{e}(t)$ *and* $\tilde{e}(t)$ *are uncorrelated* (by assumption), we have

$$\boxed{\Gamma \triangleq E\left\{\bar{y}(t)\tilde{y}^*(t)\right\} = \bar{A}P\tilde{A}^*} \tag{6.4.9}$$

which is assumed to be of dimension $M \times L$ (with $M + L = m$). For

$$\boxed{M > n, \qquad L > n} \tag{6.4.10}$$

(which cannot hold unless $m > 2n$), the rank of Γ is equal to n (under weak conditions) and the $(L - n)$–dimensional null space of this matrix contains complete information about the DOAs. To see this, let G be an $L \times (L - n)$ matrix whose columns form a basis of $\mathcal{N}(\Gamma)$ (G can be obtained from the SVD of Γ; see Result R15 in Appendix A). Then we have $\Gamma G = 0$, which implies (using the fact that $\text{rank}(\bar{A}P) = n$):

$$\tilde{A}^*G = 0$$

This observation can be used, in the manner of Sections 4.4 (YW) and 4.5 (MUSIC), to estimate the DOAs from a sample estimate of Γ such as

$$\hat{\Gamma} = \frac{1}{N}\sum_{t=1}^{N} \bar{y}(t)\tilde{y}^*(t) \tag{6.4.11}$$

Unlike all the other methods discussed in the following, *the Yule–Walker method does not impose the rather stringent condition (6.4.1).* The Yule–Walker method requires only that $E\left\{\bar{e}(t)\tilde{e}^*(t)\right\} = 0$, which is a much weaker assumption. This is a distinct advantage of the Yule–Walker method (see [VIBERG, STOICA, AND OTTERSTEN 1995] for details). Its relative drawback is that it can only be used if $m > 2n$ (all the other methods require only that $m > n$); in general, it has been found to provide accurate DOA estimates only in those applications involving large–aperture arrays.

Interestingly enough, whenever the condition (6.4.1) holds (*i.e.*, the noise at the array output is spatially white) we can use a modification of the above technique that does not require that $m > 2n$ [FUCHS 1996]. To see this, let

$$\tilde{\Gamma} \triangleq E\left\{y(t)\tilde{y}^*(t)\right\} = R\begin{bmatrix} 0 \\ I_L \end{bmatrix} \qquad (m \times L)$$

where $\tilde{y}(t)$ is as defined in (6.4.8); hence $\tilde{\Gamma}$ is made from the last L columns of R. By making use of the expression (6.4.3) for R, we obtain

$$\tilde{\Gamma} = AP\tilde{A}^* + \sigma^2 \begin{bmatrix} 0 \\ I_L \end{bmatrix} \tag{6.4.12}$$

Because the noise terms in $y(t)$ and $\tilde{y}(t)$ are correlated, the noise is still present in $\tilde{\Gamma}$ (as can be seen from (6.4.12)), and hence $\tilde{\Gamma}$ is not really a YW matrix. Nevertheless, $\tilde{\Gamma}$ has a property similar to that of the YW matrix Γ above, as we now show.

First observe that

$$\tilde{\Gamma}^*\tilde{\Gamma} = \tilde{A}(2\sigma^2 P + PA^*AP)\tilde{A}^* + \sigma^4 I$$

The matrix $2\sigma^2 P + PA^*AP$ is readily shown to be nonsingular if and only if P is nonsingular. As $\tilde{\Gamma}^*\tilde{\Gamma}$ has the same form as R in (6.4.3), we conclude that (for $m \geq L > n$) the $L \times (L-n)$ matrix $\tilde{G}$, whose columns are the eigenvectors of $\tilde{\Gamma}^*\tilde{\Gamma}$ that correspond to the multiple minimum eigenvalue of σ^4, satisfies

$$\tilde{A}^*\tilde{G} = 0 \tag{6.4.13}$$

The columns of $\tilde{G}$ are also equal to the $(L-n)$ right singular vectors of $\tilde{\Gamma}$ corresponding to the multiple minimum singular value of σ^2. For numerical precision reasons $\tilde{G}$ should be computed from the singular vectors of $\tilde{\Gamma}$ rather than from the eigenvectors of $\tilde{\Gamma}^*\tilde{\Gamma}$ (see Section A.8.2).

Because (6.4.13) has the same form as $\tilde{A}^*G = 0$, we can use (6.4.13) for subspace–based DOA estimation in exactly the same way as we used $\tilde{A}^*G = 0$ (see equation (4.5.6) and the discussion following it in Chapter 4). Note that for the method based on $\tilde{\Gamma}$ to be usable, we require only that

$$m \geq L > n \tag{6.4.14}$$

instead of the more restrictive conditions $\{m-L > n,\ L > n\}$ (see (6.4.10)) required in the YW method based on Γ. Observe that (6.4.14) can always be satisfied if $m > n$, whereas (6.4.10) requires that $m > 2n$. Finally, note that Γ is made from the first $m - L$ rows of $\tilde{\Gamma}$, and hence Γ contains "less information" than $\tilde{\Gamma}$; this provides a quick intuitive explanation why the method based on Γ requires more sensors to be applicable than does the method based on $\tilde{\Gamma}$.

6.4.3 Pisarenko and MUSIC Methods

The MUSIC algorithm (with Pisarenko as a special case), developed in Section 4.5 for the frequency estimation application, can be used without modification for DOA estimation [Bienvenu 1979; Schmidt 1979; Barabell 1983]. There are only minor differences between the DOA and the frequency estimation applications of MUSIC, as pointed out below.

First, in the spatial application we can choose between the Spectral and Root MUSIC estimators only in the case of a ULA. For most of the other array geometries, *only Spectral MUSIC is applicable.*

Second, *the standard MUSIC algorithm (4.5.15) breaks down in the case of coherent signals*, as in that case the rank condition (4.5.1) no longer holds. (Such a situation cannot happen in the frequency estimation application, because P is always (diagonal and) nonsingular there.) However, the *modified MUSIC algorithm (outlined at the end of Section 4.5) can be used when the signals are coherent provided that the array is uniform and linear.* This is so because the property (4.5.23), on which the modified MUSIC algorithm is based, continues to hold even if P is singular (see Exercise 6.11).

6.4.4 Min–Norm Method

There is no essential difference between the use of the Min–Norm method for frequency estimation and for DOA finding in the noncoherent case. As for MUSIC, in the DOA estimation application the Min–Norm method should not be used in scenarios with coherent signals, and the Root Min–Norm algorithm can only be used in the ULA case [KUMARESAN AND TUFTS 1983]. In addition, the key property that the true DOAs are asymptotically the *unique* solutions of the Min–Norm estimation problem holds in the ULA case (see the Complement in Section 6.5.1) but not necessarily for other array geometries.

6.4.5 ESPRIT Method

In the ULA case, ESPRIT can be used for DOA estimation exactly as it was for frequency estimation (see Section 4.7). In the non–ULA case ESPRIT can be used only in certain situations. More precisely, and unlike the other algorithms in this section, ESPRIT can be used for DOA finding only if *the array at hand contains two identical subarrays which are displaced by a known displacement vector* [ROY AND KAILATH 1989; STOICA AND NEHORAI 1991]. Mathematically, this condition can be formulated as follows. Let $\bar{m}$ denote the number of sensors in the two twin subarrays, and let A_1 and A_2 denote the sub–matrices of A corresponding to these subarrays. Since the sensors in the array are arbitrarily numbered, there is no restriction to assume that A_1 is made from the first $\bar{m}$ rows in A and A_2 from the last $\bar{m}$:

$$A_1 = [I_{\bar{m}} \;\; 0]A \qquad (\bar{m} \times n) \tag{6.4.15}$$

$$A_2 = [0 \;\; I_{\bar{m}}]A \qquad (\bar{m} \times n) \tag{6.4.16}$$

(here $I_{\bar{m}}$ denotes the $\bar{m} \times \bar{m}$ identity matrix). Note that the two subarrays overlap if $\bar{m} > m/2$; otherwise, they might not overlap. If the array is purposely built to meet ESPRIT's subarray condition, then normally $\bar{m} = m/2$ and the two subarrays are nonoverlapping.

Mathematically, the ESPRIT requirement means that

$$A_2 = A_1 D \tag{6.4.17}$$

where

$$D = \begin{bmatrix} e^{-i\omega_c \tau(\theta_1)} & & 0 \\ & \ddots & \\ 0 & & e^{-i\omega_c \tau(\theta_n)} \end{bmatrix} \tag{6.4.18}$$

and where $\tau(\theta)$ denotes the time needed by a wavefront impinging upon the array from the direction θ to travel between (the "reference points" of) the two twin subarrays. If the angle of arrival θ is measured with respect to the perpendicular of the line between the subarrays' center points, then a calculation similar to the one that led to (6.2.22) shows that:

$$\tau(\theta) = d \sin(\theta)/c \tag{6.4.19}$$

where d is the distance between the two subarrays. Hence, estimates of the DOAs can readily be derived from estimates of the diagonal elements of D in (6.4.18).

Equations (6.4.17) and (6.4.18) are basically equivalent to (4.7.3) and (4.7.4) in Section 4.7, and hence the ESPRIT DOA estimation method is analogous to the ESPRIT frequency estimator.

The ESPRIT DOA estimation method, like the ESPRIT frequency estimator, determines the DOA estimates by solving an $n \times n$ eigenvalue problem. *There is no search involved,* in contrast to the previous methods; in addition, *there is no problem of separating the "signal DOAs" from the "noise DOAs"*, once again in contrast to the Yule–Walker, MUSIC and Min–Norm methods. However, unlike these other methods, *ESPRIT can only be used with the special array configuration described earlier.* In particular, this requirement limits the number of resolvable sources at $n < \bar{m}$ (as both A_1 and A_2 must have full column rank). Note that *the two subarrays do not need to be calibrated* although they need to be identical, and *ESPRIT may be sensitive to differences between the two subarrays* in the same way as Yule–Walker, MUSIC, and Min–Norm are sensitive to imperfections in array calibration. Finally, note that similar to the other DOA finding algorithms presented in this section (with the exception of the NLS method), ESPRIT is not usable in the case of coherent signals.

6.5 Complements

6.5.1 On the Minimum Norm Constraint

As explained in Section 6.4.4 the Root Min–Norm (temporal) frequency estimator, introduced in Section 4.6, can without modification be used for DOA estimation with a uniform linear array. Using the definitions and notation in Section 4.6, let $\hat{g} = [1\ \hat{g}_1 \dots \hat{g}_{m-1}]^T$ denote the vector in $\mathcal{R}(\hat{G})$ that has first element equal to one and minimum Euclidean norm. Then, the Root Min–Norm DOA estimates are obtained from the roots of the polynomial

$$\hat{g}(z) = 1 + \hat{g}_1 z^{-1} + \dots + \hat{g}_{m-1} z^{-(m-1)} \tag{6.5.1}$$

which are located nearest the unit circle. (See the description of Min–Norm in Section 4.6.) As N increases, the polynomial in (6.5.1) approaches

$$g(z) = 1 + g_1 z^{-1} + \dots + g_{m-1} z^{-(m-1)} \tag{6.5.2}$$

where $g = [1\ g_1 \dots g_{m-1}]^T$ is the minimum–norm vector in $\mathcal{R}(G)$. Show that (6.5.2) has n zeroes at $\{e^{-i\omega_k}\}_{k=1}^n$ (the so–called "signal zeroes") and $(m-n-1)$ extraneous zeroes situated *strictly inside* the unit circle (the latter are normally called "noise zeroes"); here $\{\omega_k\}_{k=1}^n$ are either temporal frequencies, or spatial frequencies as in (6.2.27). To prove that the extraneous zeroes are located inside the unit disc, rely on the *minimum norm constraint* that is satisfied by g. Conclude that the choice of $\hat{g}$ in the Min–Norm algorithm makes it possible to separate the signal zeroes from the noise zeroes, at least for data samples that are sufficiently long. (For small or medium–sized samples, it might happen that noise zeroes get closer to the unit circle than signal zeroes, which would lead to spurious frequency or DOA estimates.) For more details on the topic of this complement, see [Tufts and Kumaresan 1982; Kumaresan 1983].

Solution: Let $g = [1, g_1, \dots, g_{m-1}]^T \in \mathcal{R}(G)$. Then (4.2.4) and (4.5.6) imply that

$$a^*(\omega_k) \begin{bmatrix} 1 \\ g_1 \\ \vdots \\ g_{m-1} \end{bmatrix} = 0 \iff$$

$$1 + g_1 e^{i\omega_k} + \dots + g_{m-1} e^{i(m-1)\omega_k} = 0 \quad (\text{for } k = 1, \dots, n) \tag{6.5.3}$$

Hence, any polynomial $g(z)$ whose coefficient vector belongs to $\mathcal{R}(G)$ must have zeroes at $\{e^{-i\omega_k}\}_{k=1}^n$, and thus it can be factored as:

$$g(z) = g_s(z) g_n(z) \tag{6.5.4}$$

where

$$g_s(z) = \prod_{k=1}^{n} (1 - e^{-i\omega_k} z^{-1})$$

The $(m-n-1)$–degree polynomial $g_n(z)$ in (6.5.4) contains the noise zeroes, and at this point is arbitrary. (As the coefficients of $g_n(z)$ vary, the vectors made from the corresponding coefficients of $g(z)$ span $\mathcal{R}(G)$.)

Next, assume that g satisfies the minimum norm constraint:

$$\sum_{k=0}^{m-1} |g_k|^2 = \min \qquad (g_0 \triangleq 1) \tag{6.5.5}$$

By using Parseval's theorem (see (1.2.6)), we can rewrite (6.5.5) as follows:

$$\frac{1}{2\pi}\int_{-\pi}^{\pi} |g(\omega)|^2 \, d\omega = \min \Longleftrightarrow \frac{1}{2\pi}\int_{-\pi}^{\pi} |g_n(\omega)|^2 \, |g_s(\omega)|^2 \, d\omega = \min \tag{6.5.6}$$

(where, by convention, $g(\omega) = g(z)\big|_{z=e^{i\omega}}$). Since $g_s(z)$ in (6.5.4) is fixed, the minimization in (6.5.6) is over $g_n(z)$.

To proceed, some additional notation is required. Let

$$g_n(z) = 1 + \alpha_1 z^{-1} + \cdots + \alpha_{m-n-1} z^{-(m-n-1)}$$

and let $y(t)$ be a signal whose PSD is equal to $|g_s(\omega)|^2$; hence, $y(t)$ is an nth–order MA process. By making use of (1.3.9) and (1.4.9), along with the above notation, we can write (6.5.6) in the following equivalent form:

$$\min_{\{\alpha_k\}} E\left\{|y(t) + \alpha_1 y(t-1) + \cdots + \alpha_{m-n-1} y(t-m+n+1)|^2\right\} \tag{6.5.7}$$

The minimizing coefficients $\{\alpha_k\}$ are given by the solution to a Yule–Walker system of equations similar to (3.4.6). (To show this, parallel the calculation leading to (3.4.8) and (3.4.12).) Since the covariance matrix, of any finite dimension, associated with a moving average signal is positive definite, it follows that:

- The coefficients $\{\alpha_k\}$, and hence $\{g_k\}$, are *uniquely* determined by the minimum norm constraint.
- The polynomial $g_n(z)$ whose coefficients are obtained from (6.5.7) has all its zeroes *strictly inside* the unit circle (*cf.* Exercise 3.8).

which was to be proven. As a final remark, note from (6.5.6) that there is little reason for $g_n(z)$ to have zeroes in the sectors where the signal zeroes are present (since the integrand in (6.5.6) is already quite small for ω values close to $\{\omega_k\}_{k=1}^n$). Hence, we can expect the extraneous zeroes to be more–or–less uniformly distributed inside the unit circle, in sectors which do not contain signal zeroes (see, *e.g.*, [Kumaresan 1983]).

6.6 Exercises

Exercise 6.1: Source Localization using a Sensor in Motion

This exercise illustrates how the directions of arrival of planar waves can be determined by using a single moving sensor. Conceptually this problem is related to that of DOA estimation by sensor array methods. Indeed, we can think of a sensor in motion as creating a *synthetic aperture* similar to the one corresponding to a physical array of spatially distributed sensors.

Assume that the sensor has a linear motion with constant speed equal to v. Also, assume that the sources are far field point emitters at fixed locations in the same plane as the sensor. Let θ_k denote the kth DOA parameter (defined as the angle between the direction of wave propagation and the normal to the sensor trajectory). Finally, assume that the sources emit sinusoidal signals $\{\alpha_k e^{i\omega t}\}_{k=1}^n$ with the same (center) frequency ω. These signals may be reflections of a probing sinusoidal signal from different point scatterers of a target, in which case it is not restrictive to assume that they all have the same frequency.

Show that, under the previous assumptions and after elimination of the high–frequency component corresponding to the frequency ω, the sensor output signal can be written as

$$s(t) = \sum_{k=1}^{n} \alpha_k e^{i\omega_k^D t} + e(t) \tag{6.6.1}$$

where $e(t)$ is measurement noise, and where ω_k^D is the kth *Doppler frequency* defined by:

$$\omega_k^D = -\frac{v}{c} \sin\theta_k$$

with c denoting the velocity of signal propagation. Conclude from (6.6.1) that the DOA estimation problem associated with the scenario under consideration can be solved by using the estimation methods discussed in this chapter and in Chapter 4 (provided that the sensor speed v can be accurately determined).

Exercise 6.2: Beamforming Resolution for Uniform Linear Arrays

Consider a ULA comprising m sensors, with interelement spacing equal to d. Let λ denote the wavelength of the signals impinging on the array. According to the discussion in Chapter 2, the *spatial frequency resolution* of the beamforming used with the above ULA is given by

$$\Delta\omega_s = \frac{2\pi}{m} \quad \Longleftrightarrow \quad \Delta f_s = \frac{1}{m} \tag{6.6.2}$$

Make use of the previous observation to show that the *DOA resolution* of beamforming for signals coming from *broadside* is

$$\boxed{\Delta\theta = \sin^{-1}(1/L)} \tag{6.6.3}$$

where L is the array's length measured in wavelengths:

$$\boxed{L = (m-1)d/\lambda.} \tag{6.6.4}$$

Explain how (6.6.3) approximately reduces to (6.3.20), for sufficiently large L.

Next, show that for signals impinging from an *arbitrary direction angle* θ, the *DOA resolution* of beamforming is approximately:

$$\boxed{\Delta\theta \simeq 1/|L\cos\theta|} \tag{6.6.5}$$

Hence, for signals coming from nearly end–fire directions, the DOA resolution is much worse than what is suggested in (6.3.20).

Exercise 6.3: Beamforming Resolution for L–Shaped Arrays

Consider an m–element array, with m odd, shaped as an "L" with element spacing d. Thus, the array elements are located at points $(0,0)$, $(0,d), \ldots, (0, d(m-1)/2)$ and $(d,0), \ldots, (d(m-1)/2, 0)$. Using the results in Exercise 6.2, find the DOA resolution of beamforming for signals coming from an angle θ. What is the minimum and maximum resolution, and for what angles are these extremal resolutions realized? Compare your results with the m–element ULA case in Exercise 6.2.

Exercise 6.4: Beamforming Resolution for Arbitrary Arrays

The *beam pattern*

$$W(\theta) = |a^*(\theta)a(\theta_0)|^2, \qquad \text{(some } \theta_0\text{)}$$

has the same shape as a spectral window: it has a peak at $\theta = \theta_0$, is symmetric about that point, and the peak is narrow (for large enough values of m). Consequently the *beamwidth* of the array with direction vector $a(\theta)$ can approximately be derived by using the window bandwidth formula proved in Exercise 2.13:

$$\boxed{\Delta\theta \simeq 2\sqrt{|W(\theta_0)/W''(\theta_0)|}} \tag{6.6.6}$$

Now, the array's beamwidth and the resolution of beamforming are closely related. To see this, consider the case where the array output covariance matrix is given by (6.4.3). Let $n = 2$, and assume that $P = I$ (for simplicity of explanation). The average Beamforming spectral function is then given by:

$$a^*(\theta)Ra(\theta) = |a^*(\theta)a(\theta_1)|^2 + |a^*(\theta)a(\theta_2)|^2 + m\sigma^2$$

which clearly shows that the sources with DOAs θ_1 and θ_2 are resolvable by Beamforming if and only if $|\theta_1 - \theta_2|$ is larger than the array's beamwidth. Consequently, we can approximately determine the beamforming resolution by using (6.6.6). Specialize equation (6.6.6) to both a ULA and the L–shaped array in Exercise 6.3. Compare the so–obtained beamforming resolution formulas with those derived in Exercises 6.2 and 6.3.

Exercise 6.5: Grating Lobes

The results of Exercise 6.2 might suggest that an m–element ULA can have very high resolution simply by using a large array element spacing d. However, there is an ambiguity associated with choosing $d > \lambda/2$; this drawback is sometimes referred to as the problem of *grating lobes*. Identify this drawback, and discuss what ambiguities exist as a function of d (refer to the discussion on ULAs in Section 6.2.2).

One potential remedy to this drawback is to use two ULAs: one with m_1 elements and element spacing $d_1 = \lambda/2$, and another with m_2 elements and element spacing d_2. Discuss how to choose m_1, m_2, and d_2 to both avoid ambiguities and increase resolution over a conventional ULA with element spacing $d = \lambda/2$ and $m_1 + m_2$ elements. Consider as an example using a 10–element ULA with $d_2 = 3\lambda/2$ for the second ULA; find m_1 to resolve ambiguities in this array. Finally, discuss any potential drawbacks of the two–array approach.

Exercise 6.6: Beamspace Processing

Consider an array comprising many sensors ($m \gg 1$). Such an array should be able to resolve sources that are quite closely spaced (*cf.* (6.3.20) and the discussion in Exercise 6.4). There is, however, a price to be paid for the high–resolution performance achieved by using many sensors: the computational burden associated with the *elementspace processing* (ESP) (*i.e.*, the direct processing of the output of all sensors) may be prohibitively high, and the involved circuitry (A–D converters, etc.) may be quite expensive.

Let B^* be an $\bar{m} \times m$ matrix with $\bar{m} < m$, and consider the transformed output vector $B^*y(t)$. The latter vector satisfies the following equation (*cf.* (6.2.21)):

$$B^*y(t) = B^*As(t) + B^*e(t) \tag{6.6.7}$$

The transformation matrix B^* above can be interpreted as a *beamformer* or *spatial filter* acting on $y(t)$. Determination of the DOAs of the signals impinging on the array using $B^*y(t)$ is called *beamspace processing* (BSP). Since $\bar{m} < m$, BSP should have a lower computational burden than ESP. The critical question is then how to choose the beamformer B so as not to significantly degrade the performance achievable by ESP.

Assume that a certain DOA sector is known to contain the source(s) of interest (whose DOAs are designated by the generic variable θ_0). By using this information, design a matrix B^* which passes the signals from direction θ^0 approximately undistorted. Choose B in such a way that the noise in beamspace, $B^*e(t)$, is still spatially white. For a given sector size, discuss the tradeoff between the computational burden associated with BSP and the distorting effect of the beamformer on the desired signals. Finally, use the results of Exercise 6.4 to show that the resolution of Beamforming in elementspace and beamspace are nearly the same, under the previous conditions.

Exercise 6.7: Beamforming and MUSIC under the Same Umbrella

Define the scalars

$$Y_t^*(\theta) = a^*(\theta)y(t), \qquad t = 1, \dots, N.$$

By using previous notation, we can write the beamforming spatial spectrum in (6.3.18) as follows:

$$Y^*(\theta)WY(\theta) \tag{6.6.8}$$

where

$$W = (1/N)I \qquad \text{(for beamforming)}$$

and

$$Y(\theta) = [Y_1(\theta) \dots Y_N(\theta)]^T$$

Show that the MUSIC spatial pseudospectrum

$$a^*(\theta)\hat{S}\hat{S}^*a(\theta) \tag{6.6.9}$$

(see Sections 4.5 and 6.4.3) can also be put in the form (6.6.8), for a certain "weighting matrix" W. The columns of the matrix $\hat{S}$ in (6.6.9) are the n principal eigenvectors of the sample covariance matrix $\hat{R}$ in (6.3.17).

Exercise 6.8: Subspace Fitting Interpretation of MUSIC

In words, the result (4.5.9) (on which MUSIC for both frequency and DOA estimation is based) says that the direction vectors $\{a(\theta_k)\}$ belong to the subspace spanned by the columns of S. Therefore, we can think of estimating the DOAs by choosing θ (a generic DOA variable) so that the distance between $a(\theta)$ and the closest vector in the span of $\hat{S}$ is minimized:

$$\min_{\beta,\theta} \|a(\theta) - \hat{S}\beta\|^2 \tag{6.6.10}$$

where $\|\cdot\|$ denotes the Euclidean vector norm. Note that the dummy vector variable β in (6.6.10) is defined in such a way so that $\hat{S}\beta$ is closest to $a(\theta)$ in Euclidean norm.

Show that the DOA estimation method derived from the subspace fitting criterion (6.6.10) is the same as MUSIC.

Exercise 6.9: Subspace Fitting Interpretation of MUSIC (cont'd.)

The result (4.5.9) can also be invoked to arrive at the following subspace fitting criterion:

$$\min_{B,\theta} \|A(\theta) - \hat{S}B\|_F^2 \tag{6.6.11}$$

where $\|\cdot\|_F$ stands for the Frobenius matrix norm, and θ is now the vector of all DOA parameters. This criterion seems to be a more general version of equation (6.6.10) in Exercise 6.8. Show that the minimization of the multidimensional subspace fitting criterion in (6.6.11), with respect to the DOA vector θ, still leads to the one–dimensional MUSIC method. **Hint:** It will be useful to refer to the type of result proven in equations (4.3.12)–(4.3.16) in Section 4.3.

Exercise 6.10: Subspace Fitting Interpretation of MUSIC (cont'd.)

The subspace fitting interpretations of the previous two exercises provide some insights into the properties of the MUSIC estimator. Assume, for instance, that two or more source signals are coherent. Make use of the subspace fitting interpretation in Exercise 6.9 to show that MUSIC cannot be expected to yield meaningful results in such a case. Follow the line of your argument explaining why MUSIC fails in the case of coherent signals, to suggest a subspace fitting criterion that works in such a case. Discuss the computational complexity of the method based on the latter criterion.

Exercise 6.11: Modified MUSIC for Coherent Signals

Consider an m–element ULA. Assume that n signals impinge on the array at angles $\{\theta_k\}_{k=1}^{n}$, and also that some signals are coherent (so that the signal covariance matrix P is singular). Derive a modified MUSIC DOA estimator for this case, analogous to the modified MUSIC frequency estimator in Section 4.5, and show that this method is capable of determining the n DOAs even in the coherent signal case.

Computer Exercises

Tools for Array Signal Processing:

The text web site `www.prenhall.com/~stoica` contains the following MATLAB functions for use in DOA estimation.

- `Y=uladata(theta,P,N,sig2,m,d)`
  Generates an $m \times N$ data matrix $Y = [y(1), \ldots, y(N)]$ for a ULA with `n` sources arriving at angles (in degrees from $-90°$ to $90°$) given by the elements of the $n \times 1$ vector `theta`. The source signals are zero mean Gaussian with covariance matrix $P = E\{s(t)s^*(t)\}$. The noise component is spatially white Gaussian with covariance $\sigma^2 I$, where σ^2 =`sig2`. The element spacing is equal to `d` in wavelengths.

- `phi=beamform(Y,L,d)`
  Implements the beamforming spatial spectral estimate in equation (6.3.18) for an m–element ULA with sensor spacing `d` in wavelengths. The $m \times N$ matrix `Y` is as defined above. The parameter `L` controls the DOA sampling, and `phi` is the spatial spectral estimate `phi`$= [\hat{\phi}(\theta_1), \ldots, \hat{\phi}(\theta_L)]$ where $\theta_k = -\frac{\pi}{2} + \frac{\pi k}{L}$.

- `phi=capon_sp(Y,L,d)`
  Implements the Capon spatial spectral estimator in equation (6.3.26); the input and output parameters are defined as those in `beamform`.

- `theta=root_music_doa(Y,n,d)`
  Implements the Root MUSIC method in Section 4.5, adapted for spatial spec-

tral estimation using a ULA. The parameters `Y` and `d` are as in `beamform`, and `theta` is the vector containing the `n` DOA estimates $[\hat{\theta}_1, \ldots, \hat{\theta}_n]^T$.

- `theta=esprit_doa(Y,n,d)`
  Implements the ESPRIT method for a ULA. The parameters `Y` and `d` are as in `beamform`, and `theta` is the vector containing the `n` DOA estimates $[\hat{\theta}_1, \ldots, \hat{\theta}_n]^T$. The two subarrays for ESPRIT are made from the first $m-1$ and last $m-1$ elements of the array.

Exercise C6.12: Comparison of Spatial Spectral Estimators

Simulate the following scenario. Two signals with wavelength λ impinge on an array of sensors from DOAs $\theta_1 = 0^\circ$ and a θ_2 that will be varied. The signals are mutually uncorrelated complex Gaussian with unit power, so that $P = E\{s(t)s^*(t)\} = I$. The array is a 10–element ULA with element spacing $d = \lambda/2$. The measurements are corrupted by additive complex Gaussian white noise with unit power. A total of $N = 100$ snapshots are collected.

(a) Let $\theta_2 = 15^\circ$. Compare the results of the Beamforming, Capon, Root MUSIC, and ESPRIT methods for this example. The results can be shown by plotting the spatial spectrum estimates from Beamforming and Capon for 50 Monte–Carlo experiments; for Root MUSIC and ESPRIT, plot vertical lines of equal height located at the DOA estimates from the 50 Monte–Carlo experiments. How do the methods compare? Are the properties of the various estimators analogous to the time series case for two sinusoids in noise?

(b) Repeat for $\theta_2 = 7.5^\circ$.

Exercise C6.13: Performance of Spatial Spectral Estimators for Coherent Source Signals

In this exercise we will see what happens when the source signals are fully correlated (or coherent). Use the same parameters and estimation methods as in Exercise C6.12 with $\theta_2 = 15^\circ$, but with

$$P = \begin{pmatrix} 1 & 1 \\ 1 & 1 \end{pmatrix}$$

Note that the sources are coherent as $\text{rank}(P) = 1$.

Compare the results of the four methods for this case, again by plotting the spatial spectrum and "DOA line spectrum" estimates (as in Exercise C6.12) for 50 Monte–Carlo experiments from each estimator. Which method appears to be the best in this case?

Exercise C6.14: Spatial Spectral Estimators applied to Measured Data

Apply the four DOA estimators from Exercise C6.12 to the real data in the file `df5c.mat`, which can be found at the text web site `www.prenhall.com/~stoica`. These data are underwater measurements collected by the Swedish Defense Agency in the Baltic Sea. The 7–element array of hydrophones used in the experiment can be assumed to be a ULA with interelement spacing equal to 0.9m. The wavelength of the signal is approximately 5.32m, and with good approximation, we can assume that there is only one source. Can you find the "submarine"?

Appendix A

LINEAR ALGEBRA AND MATRIX ANALYSIS TOOLS

A.1 Introduction

In this appendix we provide a review of the linear algebra terms and matrix properties used in the text. For the sake of brevity we do not present proofs for all results stated in the following, nor do we discuss related results which are not needed in the previous chapters. For most of the results included, however, we do provide proofs and motivation. The reader interested in finding out more about the topic of this appendix can consult the books [STEWART 1973; HORN AND JOHNSON 1985; STRANG 1988; HORN AND JOHNSON 1989; GOLUB AND VAN LOAN 1989] to which we also refer for the proofs omitted here.

A.2 Range Space, Null Space, and Matrix Rank

Let A be an $m \times n$ matrix with possibly complex–valued elements, $A \in \mathcal{C}^{m\times n}$, and let $(\cdot)^T$ and $(\cdot)^*$ denote the *transpose* and the *conjugate transpose* operators, respectively.

Definition D1: *The* ***range space*** *of A, also called the* ***column space***, *is the subspace spanned by (all linear combinations of) the columns of A:*

$$\mathcal{R}(A) = \{\alpha \in \mathcal{C}^{m\times 1} | \alpha = A\beta \quad \text{for} \quad \beta \in \mathcal{C}^{n\times 1}\} \tag{A.2.1}$$

The range space of A^T is usually called the ***row space*** *of A, for obvious reasons.*

Definition D2: *The* ***null space*** *of A, also called* ***kernel***, *is the following subspace:*

$$\mathcal{N}(A) = \{\beta \in \mathcal{C}^{n\times 1} | A\beta = 0\} \tag{A.2.2}$$

The previous definitions are all that we need to introduce the matrix rank and its basic properties. We return to the range and null subspaces in Section A.4

where we discuss the singular value decomposition. In particular, we derive there some convenient bases and useful projectors associated with the previous matrix subspaces.

Definition D3: *The following are equivalent definitions of the **rank** of A,*

$$r \triangleq \text{rank}(A).$$

(i) r is equal to the maximum number of linearly independent columns of A. The latter number is by definition the dimension of the $\mathcal{R}(A)$; hence

$$r = \dim \mathcal{R}(A) \tag{A.2.3}$$

(ii) r is equal to the maximum number of linearly independent rows of A,

$$r = \dim \mathcal{R}(A^T) = \dim \mathcal{R}(A^*) \tag{A.2.4}$$

(iii) r is the dimension of the nonzero determinant of maximum size that can be built from the elements of A.

The equivalence between the definitions (i) and (ii) above is an important and pleasing result (without which one should have considered the row rank and column rank of a matrix separately!).

Definition D4: *A is said to be:*

- ***Rank deficient*** *whenever $r < \min(m, n)$.*
- ***Full column rank*** *if $r = n \leq m$.*
- ***Full row rank*** *if $r = m \leq n$*
- ***Nonsingular*** *whenever $r = m = n$.*

Result R1: *Premultiplication or postmultiplication of A by a nonsingular matrix does not change the rank of A.*

Proof: This fact directly follows from the definition of rank(A) because the aforementioned multiplications do not change the number of linearly independent columns (or rows) of A. ■

Result R2: *Let $A \in \mathcal{C}^{m\times n}$ and $B \in \mathcal{C}^{n\times p}$ be two conformable matrices of rank r_A and r_B, respectively. Then:*

$$\text{rank}(AB) \leq \min(r_A,\ r_B) \tag{A.2.5}$$

Proof: We can prove the previous assertion by using the definition of the rank once again. Indeed, premultiplication of B by A cannot increase the number of linearly independent columns of B, hence $\text{rank}(AB) \leq r_B$. Similarly, post–multiplication of A by B cannot increase the number of linearly independent columns of A^T, which means that $\text{rank}(AB) \leq r_A$. ■

Result R3: *Let $A \in \mathcal{C}^{m\times m}$ be given by*

$$A = \sum_{k=1}^{N} x_k\, y_k^*$$

where $x_k,\ y_k \in \mathcal{C}^{m\times 1}$. Then,

$$\text{rank}(A) \leq \min(m, N)$$

Proof: Since A can be rewritten as

$$A = [x_1 \dots x_N] \begin{bmatrix} y_1^* \\ \vdots \\ y_N^* \end{bmatrix}$$

the result follows from R2. ■

Result R4: *Let $A \in \mathcal{C}^{m\times n}$ with $n \leq m$, let $B \in \mathcal{C}^{n\times p}$, and let*

$$\text{rank}(A) = n \tag{A.2.6}$$

Then

$$\text{rank}(AB) = \text{rank}(B) \tag{A.2.7}$$

Proof: Assumption (A.2.6) implies that A contains a nonsingular $n \times n$ submatrix, the post–multiplication of which by B gives a block of rank equal to $\text{rank}(B)$ (*cf.* R1). Hence,

$$\text{rank}(AB) \geq \text{rank}(B)$$

However, by R2, $\text{rank}(AB) \leq \text{rank}(B)$ and hence (A.2.7) follows. ■

A.3 Eigenvalue Decomposition

Definition D5: *We say that the matrix $A \in \mathcal{C}^{m\times m}$ is* ***Hermitian*** *if $A^* = A$. In the real–valued case, such an A is said to be* ***symmetric****.*

Definition D6: *A matrix $U \in \mathcal{C}^{m\times m}$ is said to be* ***unitary*** *(****orthogonal*** *if U is real–valued) whenever*

$$U^*U = UU^* = I$$

If $U \in \mathcal{C}^{m\times n}$, with $m > n$, is such that $U^*U = I$ then we say that U is ***semiunitary****.*

Next, we present a number of definitions and results pertaining to the matrix eigenvalue decomposition (EVD), first for general matrices and then for Hermitian ones.

A.3.1 General Matrices

Definition D7: *A scalar $\lambda \in \mathcal{C}$ and a (nonzero) vector $x \in \mathcal{C}^{m\times 1}$ are an* ***eigenvalue*** *and its associated* ***eigenvector*** *of a matrix $A \in \mathcal{C}^{m\times m}$ if*

$$Ax = \lambda x \tag{A.3.1}$$

In particular, an eigenvalue λ is a solution of the so–called ***characteristic equation*** *of A:*

$$|A - \lambda I| = 0 \tag{A.3.2}$$

and x is a vector in $\mathcal{N}(A - \lambda I)$. The pair (λ, x) is called an ***eigenpair****.*

Observe that if $\{(\lambda_i, x_i)\}_{i=1}^{p}$ are p eigenpairs of A (with $p \le m$) then we can write the defining equations $Ax_i = \lambda x_i$ $(i = 1, \ldots, p)$ in the following compact form:

$$AX = X\Lambda \tag{A.3.3}$$

where

$$X = [x_1 \ldots x_p]$$

and

$$\Lambda = \begin{bmatrix} \lambda_1 & & 0 \\ & \vdots & \\ 0 & & \lambda_p \end{bmatrix}$$

Result R5: *Let (λ, x) be an eigenpair of $A \in \mathcal{C}^{m\times m}$. If $B = A + \alpha I$, with $\alpha \in \mathcal{C}$, then $(\lambda + \alpha, x)$ is an eigenpair of B.*

Proof: The result follows from the fact that

$$Ax = \lambda x \implies (A + \alpha I)x = (\lambda + \alpha)x.$$

■

Result R6: *The matrices A and $B \triangleq Q^{-1}AQ$, where Q is any nonsingular matrix, share the same eigenvalues. (B is said to be related to A by a* ***similarity transformation****).*

Proof: Indeed, the equation

$$|B - \lambda I| = |Q^{-1}(A - \lambda I)Q| = |Q^{-1}||A - \lambda I||Q| = 0$$

is equivalent to $|A - \lambda I| = 0$. ∎

In general there is no simple relationship between the elements $\{A_{ij}\}$ of A and its eigenvalues $\{\lambda_k\}$. However, the *trace* of A, which is the sum of the diagonal elements of A, is related in a simple way to the eigenvalues, as described next.

Definition D8: *The* ***trace*** *of a square matrix $A \in \mathcal{C}^{m\times m}$ is defined as*

$$\operatorname{tr}(A) = \sum_{i=1}^{m} A_{ii} \tag{A.3.4}$$

Result R7: *If $\{\lambda_i\}_{i=1}^{m}$ are the eigenvalues of $A \in \mathcal{C}^{m\times m}$, then*

$$\operatorname{tr}(A) = \sum_{i=1}^{m} \lambda_i \tag{A.3.5}$$

Proof: We can write

$$|\lambda I - A| = \prod_{i=1}^{n} (\lambda - \lambda_i) \tag{A.3.6}$$

The right hand side of (A.3.6) is a polynomial in λ whose λ^{n-1} coefficient is $\sum_{i=1}^{n} \lambda_i$. From the definition of the determinant (see, *e.g.*, [STRANG 1988]) we find that the left hand side of (A.3.6) is a polynomial whose λ^{n-1} coefficient is $\sum_{i=1}^{n} A_{ii} = \operatorname{tr}(A)$. This proves the result. ∎

Interestingly, while the matrix product is not commutative, the trace is invariant to commuting the factors in a matrix product, as shown next.

Result R8: *Let $A \in \mathcal{C}^{m\times n}$ and $B \in \mathcal{C}^{n\times m}$. Then:*

$$\operatorname{tr}(AB) = \operatorname{tr}(BA) \tag{A.3.7}$$

Proof: A straightforward calculation, based on the definition of tr(·) in (A.3.4), shows that

$$\operatorname{tr}(AB) = \sum_{i=1}^{m} \sum_{j=1}^{n} A_{ij} B_{ji}$$

$$= \sum_{j=1}^{n} \sum_{i=1}^{m} B_{ji} A_{ij} = \sum_{j=1}^{n} [BA]_{jj} = \operatorname{tr}(BA)$$

■

We can also prove (A.3.7) by using Result R7. Along the way we will obtain some other useful results. First we note the following.

Result R9: *Let $A, B \in \mathcal{C}^{m\times m}$ and let $\alpha \in \mathcal{C}$. Then*

$$|AB| = |A|\,|B|$$
$$|\alpha A| = \alpha^m |A|$$

Proof: The identities follow directly from the definition of the determinant; see, *e.g.*, [STRANG 1988]. ■

Next we prove the following results.

Result R10: *Let $A \in \mathcal{C}^{m\times n}$ and $B \in \mathcal{C}^{n\times m}$. Then:*

$$|I - AB| = |I - BA|. \tag{A.3.8}$$

Proof: It is straightforward to verify that:

$$\begin{bmatrix} I & A \\ 0 & I \end{bmatrix} \begin{bmatrix} I & -A \\ -B & I \end{bmatrix} \begin{bmatrix} I & 0 \\ B & I \end{bmatrix} = \begin{bmatrix} I - AB & 0 \\ 0 & I \end{bmatrix} \tag{A.3.9}$$

and

$$\begin{bmatrix} I & 0 \\ B & I \end{bmatrix} \begin{bmatrix} I & -A \\ -B & I \end{bmatrix} \begin{bmatrix} I & A \\ 0 & I \end{bmatrix} = \begin{bmatrix} I & 0 \\ 0 & I - BA \end{bmatrix} \tag{A.3.10}$$

Because the matrices in the left–hand sides of (A.3.9) and (A.3.10) have the same determinant, equal to $\begin{vmatrix} I & -A \\ -B & I \end{vmatrix}$, it follows that the right–hand sides must also have the same determinant, which concludes the proof. ■

Result R11: *Let $A \in \mathcal{C}^{m\times n}$ and $B \in \mathcal{C}^{n\times m}$. The nonzero eigenvalues of AB and of BA are identical.*

Proof: Let $\lambda \neq 0$ be an eigenvalue of AB. Then,

$$0 = |AB - \lambda I| = \lambda^m |AB/\lambda - I| = \lambda^m |BA/\lambda - I| = \lambda^{m-n} |BA - \lambda I|$$

where the third equality follows from R10. Hence, λ is also an eigenvalue of BA. ■

We can now obtain R8 as a simple corollary of R11, by using the property (A.3.5) of the trace operator.

A.3.2 Hermitian Matrices

An important property of the class of Hermitian matrices, which does not necessarily hold for general matrices, is the following.

Result R12:

*(i) All eigenvalues of $A = A^* \in \mathcal{C}^{m\times m}$ are **real–valued**.*

*(ii) The m eigenvectors of $A = A^* \in \mathcal{C}^{m\times m}$ form an **orthonormal set**. In other words, the matrix whose columns are the eigenvectors of A is **unitary**.*

It follows from (i) and (ii) and from (A.3.3) that for a Hermitian matrix we can write:

$$AU = U\Lambda$$

where $U^*U = UU^* = I$ and the diagonal elements of Λ are real numbers. Equivalently,

$$A = U\Lambda U^* \tag{A.3.11}$$

which is the so–called eigenvalue decomposition (EVD) of $A = A^*$. The EVD of a Hermitian matrix is a special case of the singular value decomposition of a general matrix discussed in the next section.

The following is a useful result associated with Hermitian matrices.

Result R13: *Let $A = A^* \in \mathcal{C}^{m\times m}$ and let $v \in \mathcal{C}^{m\times 1}$ ($v \neq 0$). Also, let the eigenvalues of A be arranged in a nonincreasing order:*

$$\lambda_1 \geq \lambda_2 \geq \cdots \geq \lambda_m.$$

Then:

$$\lambda_m \leq \frac{v^*Av}{v^*v} \leq \lambda_1 \tag{A.3.12}$$

*The ratio in (A.3.12) is called the **Rayleigh quotient**. As this ratio is invariant to the multiplication of v by any complex number, we can rewrite (A.3.12) in the form:*

$$\lambda_m \leq v^*Av \leq \lambda_1 \quad \textit{for any } v \in \mathcal{C}^{m\times 1} \textit{ with } v^*v = 1 \tag{A.3.13}$$

The equalities in (A.3.13) are evidently achieved when v is equal to the eigenvector of A associated with λ_m and λ_1, respectively.

Proof: Let the EVD of A be given by (A.3.11), and let

$$w = U^* v = \begin{bmatrix} w_1 \\ \vdots \\ w_m \end{bmatrix}$$

We need to prove that

$$\lambda_m \leq w^* \Lambda w = \sum_{k=1}^{m} \lambda_k |w_k|^2 \leq \lambda_1$$

for any $w \in \mathcal{C}^{m \times 1}$ satisfying

$$w^* w = \sum_{k=1}^{m} |w_k|^2 = 1.$$

However, this is readily verified as follows:

$$\lambda_1 - \sum_{k=1}^{m} \lambda_k |w_k|^2 = \sum_{k=1}^{m} (\lambda_1 - \lambda_k)|w_k|^2 \geq 0$$

and

$$\sum_{k=1}^{m} \lambda_k |w_k|^2 - \lambda_m = \sum_{k=1}^{m} (\lambda_k - \lambda_m)|w_k|^2 \geq 0$$

and the proof is concluded. ■

The following result is an extension of R13.

Result R14: *Let $V \in \mathcal{C}^{m \times n}$, with $m > n$, be a semiunitary matrix (i.e., $V^* V = I$), and let $A = A^* \in \mathcal{C}^{m \times m}$ have its eigenvalues ordered as in R13. Then:*

$$\sum_{k=m-n+1}^{m} \lambda_k \leq \operatorname{tr}(V^* A V) \leq \sum_{k=1}^{n} \lambda_k \tag{A.3.14}$$

where the equalities are achieved, for instance, when the columns of V are the eigenvectors of A corresponding to $(\lambda_{m-n+1}, \ldots, \lambda_m)$ and, respectively, to $(\lambda_1, \ldots, \lambda_n)$. The ratio

$$\frac{\operatorname{tr}(V^* A V)}{\operatorname{tr}(V^* V)} = \frac{\operatorname{tr}(V^* A V)}{n}$$

is sometimes called the ***extended Rayleigh quotient****.*

Proof: Let

$$A = U\Lambda U^*$$

(*cf.* (A.3.11)), and let

$$S = U^*V \triangleq \begin{bmatrix} s_1^* \\ \vdots \\ s_m^* \end{bmatrix} \qquad (m \times n)$$

(hence s_k^* is the kth row of S). By making use of the above notation, we can write:

$$\mathrm{tr}(V^*AV) = \mathrm{tr}(V^*U\Lambda U^*V) = \mathrm{tr}(S^*\Lambda S) = \mathrm{tr}(\Lambda SS^*) = \sum_{k=1}^{m} \lambda_k c_k \qquad \text{(A.3.15)}$$

where

$$c_k \triangleq s_k^* s_k, \qquad k = 1, \ldots m \qquad \text{(A.3.16)}$$

Clearly,

$$c_k \geq 0, \qquad k = 1, \ldots, m \qquad \text{(A.3.17)}$$

and

$$\sum_{k=1}^{m} c_k = \mathrm{tr}(SS^*) = \mathrm{tr}(S^*S) = \mathrm{tr}(V^*UU^*V) = \mathrm{tr}(V^*V) = \mathrm{tr}(I) = n \qquad \text{(A.3.18)}$$

Furthermore,

$$c_k \leq 1, \qquad k = 1, \ldots, m. \qquad \text{(A.3.19)}$$

To see this, let $G \in \mathcal{C}^{m \times (m-n)}$ be such that the matrix $[S\ G]$ is unitary; and let g_k^* denote the kth row of G. Then, by construction,

$$[s_k^*\ g_k^*] \begin{bmatrix} s_k \\ g_k \end{bmatrix} = c_k + g_k^* g_k = 1 \implies c_k = 1 - g_k^* g_k \leq 1$$

which is (A.3.19).

Finally, by combining (A.3.15) with (A.3.17)–(A.3.19) we can readily verify that $\mathrm{tr}(V^*AV)$ satisfies (A.3.14), where the equalities are achieved for

$$c_1 = \cdots = c_{m-n} = 0; \quad c_{m-n+1} = \cdots = c_m = 1$$

and, respectively,

$$c_1 = \cdots = c_n = 1; \quad c_{n+1} = \cdots = c_m = 0$$

These conditions on $\{c_k\}$ are satisfied if, for example, S is equal to $[0\ I]^T$ and $[I\ 0]^T$, respectively. With this observation, the proof is concluded. ∎

Result R13 is clearly a special case of Result R14. The only reason for considering R13 separately was that the simpler result R13 is more often used in the text than R14.

A.4 Singular Value Decomposition and Projection Operators

For any matrix $A \in \mathcal{C}^{m\times n}$ there exist unitary matrices $U \in \mathcal{C}^{m\times m}$ and $V \in \mathcal{C}^{n\times n}$ and a diagonal matrix $\Sigma \in \mathcal{R}^{m\times n}$ with nonnegative diagonal elements, such that

$$A = U\Sigma V^* \tag{A.4.1}$$

By appropriate permutation, the diagonal elements of Σ can be arranged in a nonincreasing order:

$$\sigma_1 \geq \sigma_2 \geq \cdots \geq \sigma_{\min(m,n)}$$

The factorization (A.4.1) is called the *singular value decomposition* (SVD) of A and its existence is a significant result from both a theoretical and practical standpoint. We reiterate that the matrices U, Σ, and V in (A.4.1) satisfy:

$$\begin{array}{ll} U^*U = UU^* = I & (m \times m) \\ V^*V = VV^* = I & (n \times n) \\ \Sigma_{ij} = \begin{cases} \sigma_i \geq 0 & \\ 0 & \end{cases} & \begin{array}{l}\text{for } i = j \\ \text{for } i \neq j\end{array} \end{array}$$

The following terminology is most commonly associated with the SVD:

- The *left singular vectors* of A are the columns of U. These singular vectors are also the eigenvectors of the matrix AA^*.
- The *right singular vectors* of A are the columns of V. These vectors are also the eigenvectors of the matrix A^*A.
- The *singular values* of A are the diagonal elements $\{\sigma_i\}$ of Σ. Note that $\{\sigma_i\}$ are the square roots of the largest $\min(m,n)$ eigenvalues of AA^* or A^*A.
- The *singular triple* of A is the triple (singular value, left singular vector, and right singular vector) (σ_k, u_k, v_k), where u_k (v_k) is the kth column of U (V).

If

$$\text{rank}(A) = r \leq \min(m,n)$$

then one can show that:

$$\begin{cases} \sigma_k > 0, & k = 1, \ldots, r \\ \sigma_k = 0, & k = r+1, \ldots, \min(m,n) \end{cases}$$

Hence, for a matrix of rank r the SVD can be written as:

$$A = [\underbrace{U_1}_{r} \;\; \underbrace{U_2}_{m-r}] \begin{bmatrix} \Sigma_1 & 0 \\ 0 & 0 \end{bmatrix} \begin{bmatrix} V_1^* \\ V_2^* \end{bmatrix} \begin{matrix} \}\, r \\ \}\, n-r \end{matrix} = U_1 \Sigma_1 V_1^* \tag{A.4.2}$$

where $\Sigma_1 \in \mathcal{R}^{r\times r}$ is nonsingular. The factorization of A in (A.4.2) has a number of important consequences.

Result R15: *Consider the SVD of $A \in \mathcal{C}^{m\times n}$ in (A.4.2), where $r \le \min(m,n)$. Then:*

(i) U_1 is an orthonormal basis of $\mathcal{R}(A)$

(ii) U_2 is an orthonormal basis of $\mathcal{N}(A^*)$

(iii) V_1 is an orthonormal basis of $\mathcal{R}(A^*)$

(iv) V_2 is an orthonormal basis of $\mathcal{N}(A)$.

Proof: We see that (iii) and (iv) follow from the properties (i) and (ii) applied to A^*. To prove (i) and (ii), we need to show that:

$$\mathcal{R}(A) = \mathcal{R}(U_1) \tag{A.4.3}$$

and, respectively,

$$\mathcal{N}(A^*) = \mathcal{R}(U_2) \tag{A.4.4}$$

To show (A.4.3), note that

$$\begin{aligned} \alpha \in \mathcal{R}(A) &\Rightarrow \text{there exists } \beta \text{ such that } \alpha = A\beta \Rightarrow \\ &\Rightarrow \alpha = U_1(\Sigma_1 V_1^* \beta) = U_1\gamma \Rightarrow \alpha \in \mathcal{R}(U_1) \end{aligned}$$

so $\mathcal{R}(A) \subset \mathcal{R}(U_1)$. Also,

$$\alpha \in \mathcal{R}(U_1) \Rightarrow \text{there exists } \beta \text{ such that } \alpha = U_1\beta$$

From (A.4.2), $U_1 = AV_1\Sigma_1^{-1}$; it follows that

$$\alpha = A(V_1\Sigma_1^{-1}\beta) = A\rho \Rightarrow \alpha \in \mathcal{R}(A)$$

which shows $\mathcal{R}(U_1) \subset \mathcal{R}(A)$. Combining $\mathcal{R}(U_1) \subset \mathcal{R}(A)$ with $\mathcal{R}(A) \subset \mathcal{R}(U_1)$ gives (A.4.3). Similarly,

$$\alpha \in \mathcal{N}(A^*) \Rightarrow A^*\alpha = 0 \Rightarrow V_1\Sigma_1 U_1^*\alpha = 0 \Rightarrow \Sigma_1^{-1}V_1^*V_1\Sigma_1U_1^*\alpha = 0 \Rightarrow U_1^*\alpha = 0$$

Now, any vector α can be written as

$$\alpha = [U_1 \; U_2]\begin{bmatrix} \gamma \\ \beta \end{bmatrix}$$

since $[U_1 \; U_2]$ is nonsingular. However, $0 = U_1^*\alpha = U_1^*U_1\gamma + U_1^*U_2\beta = \gamma$, so $\gamma = 0$, and thus $\alpha = U_2\beta$. Thus, $\mathcal{N}(A^*) \subset \mathcal{R}(U_2)$. Finally,

$$\alpha \in \mathcal{R}(U_2) \Rightarrow \text{there exists } \beta \text{ such that } \alpha = U_2\beta$$

Then

$$A^*\alpha = V_1\Sigma_1U_1^*U_2\beta = 0 \Rightarrow \alpha \in \mathcal{N}(A^*)$$

which leads to (A.4.4). ■

The previous result, readily derived by using the SVD, has a number of interesting corollaries which complement the discussion on range and null subspaces in Section A.2.

Result R16: *For any $A \in \mathcal{C}^{m\times n}$ the subspaces $\mathcal{R}(A)$ and $\mathcal{N}(A^*)$ are orthogonal to each other and they together span $\mathcal{C}^m$. Consequently, we say that $\mathcal{N}(A^*)$ is the* ***orthogonal complement*** *of $\mathcal{R}(A)$ in $\mathcal{C}^m$, and vice versa. In particular, we have:*

$$\dim \mathcal{N}(A^*) = m - r \tag{A.4.5}$$

$$\dim \mathcal{N}(A) = n - r \tag{A.4.6}$$

(Recall that $\dim \mathcal{R}(A) = \dim \mathcal{R}(A^*) = r$.*)*

Proof: This result is a direct corollary of R15. ■

The SVD of a matrix also provides a convenient representation for the projectors onto the range and null spaces of A and A^*.

Definition D9: *Let $y \in \mathcal{C}^{m\times 1}$ be an arbitrary vector. By definition the* ***orthogonal projector*** *onto $\mathcal{R}(A)$ is the matrix Π, which is such that (i) $\mathcal{R}(\Pi) = \mathcal{R}(A)$ and (ii) the Euclidean distance between y and $\Pi y \in \mathcal{R}(A)$ is minimum:*

$$\|y - \Pi y\|^2 = \min \quad \text{over } \mathcal{R}(A)$$

Hereafter, $\|x\|^2 = x^* x$ denotes the ***Euclidean vector norm****.*

Result R17: *Let $A \in \mathcal{C}^{m\times n}$. The orthogonal projector onto $\mathcal{R}(A)$ is given by*

$$\Pi = U_1 U_1^* \tag{A.4.7}$$

whereas the orthogonal projector onto $\mathcal{N}(A^*)$ is

$$\Pi^{\perp} = I - U_1 U_1^* = U_2 U_2^* \tag{A.4.8}$$

Proof: Let $y \in \mathcal{C}^{m\times 1}$ be an arbitrary vector. As $\mathcal{R}(A) = \mathcal{R}(U_1)$, according to R15, we can find the vector in $\mathcal{R}(A)$ that is of minimal distance from y by solving the problem:

$$\min_{\beta} \|y - U_1\beta\|^2 \tag{A.4.9}$$

Because

$$\begin{aligned}\|y - U_1\beta\|^2 &= (\beta^* - y^* U_1)(\beta - U_1^* y) + y^*(I - U_1 U_1^*)y \\ &= \|\beta - U_1^* y\|^2 + \|U_2^* y\|^2\end{aligned}$$

it readily follows that the solution to the minimization problem (A.4.9) is given by $\beta = U_1^* y$. Hence the vector $U_1 U_1^* y$ is the orthogonal projection of y onto $\mathcal{R}(A)$ and the minimum distance from y to $\mathcal{R}(A)$ is $\|U_2^* y\|$. This proves (A.4.7). Then (A.4.8) follows immediately from (A.4.7) and the fact that $\mathcal{N}(A^*) = \mathcal{R}(U_2)$. ■

Note, for instance, that for the projection of y onto $\mathcal{R}(A)$ the error vector is $y - U_1U_1^*y = U_2U_2^*y$, which is in $\mathcal{R}(U_2)$ and is therefore orthogonal to $\mathcal{R}(A)$ by R15. For this reason, Π is given the name "orthogonal projector" in D9 and R17.

As an aside, we remark that the orthogonal projectors in (A.4.7) and (A.4.8) are *idempotent matrices*; see the next definition.

Definition D10: *The matrix* $A \in \mathcal{C}^{m\times m}$ *is* ***idempotent*** *if*

$$A^2 = A \tag{A.4.10}$$

Furthermore, observe by making use of R11 that the idempotent matrix in (A.4.7), for example, has r eigenvalues equal to 1 and $(m - r)$ eigenvalues equal to zero. This is a general property of idempotent matrices: their eigenvalues are either zero or one.

Finally we present a result that even alone would be enough to make the SVD an essential matrix analysis tool.

Result R18: *Let* $A \in \mathcal{C}^{m\times n}$, *with elements* A_{ij}. *Let the SVD of* A *(with the singular values arranged in a nonincreasing order) be given by:*

$$A = [\underbrace{U_1}_{p} \;\; \underbrace{U_2}_{m-p}] \begin{bmatrix} \Sigma_1 & 0 \\ 0 & \Sigma_2 \end{bmatrix} \begin{bmatrix} V_1^* \\ V_2^* \end{bmatrix} \begin{matrix} \}\, p \\ \}\, n-p \end{matrix} \tag{A.4.11}$$

where $p \le \min(m, n)$ *is an integer. Let*

$$\|A\|^2 = \mathrm{tr}(A^*A) = \sum_{i=1}^{m}\sum_{j=1}^{n} |A_{ij}|^2 = \sum_{k=1}^{\min(m,n)} \sigma_k^2 \tag{A.4.12}$$

denote the square of the so-called ***Frobenius norm****. Then* ***the best rank–p approximant of*** A *in the Frobenius norm metric, that is, the solution to*

$$\min_B \; \|A - B\|^2 \quad \text{subject to } \mathrm{rank}(B) = p \; , \tag{A.4.13}$$

is given by

$$B_0 = U_1\Sigma_1V_1^* \tag{A.4.14}$$

Furthermore, B_0 *above is the unique solution to the approximation problem (A.4.13) if and only if* $\sigma_p > \sigma_{p+1}$.

Proof: It follows from R4 and (A.4.2) that we can parameterize B in (A.4.13) as:

$$B = CD^* \tag{A.4.15}$$

where $C \in \mathcal{C}^{m\times p}$ and $D \in \mathcal{C}^{n\times p}$ are full column rank matrices. The previous parameterization of B is of course nonunique but, as we will see, this fact does not introduce any problem. By making use of (A.4.15) we can rewrite the problem (A.4.13) in the following form:

$$\min_{C,D} \|A - CD^*\|^2 \quad \text{rank}(C) = \text{rank}(D) = p \tag{A.4.16}$$

The reparameterized problem is essentially constraint free. Indeed, the full column rank condition that must be satisfied by C and D can be easily handled, see below.

First, we minimize (A.4.16) with respect to D, for a given C. To that end, observe that:

$$\begin{aligned}\|A - CD^*\|^2 = \text{tr}\{&[D - A^*C(C^*C)^{-1}](C^*C)[D^* - (C^*C)^{-1}C^*A] \\ &+A^*[I - C(C^*C)^{-1}C^*]A\}\end{aligned} \tag{A.4.17}$$

By result (iii) in Definition D11 in the next section, the matrix $[D - A^*C(C^*C)^{-1}]\cdot (C^*C)[D^* - (C^*C)^{-1}C^*A]$ is positive semidefinite for any D. This observation implies that (A.4.17) is minimized with respect to D for

$$D_0 = A^*C(C^*C)^{-1} \tag{A.4.18}$$

and the corresponding minimum value of (A.4.17) is given by

$$\text{tr}\{A^*[I - C(C^*C)^{-1}C^*]A\} \tag{A.4.19}$$

Next we minimize (A.4.19) with respect to C. Let $S \in \mathcal{C}^{m\times p}$ denote an orthogonal basis of $\mathcal{R}(C)$; that is, $S^*S = I$ and

$$S = C\Gamma$$

for some nonsingular $p \times p$ matrix Γ. It is then straightforward to verify that

$$I - C(C^*C)^{-1}C^* = I - SS^* \tag{A.4.20}$$

By combining (A.4.19) and (A.4.20) we can restate the problem of minimizing (A.4.19) with respect to C as:

$$\max_{S;\ S^*S=I} \text{tr}[S^*(AA^*)S] \tag{A.4.21}$$

The solution to (A.4.21) follows from R14: the maximizing S is given by

$$S_0 = U_1$$

which yields

$$C_0 = U_1\Gamma^{-1} \tag{A.4.22}$$

It follows that:

$$\begin{aligned} B_0 &= C_0 D_0^* = C_0(C_0^* C_0)^{-1} C_0^* A = S_0 S_0^* A \\ &= U_1 U_1^* (U_1 \Sigma_1 V_1^* + U_2 \Sigma_2 V_2^*) \\ &= U_1 \Sigma_1 V_1^*. \end{aligned}$$

Furthermore, we observe that the minimum value of the Frobenius distance in (A.4.13) is given by

$$\|A - B_0\|^2 = \|U_2 \Sigma_2 V_2^*\|^2 = \sum_{k=p+1}^{\min(m,n)} \sigma_k^2$$

If $\sigma_p > \sigma_{p+1}$ then the best rank–p approximant B_0 derived above is unique. Otherwise it is not unique. Indeed, whenever $\sigma_p = \sigma_{p+1}$ we can obtain B_0 by using either the singular vectors associated with σ_p or those corresponding to σ_{p+1}, which will generally lead to different solutions. ∎

A.5 Positive (Semi)Definite Matrices

Let $A = A^* \in \mathcal{C}^{m \times m}$ be a Hermitian matrix, and let $\{\lambda_k\}_{k=1}^m$ denote its eigenvalues.

Definition D11: *We say that A is* ***positive semidefinite*** *(psd) or* ***positive definite*** *(pd) if any of the following equivalent conditions holds true.*

(i) $\lambda_k \geq 0$ *($\lambda_k > 0$ for pd) for* $k = 1, \ldots, m$.

(ii) $\alpha^* A \alpha \geq 0$ *($\alpha^* A \alpha > 0$ for pd) for any nonzero vector* $\alpha \in \mathcal{C}^{m \times 1}$

(iii) There exists a matrix C such that

$$A = CC^* \tag{A.5.1}$$

(with $\operatorname{rank}(C) = m$ *for pd)*

(iv) $|A(i_1, \ldots, i_k)| \geq 0$ *(> 0 for pd) for all $k = 1, \ldots, m$ and all indices $i_1, \ldots, i_k \in [1, m]$, where $A(i_1, \ldots, i_k)$ is the submatrix formed from A by eliminating the $i_1, \ldots, i_k$ rows and columns of A. ($A(i_1, \ldots, i_k)$ is called a* ***principal submatrix*** *of A). The condition for A to be positive definite can be simplified to requiring that $|A(k+1, \ldots, m)| > 0$ (for $k = 1, \ldots, m-1$) and $|A| > 0$. ($A(k+1, \ldots, m)$ is called a* ***leading submatrix*** *of A).*

The notation $A > 0$ ($A \geq 0$) is commonly used to denote that A is pd (psd).

Of the previous defining conditions, (iv) is apparently more involved. The necessity of (iv) can be proven as follows. Let α be a vector in $\mathcal{C}^m$ with zeroes at the positions $\{i_1, \ldots, i_k\}$ and arbitrary elements elsewhere. Then, by using (ii) we readily see that $A \geq 0$ (> 0) implies $A(i_1, \ldots, i_k) \geq 0$ (> 0) which, in turn, implies (iv) by making use of (i) and the fact that the determinant of a matrix equals the product of its eigenvalues. The sufficiency of (iv) is shown in [STRANG 1988].

The equivalence of the remaining conditions, (i), (ii), and (iii), is easily proven by making use of the EVD of A: $A = U\Lambda U^*$. To show that (i) $\Leftrightarrow$ (ii), assume first that (i) holds and let $\beta = U^*\alpha$. Then:

$$\alpha^* A\alpha = \beta^*\Lambda\beta = \sum_{k=1}^{m} \lambda_k |\beta_k|^2 \geq 0 \tag{A.5.2}$$

and hence, (ii) holds as well. Conversely, since U is invertible it follows from (A.5.2) that (ii) can hold only if (i) holds; indeed, if (i) does not hold one can choose β to make (A.5.2) negative; thus there exists an $\alpha = U\beta$ such that $\alpha^* A\alpha < 0$, which contradicts the assumption that (ii) holds. Hence (i) and (ii) are equivalent. To show that (iii) $\Rightarrow$ (ii), note that

$$\alpha^* A\alpha = \alpha^* CC^*\alpha = \|C^*\alpha\|^2 \geq 0$$

and hence (ii) holds as well. Since (iii) $\Rightarrow$ (ii) and (ii) $\Rightarrow$ (i), we have (iii) $\Rightarrow$ (i). To show that (i) $\Rightarrow$ (iii), we assume (i) and write

$$A = U\Lambda U^* = (U\Lambda^{1/2}\Lambda^{1/2}U^*) = (U\Lambda^{1/2}U^*)(U\Lambda^{1/2}U^*) \triangleq CC^* \tag{A.5.3}$$

and hence (iii) is also satisfied. In (A.5.3), $\Lambda^{1/2}$ is a diagonal matrix the diagonal elements of which are equal to $\{\lambda_k^{1/2}\}$. In other words, $\Lambda^{1/2}$ is the "square root" of Λ.

In a general context, the square root of a positive semidefinite matrix is defined as follows.

Definition D12: *Let $A = A^*$ be a positive semidefinite matrix. Then any matrix C that satisfies*

$$A = CC^* \tag{A.5.4}$$

*is called a **square root** of A. Sometimes such a C is denoted by $A^{1/2}$.*

If C is a square root of A, then so is CB for any unitary matrix B, and hence there are an infinite number of square roots of a given positive semidefinite matrix. Two often–used particular choices for square roots are:

(i) *Hermitian square root*: $C = C^*$. In this case we can simply write (A.5.4) as $A = C^2$. Note that we have already obtained such a square root of A in (A.5.3):

$$C = U\Lambda^{1/2}U^* \tag{A.5.5}$$

The Hermitian square root is unique.

(ii) *Cholesky factor*. If C is lower triangular with nonnegative diagonal elements, then C is called the *Cholesky factor* of A. In computational exercises, the triangular form of the square–root matrix is often preferred to other forms. If A is positive definite, the Cholesky factor is unique.

We also note that equation (A.5.4) implies that A and C *have the same rank as well as the same range space.* This follows easily, for example, by inserting the SVD of C into (A.5.4).

Next we prove three specialized results on positive semidefinite matrices required in Section 2.5 and in Appendix B.

Result R19: *Let $A \in \mathcal{C}^{m\times m}$ and $B \in \mathcal{C}^{m\times m}$ be positive semidefinite matrices. Then the matrix $A\odot B$ is also positive semidefinite, where $\odot$ denotes the* ***Hadamard matrix product*** *(also called* ***elementwise multiplication****: $[A\odot B]_{ij} = A_{ij}B_{ij}$).*

Proof: Because B is positive semidefinite it can be written as $B = CC^*$ for some matrix $C \in \mathcal{C}^{m\times m}$. Let c_k^* denote the kth row of C. Then,

$$[A \odot B]_{ij} = A_{ij}B_{ij} = A_{ij}\ c_i^* c_j$$

and hence, for any $\alpha \in \mathcal{C}^{m\times 1}$,

$$\alpha^*(A \odot B)\alpha = \sum_{i=1}^{m}\sum_{j=1}^{m} \alpha_i^* A_{ij} c_i^* c_j \alpha_j \tag{A.5.6}$$

By letting $\{c_{jk}\}_{k=1}^m$ denote the elements of the vector c_j, we can rewrite (A.5.6) as:

$$\alpha^*(A \odot B)\alpha = \sum_{k=1}^{m}\sum_{i=1}^{m}\sum_{j=1}^{m} \alpha_i^* c_{ik}^* A_{ij}\alpha_j c_{jk} = \sum_{k=1}^{m} \beta_k^* A \beta_k \tag{A.5.7}$$

where

$$\beta_k \triangleq [\alpha_1 c_{1k} \cdots \alpha_m c_{mk}]^T$$

As A is positive semidefinite by assumption, $\beta_k^* A \beta_k \geq 0$ for each k, and it follows from (A.5.7) that $A \odot B$ must be positive semidefinite as well. ■

Result R20: *Let $A \in \mathcal{C}^{m\times m}$ and $B \in \mathcal{C}^{m\times m}$ be Hermitian matrices. Assume that B is nonsingular and that the partitioned matrix*

$$\begin{bmatrix} A & I \\ I & B \end{bmatrix}$$

is positive semidefinite. Then the matrix $(A - B^{-1})$ is also positive semidefinite,

$$A \geq B^{-1}$$

Proof: By Definition D11, part (ii),

$$\begin{bmatrix} \alpha_1^* & \alpha_2^* \end{bmatrix} \begin{bmatrix} A & I \\ I & B \end{bmatrix} \begin{bmatrix} \alpha_1 \\ \alpha_2 \end{bmatrix} \geq 0 \tag{A.5.8}$$

for any vectors $\alpha_1, \alpha_2 \in \mathcal{C}^{m\times 1}$. Let

$$\alpha_2 = -B^{-1}\alpha_1$$

Then (A.5.8) becomes:

$$\alpha_1^*(A - B^{-1})\alpha_1 \ \geq \ 0$$

As the above inequality must hold for any $\alpha_1 \in \mathcal{C}^{m\times 1}$, the proof is concluded. ∎

Result R21: *Let $C \in \mathcal{C}^{m\times m}$ be a (Hermitian) positive definite matrix depending on a real-valued parameter α. Assume that C is a differentiable function of α. Then*

$$\frac{\partial}{\partial \alpha}[\ln |C|] = \operatorname{tr}\left[C^{-1}\frac{\partial C}{\partial \alpha}\right]$$

Proof: Let $\{\lambda_i\} \in \mathcal{R}$ $(i = 1, \ldots, m)$ denote the eigenvalues of C. Then

$$\begin{aligned}
\frac{\partial}{\partial \alpha}[\ln |C|] = \frac{\partial}{\partial \alpha}\left[\ln \prod_{k=1}^{m} \lambda_k\right] &= \sum_{k=1}^{m} \frac{\partial}{\partial \alpha}(\ln \lambda_k) \\
&= \sum_{k=1}^{m} \frac{1}{\lambda_k}\frac{\partial \lambda_k}{\partial \alpha} \\
&= \operatorname{tr}\left[\Lambda^{-1}\frac{\partial \Lambda}{\partial \alpha}\right]
\end{aligned}$$

where $\Lambda = \operatorname{diag}(\lambda_1, \ldots, \lambda_m)$. Let Q be a unitary matrix such that $Q^*\Lambda Q = C$ (which is the EVD of C). Since Q is unitary, $Q^*Q = I$, we obtain

$$\frac{\partial Q^*}{\partial \alpha}Q + Q^*\frac{\partial Q}{\partial \alpha} = 0$$

Thus, we get

$$\begin{aligned}
\operatorname{tr}\left[\Lambda^{-1}\frac{\partial \Lambda}{\partial \alpha}\right] &= \operatorname{tr}\left[(Q^*\Lambda^{-1}Q)\left(Q^*\frac{\partial \Lambda}{\partial \alpha}Q\right)\right] \\
&= \operatorname{tr}\left[C^{-1}\left(\frac{\partial}{\partial \alpha}(Q^*\Lambda Q) - \frac{\partial Q^*}{\partial \alpha}\Lambda Q - Q^*\Lambda\frac{\partial Q}{\partial \alpha}\right)\right] \\
&= \operatorname{tr}\left[C^{-1}\frac{\partial C}{\partial \alpha}\right] - \operatorname{tr}\left[Q^*\Lambda^{-1}Q\left(\frac{\partial Q^*}{\partial \alpha}\Lambda Q + Q^*\Lambda\frac{\partial Q}{\partial \alpha}\right)\right]
\end{aligned}$$

$$= \operatorname{tr}\left[C^{-1}\frac{\partial C}{\partial \alpha}\right] - \operatorname{tr}\left[\frac{\partial Q^*}{\partial \alpha}Q + Q^*\frac{\partial Q}{\partial \alpha}\right]$$

$$= \operatorname{tr}\left[C^{-1}\frac{\partial C}{\partial \alpha}\right]$$

which is the result stated. ■

Finally we make use of a simple property of positive semidefinite matrices to prove the *Cauchy–Schwartz inequality* for vectors and for functions.

Result R22: *(Cauchy–Schwartz inequality for vectors). Let $x,\ y \in \mathcal{C}^{m\times 1}$. Then:*

$$|x^*y|^2 \leq \|x\|^2\ \|y\|^2 \tag{A.5.9}$$

where $|\cdot|$ denotes the modulus of a possibly complex–valued number, and $\|\cdot\|$ denotes the Euclidean vector norm ($\|x\|^2 = x^*x$). Equality in (A.5.9) is achieved if and only if x is proportional to y.

Proof: The (2×2) matrix

$$\begin{bmatrix} \|x\|^2 & x^*y \\ y^*x & \|y\|^2 \end{bmatrix} = \begin{bmatrix} x^* \\ y^* \end{bmatrix} \begin{bmatrix} x & y \end{bmatrix} \tag{A.5.10}$$

is clearly positive semidefinite (observe that condition (iii) in D11 is satisfied). It follows from condition (iv) in D11 that the determinant of the above matrix must be nonnegative:

$$\|x\|^2\ \|y\|^2 - |x^*y|^2 \geq 0$$

which gives (A.5.9). Equality in (A.5.9) holds if and only if the determinant of (A.5.10) is equal to zero. The latter condition is equivalent to requiring that x is proportional to y (*cf.* D3: the columns of the matrix $[x\ y]$ will then be linearly dependent). ■

Result R23: *(Cauchy–Schwartz inequality for functions). Let $f(x)$ and $g(x)$ be two complex–valued functions defined for real–valued argument x. Then, assuming that the integrals below exist,*

$$\left|\int_I f(x)g^*(x)dx\right|^2 \leq \left[\int_I |f(x)|^2 dx\right]\left[\int_I |g(x)|^2 dx\right]$$

where $I \subset \mathcal{R}$ is an integration interval. The inequality above becomes an equality if and only if $f(x)$ is proportional to $g(x)$ on I.

Proof: The following matrix

$$\int_I \begin{bmatrix} f(x) \\ g(x) \end{bmatrix} [f^*(x)\ g^*(x)]\, dx$$

is seen to be positive semidefinite (since the integrand is a positive semidefinite matrix for every $x \in I$). Hence the stated result follows from the type of argument used in the proof of Result R22. ■

A.6 Matrices with Special Structure

In this section we consider several types of matrices with a special structure, for which we prove some basic properties used in the text.

Definition D13: *A matrix $A \in C^{m \times n}$ is called* ***Vandermonde*** *if it has the following structure:*

$$A = \begin{bmatrix} 1 & \cdots & 1 \\ z_1 & & z_n \\ \vdots & & \vdots \\ z_1^{m-1} & \cdots & z_n^{m-1} \end{bmatrix} \tag{A.6.1}$$

where $z_k \in C$ are usually assumed to be distinct.

Result R24: *Consider the matrix A in (A.6.1) with $z_k \neq z_p$ for $k, p = 1, \ldots, n$ and $k \neq p$. Also let $m \geq n$ and assume that $z_k \neq 0$ for all k. Then any n consecutive rows of A are linearly independent.*

Proof: To prove the assertion, it is sufficient to show that the following $n \times n$ Vandermonde matrix is nonsingular:

$$\bar{A} = \begin{bmatrix} 1 & \cdots & 1 \\ z_1 & & z_n \\ \vdots & & \vdots \\ z_1^{n-1} & \cdots & z_n^{n-1} \end{bmatrix}$$

Let $\beta = [\beta_0 \cdots \beta_{n-1}]^* \neq 0$. The equation $\beta^* \bar{A} = 0$ is equivalent to

$$\beta_0 + \beta_1 z + \cdots + \beta_{n-1} z^{n-1} = 0 \quad \text{at} \quad z = z_k \quad (k = 1, \ldots, n) \tag{A.6.2}$$

However, (A.6.2) is impossible as a $(n-1)$-degree polynomial cannot have n zeroes. Hence, $\bar{A}$ has full rank. ∎

Definition D14: *A matrix $A \in C^{m \times n}$ is called:*

- ***Toeplitz*** *when $A_{ij} = A_{i-j}$*

- ***Hankel*** *when $A_{ij} = A_{i+j}$*

Observe that a Toeplitz matrix has the same element along each diagonal, whereas a Hankel matrix has identical elements on each of the antidiagonals.

Result R25: *The eigenvectors of a symmetric Toeplitz matrix $A \in \mathcal{R}^{m\times m}$ are either symmetric or skew–symmetric. More precisely, if J denotes the exchange (or reversal) matrix*

$$J = \begin{bmatrix} 0 & & 1 \\ & \cdot^{\cdot^{\cdot}} & \\ 1 & & 0 \end{bmatrix}$$

and if x is an eigenvector of A, then either $x = Jx$ or $x = -Jx$.

Proof: By the property (3.5.3) proven in Section 3.5, A satisfies

$$AJx = JAx$$

or equivalently

$$(JAJ)x = Ax$$

for any $x \in \mathcal{C}^{m\times 1}$. Hence, we must have:

$$JAJ = A \tag{A.6.3}$$

Let (λ, x) denote an eigenpair of A:

$$Ax = \lambda x \tag{A.6.4}$$

Combining (A.6.3) and (A.6.4) yields:

$$\lambda Jx = JAx = J(JAJ)x = A(Jx) \tag{A.6.5}$$

Because the eigenvectors of a symmetric matrix are unique modulo multiplication by a scalar, it follows from (A.6.5) that:

$$x = \alpha Jx \quad \text{for some } \alpha \in \mathcal{R}$$

As x and hence Jx must have unit norm, α must satisfy $\alpha^2 = 1 \Rightarrow \alpha = \pm 1$; thus, either $x = Jx$ (x is symmetric) or $x = -Jx$ (x is skew–symmetric). ■

One can show that for m even, the number of symmetric eigenvectors is $m/2$, as is the number of skew–symmetric eigenvectors; for odd m the number of symmetric eigenvectors is $(m+1)/2$ and the number of skew–symmetric eigenvectors is $(m-1)/2$ (see [CANTONI AND BUTLER 1976]).

For many additional results on Toeplitz matrices, the reader can consult [IOHVIDOV 1982; BÖTTCHER AND SILBERMANN 1983].

A.7 Matrix Inversion Lemmas

The following formulas for *the inverse of a partitioned matrix* are used in the text.

Result R26: *Let $A \in \mathcal{C}^{m\times m}$, $B \in \mathcal{C}^{n\times n}$, $C \in \mathcal{C}^{m\times n}$ and $D \in \mathcal{C}^{n\times m}$. Then, provided that the matrix inverses appearing below exist,*

$$\begin{bmatrix} A & C \\ D & B \end{bmatrix} = \begin{bmatrix} I \\ 0 \end{bmatrix} A^{-1} \begin{bmatrix} I & 0 \end{bmatrix} + \begin{bmatrix} -A^{-1}C \\ I \end{bmatrix} (B - DA^{-1}C)^{-1} [-DA^{-1} \;\; I]$$

$$= \begin{bmatrix} 0 \\ I \end{bmatrix} B^{-1} \begin{bmatrix} 0 & I \end{bmatrix} + \begin{bmatrix} I \\ -B^{-1}D \end{bmatrix} (A - CB^{-1}D)^{-1} [I \;\; -CB^{-1}]$$

Proof: By direct verification. ■

By equating the top–left blocks in the above two equations we obtain the so–called *Matrix Inversion Lemma.*

Result R27: *(Matrix Inversion Lemma) Let A, B, C and D be as in R26. Then, assuming that the matrix inverses appearing below exist,*

$$(A - CB^{-1}D)^{-1} = A^{-1} + A^{-1}C(B - DA^{-1}C)^{-1}DA^{-1}$$

A.8 Systems of Linear Equations

Let $A \in \mathcal{C}^{m\times n}$, $B \in \mathcal{C}^{m\times p}$, and $X \in \mathcal{C}^{n\times p}$. A general *system of linear equations* in X can be written as:

$$AX = B \tag{A.8.1}$$

where A and B are given and X is the unknown matrix. The special case of (A.8.1) corresponding to $p = 1$ (for which X and B are vectors) is perhaps the most common one in applications. For the sake of generality, we consider the system (A.8.1) with $p \geq 1$. (The ESPRIT system of equations encountered in Section 4.7 is of the form of (A.8.1) with $p > 1$.) We say that (A.8.1) is *exactly determined* whenever $m = n$, *overdetermined* if $m > n$ and *underdetermined* if $m < n$. In the following discussion, we first address the case where (A.8.1) has an exact solution and then the case where (A.8.1) cannot be exactly satisfied.

A.8.1 Consistent Systems

Result R28: *The linear system (A.8.1) is* ***consistent****, that is it admits an exact solution X, if and only if $\mathcal{R}(B) \subset \mathcal{R}(A)$ or equivalently*

$$\text{rank}([A \;\; B]) = \text{rank}(A) \tag{A.8.2}$$

Proof: The result is readily shown by using simple rank and range properties. ■

Result R29: *Let X_0 be a particular solution to (A.8.1). Then* ***the set of all solutions*** *to (A.8.1) is given by:*

$$X = X_0 + \Delta \tag{A.8.3}$$

where $\Delta \in \mathcal{C}^{n\times p}$ is any matrix whose columns are in $\mathcal{N}(A)$.

Proof: Obviously (A.8.3) satisfies (A.8.1). To show that no solution outside the set (A.8.3) exists, let $\Omega \in \mathcal{C}^{n\times p}$ be a matrix whose columns do not all belong to $\mathcal{N}(A)$. Then $A\Omega \neq 0$ and

$$A(X_0 + \Delta + \Omega) = A\Omega + B \neq B$$

and hence $X_0 + \Delta + \Omega$ is not a solution to $AX = B$. ∎

Result R30: *The system of linear equations (A.8.1) has a* ***unique solution*** *if and only if (A.8.2) holds and A has full column rank:*

$$\text{rank}(A) = n \leq m \tag{A.8.4}$$

Proof: The assertion follows from R28 and R29. ∎

Next let us assume that (A.8.1) is consistent but A does *not* satisfy (A.8.4) (hence $\dim \mathcal{N}(A) \geq 1$). Then, according to R29 there are an infinite set of solutions. In what follows we obtain the unique solution X_0, which has *minimum norm.*

Result R31: *Consider a linear system that satisfies the consistency condition in (A.8.2). Let A have rank $r \leq \min(m,n)$, and let*

$$A = [\underbrace{U_1}_{r} \;\; \underbrace{U_2}_{m-r}] \begin{bmatrix} \Sigma_1 & 0 \\ 0 & 0 \end{bmatrix} \begin{bmatrix} V_1^* \\ V_2^* \end{bmatrix} \begin{matrix} \}\, r \\ \}\, n-r \end{matrix} = U_1 \Sigma_1 V_1^*$$

denote the SVD of A. (Here Σ_1 is nonsingular, cf. the discussion in Section A.4). Then:

$$X_0 = V_1 \Sigma_1^{-1} U_1^* B \tag{A.8.5}$$

is ***the minimum Frobenius norm solution*** *of (A.8.1) in the sense that*

$$\|X_0\|^2 < \|X\|^2 \tag{A.8.6}$$

for any other solution $X \neq X_0$.

Proof: First we verify that X_0 satisfies (A.8.1). We have

$$AX_0 = U_1 U_1^* B \tag{A.8.7}$$

In (A.8.7) $U_1U_1^*$ is the orthogonal projector onto $\mathcal{R}(A)$ (*cf.* R17). Because B must belong to $\mathcal{R}(A)$ (see R28), we conclude that $U_1U_1^*B = B$ and hence that X_0 is indeed a solution.

Next note that, according to R15,

$$\mathcal{N}(A) = \mathcal{R}(V_2)$$

Consequently, the general solution (A.8.3) can be written as (*cf.* R29)

$$X = X_0 + V_2Q \; ; \qquad Q \in \mathcal{C}^{(n-r)\times p}$$

from which we obtain:

$$\begin{aligned} \|X\|^2 &= \mathrm{tr}[(X_0^* + Q^*V_2^*)(X_0 + V_2Q)] \\ &= \|X_0\|^2 + \|V_2Q\|^2 > \|X_0\|^2 \quad \text{for } X \neq X_0 \end{aligned}$$

■

Definition D15: *The matrix*

$$A^\dagger \triangleq V_1\Sigma_1^{-1}U_1^* \tag{A.8.8}$$

in (A.8.5) is the so-called ***Moore–Penrose pseudoinverse*** *(or* ***generalized inverse****) of A.*

It can be shown that $A^\dagger$ is the unique solution to the following set of equations:

$$\begin{cases} AA^\dagger A = A \\ A^\dagger AA^\dagger = A^\dagger \\ A^\dagger A \text{ and } AA^\dagger \text{ are Hermitian} \end{cases}$$

Evidently whenever A is square and nonsingular we have $A^\dagger = A^{-1}$, which partly motivates the name of "generalized inverse" (or "pseudoinverse") given to $A^\dagger$ in the general case.

The computation of a solution to (A.8.1), whenever one exists, is an important issue which we address briefly in the following. We begin by noting that in the general case there is of course no computer algorithm which can compute a solution to (A.8.1) *exactly* (*i.e.*, without any numerical errors). In effect, the best we can hope for is to compute the exact solution to a slightly perturbed (fictitious) system of linear equations:

$$(A + \Delta_A)(X + \Delta_X) = B + \Delta_B \tag{A.8.9}$$

where Δ_A and Δ_B are small perturbation terms, the magnitude of which depends on the algorithm and the length of the computer word, and where Δ_X is the solution perturbation induced. An algorithm which, when applied to (A.8.1), provides a solution to (A.8.9) corresponding to perturbation terms (Δ_A, Δ_B) whose magnitude is of the order afforded by the "machine epsilon" is said to be *numerically stable*. Now, assuming that (A.8.1) has a unique solution (and hence that A satisfies (A.8.4)), one can show that the perturbations in A and B in (A.8.9) are retrieved in Δ_X multiplied by a proportionality factor given by

$$\text{cond}(A) = \sigma_1/\sigma_n \tag{A.8.10}$$

where σ_1 and σ_n are the largest and smallest singular values of A, respectively, and where "cond" is short for "condition". The system (A.8.1) is said to be *well-conditioned* if the corresponding ratio (A.8.10) is "small" (that is, not much larger than one). The ratio in (A.8.10) is called the *condition number* of the matrix A and is an important parameter of a given system of linear equations. Note from the previous discussion that even a numerically stable algorithm (*i.e.*, one that induces quite small Δ_A and Δ_B) can yield an inaccurate solution X when applied to an ill-conditioned system of linear equations (*i.e.*, a system with a very large cond(A)). For more details on the topic of this paragraph, including specific algorithms for solving linear systems, we refer the reader to [STEWART 1973; GOLUB AND VAN LOAN 1989].

A.8.2 Inconsistent Systems

The systems of linear equations that appear in applications (such as those in the text) are quite often perturbed versions of a "nominal system" and usually they do *not* admit any exact solution. Such systems are said to be *inconsistent*, and frequently they are overdetermined and have a matrix A that has full column rank:

$$\text{rank}(A) = n \le m \tag{A.8.11}$$

In what follows, we present two approaches to obtain an approximate solution to an inconsistent system of linear equations

$$AX \simeq B \tag{A.8.12}$$

under the condition (A.8.11).

Definition D16: *The* ***least squares*** *(LS) approximate solution to (A.8.12) is given by the minimizer X_{LS} of the following criterion:*

$$\|AX - B\|^2$$

Equivalently, X_{LS} can be defined as follows. Obtain the minimal perturbation Δ_B that makes the system (A.8.12) consistent:

$$\min \|\Delta_B\|^2 \quad \text{subject to} \quad AX = B + \Delta_B \tag{A.8.13}$$

Then derive X_{LS} by solving the system in (A.8.13) corresponding to the optimal perturbation Δ_B.

The *LS* solution introduced above can be obtained in several ways. A simple way is as follows.

Result R32: *The LS solution to (A.8.12) is given by:*

$$X_{LS} = (A^*A)^{-1}A^*B \tag{A.8.14}$$

The inverse matrix in the above equation exists in view of (A.8.11).

Proof: The matrix B_0 that makes the system consistent and which is of minimal distance (in the Frobenius norm metric) from B is given by the orthogonal projection of (the columns of) B onto $\mathcal{R}(A)$:

$$B_0 = A(A^*A)^{-1}A^*B \tag{A.8.15}$$

To motivate (A.8.15) by using only the results proven so far in this appendix, we digress from the main proof and let U_1 denote an orthogonal basis of $\mathcal{R}(A)$. Then R17 implies that $B_0 = U_1U_1^*B$. However, U_1 and A span the same subspace and hence they must be related to one another by a nonsingular linear transformation: $U_1 = AQ$ ($|Q| \neq 0$). It follows from this observation that $U_1U_1^* = AQQ^*A^*$ and also that $Q^*A^*AQ = I$, which lead to the following projector formula: $U_1U_1^* = A(A^*A)^{-1}A^*$ (as used in (A.8.15)).

Next, we return to the proof of (A.8.14). The unique solution to

$$AX - B_0 = A[X - (A^*A)^{-1}A^*B]$$

is obviously (A.8.14) since $\dim \mathcal{N}(A) = 0$ by assumption. ■

The *LS* solution X_{LS} can be computed by means of the SVD of the $m \times n$ matrix A. The X_{LS} can, however, be obtained in a computationally more efficient way as briefly described below. Note that X_{LS} should *not* be computed by directly evaluating the formula in (A.8.14) as it stands. Briefly stated, the reason is as follows. Recall from (A.8.10) that the condition number of A is given by:

$$\text{cond}(A) = \sigma_1/\sigma_n \tag{A.8.16}$$

(note that $\sigma_n \neq 0$ under (A.8.11)). When working directly on A, the numerical errors made in the computation of X_{LS} can be shown to be proportional to (A.8.16). However, in (A.8.14) one would need to invert the matrix A^*A whose condition number is:

$$\text{cond}(A^*A) = \sigma_1^2/\sigma_n^2 = [\text{cond}(A)]^2 \tag{A.8.17}$$

Working with (A^*A) may hence induce much larger numerical errors during the computation of X_{LS} and is therefore not advisable. The algorithm sketched in what follows derives X_{LS} by operating on A directly.

For any matrix A satisfying (A.8.11) there exist a unitary matrix $Q \in \mathcal{C}^{m\times m}$ and nonsingular upper–triangular matrix $R \in \mathcal{C}^{n\times n}$ such that

$$A = Q \begin{bmatrix} R \\ 0 \end{bmatrix} \triangleq [\underbrace{Q_1}_{n} \quad \underbrace{Q_2}_{m-n}] \begin{bmatrix} R \\ 0 \end{bmatrix} \tag{A.8.18}$$

The previous factorization of A is called the *QR decomposition* (QRD). Inserting (A.8.18) into (A.8.14) we obtain

$$X_{LS} = R^{-1}Q_1^*B$$

Hence, once the QRD of A has been performed, X_{LS} can be conveniently obtained as the solution of a triangular system of linear equations:

$$RX_{LS} = Q_1^*B \tag{A.8.19}$$

We note that the computation of the QRD is faster than that of the SVD (see, *e.g.*, [STEWART 1973; GOLUB AND VAN LOAN 1989]).

The previous definition and derivation of X_{LS} make it clear that the LS approach derives an approximate solution to (A.8.12) by implicitly assuming that only the right–hand side matrix B was perturbed. In applications quite frequently *both A and B* can be considered to be perturbed versions of some nominal (and unknown) matrices. In such cases we may think of determining an approximate solution to (A.8.12) by explicitly recognizing the fact that neither A nor B are perturbation free. An approach based on this idea is described next (see, *e.g.*, [VAN HUFFEL AND VANDEWALLE 1991]).

Definition D17: *The* ***total least squares*** *(TLS) approximate solution to (A.8.12) is defined as follows. First derive the minimal perturbations Δ_A and Δ_B that make the system consistent:*

$$\min \|[\Delta_A \ \Delta_B]\|^2 \quad \text{subject to} \quad (A + \Delta_A)X = B + \Delta_B \tag{A.8.20}$$

Then obtain X_{TLS} by solving the system in (A.8.20) corresponding to the optimal perturbations $(\Delta_A, \ \Delta_B)$.

A simple way to derive a more explicit formula for calculating the X_{TLS} runs as follows.

Result R33: *Let*

$$[A \ B] = [\underbrace{\tilde{U}_1}_{n} \quad \underbrace{\tilde{U}_2}_{m-n}] \begin{bmatrix} \tilde{\Sigma}_1 & 0 \\ 0 & \tilde{\Sigma}_2 \end{bmatrix} \begin{bmatrix} \tilde{V}_1^* \\ \tilde{V}_2^* \end{bmatrix} \begin{matrix} \}\, n \\ \}\, p \end{matrix} \tag{A.8.21}$$

denote the SVD of the matrix $[A \ B]$. Furthermore, partition $\tilde{V}_2^*$ as

$$\tilde{V}_2^* = [\underbrace{\tilde{V}_{21}^*}_{n} \quad \underbrace{\tilde{V}_{22}^*}_{p}] \tag{A.8.22}$$

Then

$$X_{TLS} = -\tilde{V}_{21}\ \tilde{V}_{22}^{-1} \tag{A.8.23}$$

if $\tilde{V}_{22}^{-1}$ *exists.*

Proof: The optimization problem with constraints in (A.8.20) can be restated in the following way: Find the minimal perturbation $[\Delta_A\ \Delta_B]$ and the corresponding matrix X such that

$$\{\ [A\ B] + [\Delta_A\ \Delta_B]\ \} \begin{bmatrix} -X \\ I \end{bmatrix} = 0 \tag{A.8.24}$$

Since rank $\begin{bmatrix} -X \\ I \end{bmatrix} = p$ it follows that $[\Delta_A\ \Delta_B]$ should be such that $\dim \mathcal{N}(\ [A\ B] + [\Delta_A\ \Delta_B]\ ) \geq p$ or, equivalently,

$$\text{rank}(\ [A\ B] + [\Delta_A\ \Delta_B]\ ) \leq n \tag{A.8.25}$$

According to R18, the minimal perturbation matrix $[\Delta_A\ \Delta_B]$ that achieves (A.8.25) is given by

$$[\Delta_A\ \Delta_B] = -\tilde{U}_2\tilde{\Sigma}_2\tilde{V}_2^* \tag{A.8.26}$$

Inserting (A.8.26) along with (A.8.21) into (A.8.24), we obtain the following matrix equation in X:

$$\tilde{U}_1\tilde{\Sigma}_1\tilde{V}_1^* \begin{bmatrix} -X \\ I \end{bmatrix} = 0$$

or, equivalently,

$$\tilde{V}_1^* \begin{bmatrix} -X \\ I \end{bmatrix} = 0 \tag{A.8.27}$$

Equation (A.8.27) implies that X must satisfy

$$\begin{bmatrix} -X \\ I \end{bmatrix} = \tilde{V}_2 Q = \begin{bmatrix} \tilde{V}_{21} \\ \tilde{V}_{22} \end{bmatrix} Q \tag{A.8.28}$$

for some nonsingular normalizing matrix Q. The expression (A.8.23) for X_{TLS} is readily obtained from (A.8.28). ■

The TLS solution in (A.8.23) is unique if and only if the singular values $\{\tilde{\sigma}_k\}$ of the matrix $[A\ B]$ are such that $\tilde{\sigma}_n > \tilde{\sigma}_{n+1}$ (this follows from R18). When $\tilde{V}_{22}$ is singular, the TLS solution does not exist; see [VAN HUFFEL AND VANDEWALLE 1991].

The computation of the X_{TLS} requires the SVD of the $m \times (n+p)$ matrix $[A\ B]$. The solution X_{TLS} can be rewritten in a slightly different form. Let $\tilde{V}_{11}$, $\tilde{V}_{12}$ be defined via the following partition of $\tilde{V}_1^*$

$$\tilde{V}_1^* = [\underbrace{\tilde{V}_{11}}_{n}\ \ \underbrace{\tilde{V}_{12}}_{p}]$$

The orthogonality condition $\tilde{V}_1^* \tilde{V}_2 = 0$ can be rewritten as

$$\tilde{V}_{11}\tilde{V}_{21} + \tilde{V}_{12}\tilde{V}_{22} = 0$$

which yields

$$X_{TLS} = -\tilde{V}_{21}\tilde{V}_{22}^{-1} = \tilde{V}_{11}^{-1}\tilde{V}_{12} \tag{A.8.29}$$

Since usually p is (much) smaller than n, the formula (A.8.23) for X_{TLS} may often be computationally more convenient than (A.8.29) (for example, in the common case of $p = 1$, (A.8.23) does not require any matrix inversion whereas (A.8.29) requires the calculation of an $n \times n$ matrix inverse).

A.9 Quadratic Minimization

Several problems in this text require the solution to *quadratic minimization problems*. In this section, we make use of matrix analysis techniques to derive two results: one on unconstrained minimization, and the other on constrained minimization.

Result R34: *Let A be an $(n \times n)$ Hermitian positive definite matrix, let X and B be $(n \times m)$ matrices, and let C be an $m \times m$ Hermitian matrix. Then the unique solution to the minimization problem*

$$\min_X F(X), \quad F(X) = X^*AX + X^*B + B^*X + C \tag{A.9.1}$$

is given by

$$X_0 = -A^{-1}B, \qquad F(X_0) = C - B^*A^{-1}B \tag{A.9.2}$$

Here, the matrix minimization means $F(X_0) \le F(X)$ for every $X \ne X_0$; that is, $F(X) - F(X_0)$ is a positive semidefinite matrix.

Proof: Let $X = X_0 + \Delta$, where Δ is an arbitrary $(n \times m)$ complex matrix. Then

$$\begin{aligned} F(X) &= (-A^{-1}B + \Delta)^*A(-A^{-1}B + \Delta) + (-A^{-1}B + \Delta)^*B \\ &\quad + B^*(-A^{-1}B + \Delta) + C \\ &= \Delta^*A\Delta + F(X_0) \end{aligned} \tag{A.9.3}$$

Since A is positive definite, $\Delta^*A\Delta \ge 0$ for all nonzero Δ; thus, the minimum value of $F(X)$ is $F(X_0)$, and the result is proven. ■

We next present a result on linearly constrained quadratic minimization.

Result R35: *Let A be an $(n \times n)$ Hermitian positive definite matrix, and let $X \in \mathcal{C}^{n\times m}$, $B \in \mathcal{C}^{n\times k}$, and $C \in \mathcal{C}^{m\times k}$. Assume that B has full column rank equal to k (hence $n \ge k$). Then the unique solution to the minimization problem*

$$\min_X X^*AX \quad \text{subject to} \quad X^*B = C \tag{A.9.4}$$

is given by

$$X_0 = A^{-1}B(B^*A^{-1}B)^{-1}C^*. \quad \text{(A.9.5)}$$

Proof: First note that $(B^*A^{-1}B)^{-1}$ exists and that $X_0^*B = C$. Let $X = X_0 + \Delta$, where $\Delta \in \mathcal{C}^{n \times m}$ satisfies $\Delta^*B = 0$ (so that X also satisfies the constraint $X^*B = C$). Then

$$X^*AX = X_0^*AX_0 + X_0^*A\Delta + \Delta^*AX_0 + \Delta^*A\Delta \quad \text{(A.9.6)}$$

where the two middle terms are equal to zero:

$$\Delta^*AX_0 = \Delta^*B(B^*A^{-1}B)^{-1}C^* = 0$$

Hence,

$$X^*AX - X_0^*AX_0 = \Delta^*A\Delta \geq 0 \quad \text{(A.9.7)}$$

as A is positive definite. It follows from (A.9.7) that the minimizing X matrix is given by X_0. ■

A common special case of Result R35 is $m = k = 1$ (so X and B are both vectors) and $C = 1$. Then

$$X_0 = \frac{A^{-1}B}{B^*A^{-1}B}$$

Appendix B

CRAMÉR–RAO BOUND TOOLS

B.1 Introduction

In the text we have kept the discussion of statistical aspects at a minimum for conciseness reasons. However, we have presented certain statistical tools and analyses that we have found useful to the understanding of the spectral analysis material discussed. In this appendix we introduce some basic facts on an important statistical tool: the Cramér–Rao bound (abbreviated as CRB). We begin our discussion by explaining the importance of the CRB for *parametric spectral analysis.*

Let $\phi(\omega, \theta)$ denote a parametric spectral model, depending on a *real–valued* vector θ, and let $\phi(\omega, \hat{\theta})$ denote the spectral density estimated from N data samples. Assume that the estimate $\hat{\theta}$ of θ is *consistent* such that the estimation error is small for large values of N. Then, by making use of a Taylor series expansion technique, we can approximately write the estimation error $[\phi(\omega, \hat{\theta}) - \phi(\omega, \theta)]$ as a linear function of $\hat{\theta} - \theta$:

$$[\phi(\omega, \hat{\theta}) - \phi(\omega, \theta)] \simeq \psi^T(\omega, \theta)(\hat{\theta} - \theta) \tag{B.1.1}$$

where the symbol $\simeq$ denotes an asymptotically (in N) valid approximation, and $\psi(\omega, \theta)$ is the gradient of $\phi(\omega, \theta)$ with respect to θ (evaluated at the true parameter values):

$$\psi(\omega, \theta) = \frac{\partial \phi(\omega, \theta)}{\partial \theta} \tag{B.1.2}$$

It follows from (B.1.1) that the mean squared error (MSE) of $\phi(\omega, \hat{\theta})$ is approximately given by

$$\text{MSE}[\phi(\omega, \hat{\theta})] \simeq \psi^T(\omega, \theta) P \psi(\omega, \theta) \qquad (\text{for } N \gg 1) \tag{B.1.3}$$

where

$$P = \text{MSE}[\hat{\theta}] = E\left\{(\hat{\theta} - \theta)(\hat{\theta} - \theta)^T\right\} \tag{B.1.4}$$

We see from (B.1.3) that the variance (or MSE) of the estimation errors in the spectral domain are linearly related to the variance (or MSE) of the parameter vector estimate $\hat{\theta}$, so that we can get an accurate spectral estimate only if we use an accurate parameter estimator. We start from this simple observation, which reduces the statistical analysis of $\phi(\omega, \hat{\theta})$ to the analysis of $\hat{\theta}$, to explain the importance of the CRB for the performance study of spectral analysis. Toward that end, we discuss several facts in the paragraphs that follow.

Assume that $\hat{\theta}$ is some *unbiased estimate* of θ (that is $E\{\hat{\theta}\} = \theta$), and let P denote the covariance matrix of $\hat{\theta}$ (*cf.* (B.1.4)):

$$P = E\left\{(\hat{\theta} - \theta)(\hat{\theta} - \theta)^T\right\} \tag{B.1.5}$$

(Note that here we do not require that N be large). Then, under quite general conditions, there is a matrix (which we denote by P_{cr}) such that

$$P \geq P_{cr} \tag{B.1.6}$$

in the sense that the difference $(P - P_{cr})$ is a positive semidefinite matrix. This is basically the celebrated Cramér–Rao bound result [CRAMÉR 1946; RAO 1945]. We will derive the inequality (B.1.6) along with an expression for the CRB in the next section.

In view of (B.1.6) we may think of assessing the performance of a given estimation method by comparing its covariance matrix P with the CRB. Such a comparison would make perfect sense whenever the CRB is *achievable*; that is, whenever there exists an estimation method such that its P equals the CRB. Unfortunately, this is rarely the case for finite N. Additionally, *biased estimators* may exist whose MSEs are smaller than the CRB under discussion (see, for example, [STOICA AND MOSES 1990; STOICA AND OTTERSTEN 1996]). Hence, in the *finite sample case* (particularly for small samples) comparing with the CRB does not really make too much sense because:

(i) There might be no unbiased estimator that attains the CRB and, consequently, a large difference $(P - P_{cr})$ may not necessarily mean bad accuracy; and

(ii) The equality $P = P_{cr}$ does not necessarily mean that we have achieved the ultimate possible performance, as there might be biased estimators with lower MSE than the CRB.

In the *large sample case*, on the other hand, the utility of the CRB result for the type of parameter estimation problems addressed in the text is significant, as explained next.

Let $y \in \mathcal{R}^{N\times 1}$ denote the sample of available observations. Any estimate $\hat{\theta}$ of θ will be a function of y. We assume that both θ and y are *real–valued*. Working with real θ and y vectors appears to be the most convenient way when discussing the CRB theory, even when the original parameters and measurements are complex–valued. (If the parameters and measurements are complex–valued, θ and y are obtained by concatenating the real and imaginary parts of the complex parameter and data vectors, respectively.) We also assume that the probability density of y, which we denote by $p(y, \theta)$, is a differentiable function of θ. An important general method for parameter estimation consists of maximizing $p(y, \theta)$ with respect to θ:

$$\hat{\theta} = \arg\max_{\theta} p(y, \theta) \tag{B.1.7}$$

The $p(y, \theta)$ in (B.1.7) with y fixed and θ variable is called the *likelihood function*, and $\hat{\theta}$ is called the *maximum likelihood* (ML) *estimate* of θ. Under regularity conditions the ML estimate (MLE) is *consistent* (*i.e.*, $\lim_{N\to\infty} \hat{\theta} = \theta$ stochastically) and its covariance matrix approaches the CRB as N increases:

$$P \simeq P_{cr} \qquad \text{for a MLE with } N \gg 1 \tag{B.1.8}$$

The aforementioned regularity conditions basically amount to requiring that the number of free parameters does not increase with N, which is true for all but one of the parametric spectral estimation problems discussed in the text. The array processing problem of Chapter 6 does not satisfy the previous requirement when the signal snapshots are assumed to be unknown deterministic variables; in such a case the number of unknown parameters grows without bound as N increases, and the equality in (B.1.8) does not hold [Stoica and Nehorai 1989a; Stoica and Nehorai 1990].

In summary, then, in *large samples* the ML method attains the ultimate performance corresponding to the CRB, under rather general conditions. Furthermore, there are no other known *practical methods* that can provide consistent estimates of θ with lower variance than the CRB[1]. Hence, the ML method can be said to be asymptotically a *statistically efficient practical estimation approach.* The accuracy achieved by any other estimation method can therefore be assessed *by comparing the (large sample) covariance matrix of that method with the CRB*, which approximately equals the covariance matrix of the MLE in large samples (*cf.* (B.1.8)). This performance comparison ability is one of the most important uses of the CRB.

With reference to the spectral estimation problem, it follows from (B.1.3) and the previous observation that we can assess the performance of a given spectral estimator by comparing its large sample MSE values with

$$\boxed{\psi^T(\omega, \theta)[P_{cr}]\psi(\omega, \theta)} \tag{B.1.9}$$

The MSE values can be obtained by the Monte–Carlo simulation of a typical scenario representative of the problem of interest, or by using analytical MSE formulas whenever they are available. In the text we have emphasized the former, more pragmatic way of determining the MSE of a given spectral estimator.

[1]Consistent estimation methods whose asymptotic variance is lower than the CRB, at certain points in the parameter set, do exist! However, such methods (which are called "asymptotically statistically super–efficient") have little practical relevance (they are mainly of a theoretical interest); see, *e.g.*, [Stoica and Ottersten 1996].

B.2 The CRB for General Distributions

Result R36: (Cramér–Rao) *Consider the likelihood function $p(y,\theta)$, introduced in the previous section, and define*

$$P_{cr} = \left(E\left\{ \left[\frac{\partial \ln p(y,\theta)}{\partial \theta} \right] \left[\frac{\partial \ln p(y,\theta)}{\partial \theta} \right]^T \right\} \right)^{-1} \tag{B.2.1}$$

where the inverse is assumed to exist. Then

$$P \geq P_{cr} \tag{B.2.2}$$

holds for any unbiased estimate of θ. Furthermore, the CRB matrix can alternatively be expressed as:

$$P_{cr} = -\left(E\left\{ \frac{\partial^2 p(y,\theta)}{\partial \theta\, \partial \theta^T} \right\} \right)^{-1} \tag{B.2.3}$$

Proof: As $p(y,\theta)$ is a probability density function,

$$\int p(y,\theta) dy = 1 \tag{B.2.4}$$

where the integration is over $\mathcal{R}^N$. The assumption that $\hat{\theta}$ is an unbiased estimate implies

$$\int \hat{\theta} p(y,\theta) dy = \theta \tag{B.2.5}$$

Differentiation of (B.2.4) and (B.2.5) with respect to θ yields, under regularity conditions,

$$\int \frac{\partial p(y,\theta)}{\partial \theta} dy = \int \frac{\partial \ln p(y,\theta)}{\partial \theta} p(y,\theta) dy = E\left\{ \frac{\partial \ln p(y,\theta)}{\partial \theta} \right\} = 0 \tag{B.2.6}$$

and

$$\int \hat{\theta} \frac{\partial p(y,\theta)}{\partial \theta} dy = \int \hat{\theta} \frac{\partial \ln p(y,\theta)}{\partial \theta} p(y,\theta) dy = E\left\{ \hat{\theta} \frac{\partial \ln p(y,\theta)}{\partial \theta} \right\} = I \tag{B.2.7}$$

It follows from (B.2.6) and (B.2.7) that

$$E\left\{ (\hat{\theta} - \theta) \frac{\partial \ln p(y,\theta)}{\partial \theta} \right\} = I \tag{B.2.8}$$

Next note that the matrix

$$E\left\{ \begin{bmatrix} (\hat{\theta} - \theta) \\ \dfrac{\partial \ln p(y,\theta)}{\partial \theta} \end{bmatrix} \begin{bmatrix} (\hat{\theta} - \theta)^T & \left(\dfrac{\partial \ln p(y,\theta)}{\partial \theta} \right)^T \end{bmatrix} \right\} = \begin{bmatrix} P & I \\ I & P_{cr}^{-1} \end{bmatrix} \tag{B.2.9}$$

is, by construction, positive semidefinite. (To obtain the equality in (B.2.9) we used (B.2.8)). This observation implies (B.2.2) (see Result R20 in Appendix A).

Next we prove the equality in (B.2.3). Differentiation of (B.2.6) gives:

$$\int \frac{\partial^2 \ln p(y,\theta)}{\partial\theta\, \partial\theta^T} p(y,\theta)dy + \int \left[\frac{\partial \ln p(y,\theta)}{\partial\theta}\right] \left[\frac{\partial \ln p(y,\theta)}{\partial\theta}\right]^T p(y,\theta)dy = 0$$

or, equivalently,

$$E\left\{\left[\frac{\partial \ln p(y,\theta)}{\partial\theta}\right] \left[\frac{\partial \ln p(y,\theta)}{\partial\theta}\right]^T\right\} = -E\left\{\frac{\partial^2 \ln p(y,\theta)}{\partial\theta\, \partial\theta^T}\right\}$$

which is precisely what we had to prove. ■

The matrix

$$\begin{aligned} J &= E\left\{\left[\frac{\partial \ln p(y,\theta)}{\partial\theta}\right] \left[\frac{\partial \ln p(y,\theta)}{\partial\theta}\right]^T\right\} \\ &= -E\left\{\frac{\partial^2 \ln p(y,\theta)}{\partial\theta\, \partial\theta^T}\right\}, \end{aligned} \tag{B.2.10}$$

the inverse of which appears in the CRB formula (B.2.1) (or (B.2.3)), is called the (Fisher) *information matrix* [FISHER 1922].

B.3 The CRB for Gaussian Distributions

The CRB matrix in (B.2.1) depends implicitly on the data properties via the probability density function $p(y,\theta)$. To obtain a more explicit expression for the CRB we should specify the data distribution. A particularly convenient CRB formula is obtained if the data vector is assumed to be Gaussian distributed:

$$p(y,\theta) = \frac{1}{(2\pi)^{N/2}|C|^{1/2}} e^{-(y-\mu)^T C^{-1}(y-\mu)/2} \tag{B.3.1}$$

where μ and C are, respectively, the mean and the covariance matrix of y (C is assumed to be invertible). In the case of (B.3.1), the log–likelihood function that appears in (B.2.1) is given by:

$$\ln p(y,\theta) = -\frac{N}{2}\ln 2\pi - \frac{1}{2}\ln|C| - \frac{1}{2}(y-\mu)^T C^{-1}(y-\mu) \tag{B.3.2}$$

Result R37: *The CRB matrix corresponding to the Gaussian data distribution in (B.3.1) is given (elementwise) by:*

$$\boxed{[P_{cr}^{-1}]_{ij} = \frac{1}{2}\operatorname{tr}\left[C^{-1}C_i' C^{-1} C_j'\right] + \left[\mu_i'^T C^{-1}\mu_j'\right]} \tag{B.3.3}$$

where C_i' denotes the derivative of C with respect to the ith element of θ (and similarly for μ_i').

Proof: By using Result R21 and the notational convention for the first–order and second–order derivatives, we obtain:

$$\begin{aligned}
2[\ln p(y,\theta)]_{ij}'' &= \frac{\partial}{\partial \theta_i}\{-\operatorname{tr}[C^{-1}C_j'] + 2\mu_j'^T C^{-1}(y-\mu) \\
&\quad +(y-\mu)^T C^{-1}C_j' C^{-1}(y-\mu)\} \\
&= \operatorname{tr}[C^{-1}C_i'C^{-1}C_j'] - \operatorname{tr}[C^{-1}C_{ij}''] \\
&\quad +2\left\{[\mu_j'^T C^{-1}]_i'(y-\mu) - \mu_j'^T C^{-1}\mu_i'\right\} \\
&\quad -2\mu_i'^T C^{-1}C_j'C^{-1}(y-\mu) \\
&\quad +\operatorname{tr}\{(y-\mu)(y-\mu)^T \\
&\quad \cdot[-C^{-1}C_i'C^{-1}C_j'C^{-1} + C^{-1}C_{ij}''C^{-1} - C^{-1}C_j'C^{-1}C_i'C^{-1}]\}
\end{aligned}$$

Taking the expectation of both sides of the equation above yields:

$$\begin{aligned}
2\left[P_{cr}^{-1}\right]_{ij} &= -\operatorname{tr}[C^{-1}C_i'C^{-1}C_j'] + \operatorname{tr}[C^{-1}C_{ij}''] + 2\mu_i'^T C^{-1}\mu_j' \\
&\quad +\operatorname{tr}[C^{-1}C_i'C^{-1}C_j'] - \operatorname{tr}[C^{-1}C_{ij}''] + \operatorname{tr}[C^{-1}C_i'C^{-1}C_j'] \\
&= \operatorname{tr}[C^{-1}C_i'C^{-1}C_j'] + 2\mu_i'^T C^{-1}\mu_j'
\end{aligned}$$

which concludes the proof. ■

The CRB expression in (B.3.3) is sometimes referred to as the *Slepian–Bangs formula.* (The second term in (B.3.3) is due to Slepian [SLEPIAN 1954] and the first to Bangs [BANGS 1971]).

Next we specialize the CRB formula (B.3.3) to a particular type of Gaussian distribution. Let $N = 2\bar{N}$ (hence, N is assumed to be even). Partition the vector y as

$$y = \begin{bmatrix} y_1 \\ y_2 \end{bmatrix} \begin{matrix} \}\bar{N} \\ \}\bar{N} \end{matrix} \tag{B.3.4}$$

Accordingly, partition μ and C as

$$\mu = \begin{bmatrix} \mu_1 \\ \mu_2 \end{bmatrix} \tag{B.3.5}$$

and

$$C = \begin{bmatrix} C_{11} & C_{12} \\ C_{12}^T & C_{22} \end{bmatrix} \tag{B.3.6}$$

The vector y is said to have a *circular* (or *circularly symmetric*) *Gaussian distribution* if

$$\boxed{C_{11} = C_{22}} \tag{B.3.7}$$

$$\boxed{C_{12}^T = -C_{12}} \tag{B.3.8}$$

Let

$$\mathbf{y} \triangleq y_1 + iy_2 \tag{B.3.9}$$

and

$$\boldsymbol{\mu} = \mu_1 + i\mu_2 \tag{B.3.10}$$

We also say that the *complex–valued random vector* $\mathbf{y}$ *has a circular Gaussian distribution* whenever the conditions (B.3.7) and (B.3.8) are satisfied. It is a straightforward exercise to verify that the aforementioned conditions can be more compactly written as:

$$\boxed{E\left\{(\mathbf{y}-\boldsymbol{\mu})(\mathbf{y}-\boldsymbol{\mu})^T\right\} = 0} \tag{B.3.11}$$

The Fourier transform, as well as the complex demodulation operation (see Chapter 6), often lead to signals satisfying (B.3.11) (see, *e.g.*, [BRILLINGER 1981]). Hence, the *circularity* is a relatively frequent property of the Gaussian random signals encountered in the spectral analysis problems discussed in this text.

Remark: If a random vector $\mathbf{y}$ satisfies the "circularity condition" (B.3.11) then it is readily verified that $\mathbf{y}$ and $\mathbf{y}e^{iz}$ have the same second–order properties for every constant z in $[-\pi, \pi]$. Hence, the second–order properties of $\mathbf{y}$ do not change if its generic element $\mathbf{y}_k$ is replaced by any other value, $\mathbf{y}_k e^{iz}$, on the *circle* with radius $|\mathbf{y}_k|$ (recall that z is nonrandom and it does not depend on k). This observation provides a motivation for the name "circularly symmetric" given to such a random vector $\mathbf{y}$. ■

Let

$$\Gamma = E\left\{(\mathbf{y}-\boldsymbol{\mu})(\mathbf{y}-\boldsymbol{\mu})^*\right\} \tag{B.3.12}$$

For circular Gaussian random vectors y (or $\mathbf{y}$), the CRB formula (B.3.3) can be rewritten in a compact form as a function of Γ and $\boldsymbol{\mu}$. (Note that the dimensions of Γ and $\boldsymbol{\mu}$ are half the dimensions of C and μ appearing in (B.3.3).) In order to show how this can be done, we need some preparations.

Let

$$\bar{C} = C_{11} = C_{22} \tag{B.3.13}$$

$$\tilde{C} = C_{12}^T = -C_{12} \tag{B.3.14}$$

Hence,

$$C = \begin{bmatrix} \bar{C} & -\tilde{C} \\ \tilde{C} & \bar{C} \end{bmatrix} \tag{B.3.15}$$

and

$$\Gamma = 2(\bar{C} + i\tilde{C}) \tag{B.3.16}$$

To any complex–valued matrix $\mathcal{C} = \bar{C} + i\tilde{C}$ we associate a real–valued matrix C as defined in (B.3.15), and vice versa. It is a simple exercise to verify that if

$$\mathcal{A} = \mathcal{B}\mathcal{C} \iff \bar{A} + i\tilde{A} = (\bar{B} + i\tilde{B})(\bar{C} + i\tilde{C}) \tag{B.3.17}$$

then the real–valued matrix associated with $\mathcal{A}$ is given by

$$A = BC \iff \begin{bmatrix} \bar{A} & -\tilde{A} \\ \tilde{A} & \bar{A} \end{bmatrix} = \begin{bmatrix} \bar{B} & -\tilde{B} \\ \tilde{B} & \bar{B} \end{bmatrix} \begin{bmatrix} \bar{C} & -\tilde{C} \\ \tilde{C} & \bar{C} \end{bmatrix} \tag{B.3.18}$$

In particular, it follows from (B.3.17) and (B.3.18) with $A = I$ (and hence $\mathcal{A} = I$) that the matrices C^{-1} and $\mathcal{C}^{-1}$ form a real–complex pair as defined above.

We deduce from the results previously derived that the matrix in the first term of (B.3.3),

$$D = C^{-1}C_i'C^{-1}C_j' \tag{B.3.19}$$

is associated with

$$\mathcal{D} = \mathcal{C}^{-1}\mathcal{C}_i'\mathcal{C}^{-1}\mathcal{C}_j' = \Gamma^{-1}\Gamma_i'\Gamma^{-1}\Gamma_j' \tag{B.3.20}$$

Furthermore, we have

$$\frac{1}{2}\operatorname{tr}(D) = \operatorname{tr}(\bar{D}) = \operatorname{tr}(\mathcal{D}) \tag{B.3.21}$$

The second equality above follows from the fact that $\mathcal{C}$ is Hermitian, and hence

$$\operatorname{tr}(\mathcal{D}^*) = \operatorname{tr}(\mathcal{C}_j'\mathcal{C}^{-1}\mathcal{C}_i'\mathcal{C}^{-1}) = \operatorname{tr}(\mathcal{C}^{-1}\mathcal{C}_i'\mathcal{C}^{-1}\mathcal{C}_j') = \operatorname{tr}(\mathcal{D})$$

which in turn implies that $\operatorname{tr}(\tilde{D}) = 0$ and therefore that $\operatorname{tr}(\mathcal{D}) = \operatorname{tr}(\bar{D})$. Combining (B.3.20) and (B.3.21) shows that the first term in (B.3.3) can be rewritten as:

$$\operatorname{tr}(\Gamma^{-1}\Gamma_i'\Gamma^{-1}\Gamma_j') \tag{B.3.22}$$

Next we consider the second term in (B.3.3). Let

$$x = \begin{bmatrix} x_1 \\ x_2 \end{bmatrix} \quad \text{and} \quad z = \begin{bmatrix} z_1 \\ z_2 \end{bmatrix}$$

be two arbitrary vectors partitioned similarly to μ, and let $\mathbf{x} = x_1 + ix_2$ and $\mathbf{z} = z_1 + iz_2$. A straightforward calculation shows that:

$$\begin{aligned} x^T Az &= x_1^T\bar{A}z_1 + x_2^T\bar{A}z_2 + x_2^T\tilde{A}z_1 - x_1^T\tilde{A}z_2 \\ &= \operatorname{Re}\{\mathbf{x}^*\mathcal{A}\mathbf{z}\} \end{aligned} \tag{B.3.23}$$

Hence,

$$\begin{aligned} \mu_i'^T C^{-1}\mu_j' &= \operatorname{Re}\{\boldsymbol{\mu}_i'^*\mathcal{C}^{-1}\boldsymbol{\mu}_j'\} \\ &= 2\operatorname{Re}\{\boldsymbol{\mu}_i'^*\Gamma^{-1}\boldsymbol{\mu}_j'\} \end{aligned} \tag{B.3.24}$$

Insertion of (B.3.22) and (B.3.24) into (B.3.3) yields the following CRB formula that holds in the case of *circularly Gaussian distributed data vectors* y (or $\mathbf{y}$):

$$[P_{cr}^{-1}]_{ij} = \operatorname{tr}\left[\Gamma^{-1}\Gamma_i'\Gamma^{-1}\Gamma_j'\right] + 2\operatorname{Re}\left[\boldsymbol{\mu}_i'^{*}\Gamma^{-1}\boldsymbol{\mu}_j'\right] \tag{B.3.25}$$

The importance of the Gaussian CRB formulas lies not only in the fact that Gaussian data are rather frequently encountered in applications, but also in a more subtle aspect explained in what follows. Briefly stated, the second reason for the importance of the CRB formulas derived in this section is that:

Under rather general conditions and (at least) in large samples, the Gaussian CRB is the largest of all CRB matrices corresponding to different congruous distributions of the data sample[2].	(B.3.26)

To motivate the previous assertion, consider the ML estimate of θ derived under the Gaussian data hypothesis, which we denote by $\hat{\theta}_G$. According to the discussion around equation (B.1.8), the large sample covariance matrix of $\hat{\theta}$ equals P_{cr}^G (similar to $\hat{\theta}_G$, we use an index G to denote the CRB matrix in the Gaussian hypothesis case). Now, under rather general conditions, the large sample properties of the Gaussian ML estimator are independent of the data distribution (see, *e.g.*, [SÖDERSTRÖM AND STOICA 1989]). In other words, the large sample covariance matrix of $\hat{\theta}_G$ is equal to P_{cr}^G for many other data distributions besides the Gaussian one. This observation, along with the general CRB inequality, implies that:

$$P_{cr}^G \geq P_{cr} \tag{B.3.27}$$

where the right–hand side is the CRB matrix corresponding to the data distribution at hand.

The inequality (B.3.27) (or, equivalently, the assertion (B.3.26)) shows that a method whose covariance matrix is much larger than P_{cr}^G cannot be a good estimation method. As a matter of fact, the "asymptotic properties" of most existing parameter estimation methods do not depend on the data distribution. This means that P_{cr}^G is a lower bound for the covariance matrices of a large class of estimation methods, regardless of the data distribution. On the other hand, the inequality (B.3.27) also shows that for non–Gaussian data it should be possible to beat the Gaussian CRB (for instance by exploiting higher–order moments of the data, beyond the first and second–order moments used in the Gaussian ML method). However, general estimation methods with covariance matrices uniformly smaller than P_{cr}^G are yet to be discovered. In summary, comparing against the P_{cr}^G makes sense in most parameter estimation exercises.

[2]A meaningful comparison of the CRBs under two different data distributions requires that the hypothesized distributional models do not contain conflicting assumptions. In particular, when one of the two distributions is the Gaussian, the mean and covariance matrix should be the same for both distributions.

In what follows, we briefly consider the application of the general Gaussian CRB formulas derived above to the two main parameter estimation problems treated in the text.

B.4 The CRB for Line Spectra

As explained in Chapter 4 the estimation of line spectra is basically a parameter estimation problem. The corresponding parameter vector is

$$\theta = \left[\alpha_1 \;\ldots\; \alpha_n,\, \varphi_1 \;\ldots\; \varphi_n,\, \omega_1 \;\ldots\; \omega_n,\, \sigma^2\right]^T \tag{B.4.1}$$

and the data vector is

$$\mathbf{y} = [y(1) \cdots y(N)]^T \tag{B.4.2}$$

or, in real–valued form,

$$y = \left[\; \mathrm{Re}[y(1)] \cdots \mathrm{Re}[y(N)] \quad \mathrm{Im}[y(1)] \cdots \mathrm{Im}[y(N)] \;\right]^T \tag{B.4.3}$$

When $\{\varphi_k\}$ are assumed to be random variables uniformly distributed on $[0, 2\pi]$ (whereas $\{\alpha_k\}$ and $\{\omega_k\}$ are deterministic constants), the distribution of $\mathbf{y}$ is *not* Gaussian and hence neither of the CRB formulas of the previous section are usable. To overcome this difficulty it is customary to consider the distribution of $\mathbf{y}$ *conditioned on* $\{\varphi_k\}$ (*i.e.*, for $\{\varphi_k\}$ fixed). This distribution is circular Gaussian, under the assumption that the (white) noise is circularly Gaussian distributed, with the following mean and covariance matrix:

$$\boldsymbol{\mu} = E\{\mathbf{y}\} = \begin{bmatrix} 1 & \cdots & 1 \\ e^{i\omega_1} & \cdots & e^{i\omega_n} \\ \vdots & & \vdots \\ e^{i(N-1)\omega_1} & \cdots & e^{i(N-1)\omega_n} \end{bmatrix} \begin{bmatrix} \alpha_1 e^{i\varphi_1} \\ \vdots \\ \alpha_n e^{i\varphi_n} \end{bmatrix} \tag{B.4.4}$$

$$\Gamma = E\{(\mathbf{y} - \boldsymbol{\mu})(\mathbf{y} - \boldsymbol{\mu})^*\} = \sigma^2 I \tag{B.4.5}$$

The differentiation of (B.4.4) and (B.4.5) with respect to the elements of the parameter vector θ can be easily done (we leave the details of this differentiation operation as an exercise to the reader). Hence, we can readily obtain all ingredients required to evaluate the CRB matrix in equation (B.3.25). If the distribution of $\mathbf{y}$ (or y) is Gaussian but not circular, we need additional parameters, besides σ^2, to characterize the matrix $E\left\{(\mathbf{y} - \boldsymbol{\mu})(\mathbf{y} - \boldsymbol{\mu})^T\right\}$. Once these parameters are introduced, the use of formula (B.3.3) to obtain the CRB is straightforward.

In Section 4.3 we have given a simple formula for the block of the CRB matrix corresponding to the frequency estimates $\{\hat{\omega}_k\}$. That formula holds asymptotically, as N increases. For finite values of N, it is a good approximation of the exact CRB whenever the minimum frequency separation is larger than $1/N$ [Stoica, Moses, Friedlander, and Söderström 1989]. In any case, the approximate (large sample) CRB formula given in Section 4.3 is computationally much simpler to implement than the exact CRB.

B.5 The CRB for Rational Spectra

For rational (or ARMA) spectra, the Cramér–Rao lower bound on the variance of any consistently estimated spectrum is asymptotically (for $N \gg 1$) given by (B.1.9). The CRB matrix for the parameter vector estimate, which appears in (B.1.9), can be evaluated as outlined in what follows.

In the case of ARMA spectral models, the parameter vector consists of the white noise power σ^2 and the polynomial coefficients $\{a_k, b_k\}$. We arrange the ARMA coefficients in the following real–valued vector:

$$\theta = [\text{Re}(a_1) \cdots \text{Re}(a_n)\ \text{Re}(b_1) \cdots \text{Re}(b_m)\, \text{Im}(a_1) \cdots \text{Im}(a_n)\ \text{Im}(b_1) \cdots \text{Im}(b_m)]^T$$

The data vector is defined as in equations (B.4.2) or (B.4.3) and has zero mean ($\mu = 0$). The calculation of the covariance matrix of the data vector reduces to the calculation of ARMA covariances:

$$r(k) = \sigma^2 E\left\{ \left[\frac{B(q)}{A(q)} w(t) \right] \left[\frac{B(q)}{A(q)} w(t-k) \right]^* \right\}$$

where the white noise sequence $\{w(t)\}$ is normalized such that its variance is one. Methods for computation of $\{r_k\}$ (for given values of σ^2 and θ) were outlined in Exercises C1.12 and 3.2. The method in Exercise C1.12 should perform reasonably well as long as the zeroes of $A(q)$ are not too close to the unit circle. If the zeroes of $A(q)$ are close to the unit circle, it is advisable to use the method in Exercise 3.2 or in [KINKEL, PERL, SCHARF, AND STUBBERUD 1979; DEMEURE AND MULLIS 1989].

The calculation of the derivatives of $\{r(k)\}$ with respect to σ^2 and the elements of θ, which appear in the CRB formulas (B.3.3) or (B.3.25), can also be reduced to ARMA (cross)covariance computation. To see this, let α and γ be the real parts of a_p and b_p, respectively. Then

$$\begin{aligned} \frac{\partial r(k)}{\partial \alpha} = -\sigma^2 E\Bigg\{ & \left[\frac{B(q)}{A^2(q)} w(t-p) \right] \left[\frac{B(q)}{A(q)} w(t-k) \right]^* \\ & + \left[\frac{B(q)}{A(q)} w(t) \right] \left[\frac{B(q)}{A^2(q)} w(t-k-p) \right]^* \Bigg\} \end{aligned}$$

and

$$\begin{aligned} \frac{\partial r(k)}{\partial \gamma} = \sigma^2 E\Bigg\{ & \left[\frac{1}{A(q)} w(t-p) \right] \left[\frac{B(q)}{A(q)} w(t-k) \right]^* \\ & + \left[\frac{B(q)}{A(q)} w(t) \right] \left[\frac{1}{A(q)} w(t-k-p) \right]^* \Bigg\} \end{aligned}$$

The derivatives of $r(k)$ with respect to the imaginary parts of a_p and b_p can be similarly obtained. The differentiation of $r(k)$ with respect to σ^2 is immediate. Hence, by making use of an algorithm for ARMA cross–covariance calculation (similar to the ones for autocovariance calculation in Exercises C1.12 and 3.2) we can readily obtain all the ingredients needed to evaluate the CRB matrix in equation (B.3.3) or (B.3.25).

Similarly to the case of line spectra, for relatively large values of N (*e.g.*, on the order of hundreds) the use of the exact CRB formula for rational spectra may be computationally burdensome (owing to the need to multiply and invert matrices of large dimensions). In such large–sample cases, we may want to use an asymptotically valid approximation of the exact CRB such as the one developed in [SÖDERSTRÖM AND STOICA 1989]. Below we present such an approximate (large sample) CRB formula for ARMA parameter estimates.

Let

$$\Lambda = E\left\{\begin{bmatrix} \text{Re}[e(t)] \\ \text{Im}[e(t)] \end{bmatrix}\begin{bmatrix} \text{Re}[e(t)] & \text{Im}[e(t)] \end{bmatrix}\right\} \tag{B.5.1}$$

Typically the real and imaginary parts of the complex–valued white noise sequence $\{e(t)\}$ are assumed to be mutually uncorrelated and have the same variance $\sigma^2/2$. In such a case, we have $\Lambda = (\sigma^2/2)I$. However, this assumption is not necessary for the result discussed below to hold, and hence we do not impose it (in other words, Λ in (B.5.1) is only constrained to be a positive definite matrix). We should also remark that, for the sake of simplicity, we assumed the ARMA signal under discussion is scalar. Nevertheless, the extension of the discussion that follows to multivariate ARMA signals is immediate. Finally, note that for real–valued signals the imaginary parts in (B.5.1) (and in equation (B.5.2)) should be omitted.

The real–valued white noise vector in (B.5.1) satisfies the following equation:

$$\underbrace{\begin{bmatrix} \text{Re}[e(t)] \\ \\ \text{Im}[e(t)] \end{bmatrix}}_{\varepsilon(t)} = \underbrace{\begin{bmatrix} \text{Re}\left[\dfrac{A(q)}{B(q)}\right] & -\text{Im}\left[\dfrac{A(q)}{B(q)}\right] \\ \text{Im}\left[\dfrac{A(q)}{B(q)}\right] & \text{Re}\left[\dfrac{A(q)}{B(q)}\right] \end{bmatrix}}_{H(q)} \underbrace{\begin{bmatrix} \text{Re}[y(t)] \\ \\ \text{Im}[y(t)] \end{bmatrix}}_{z(t)} \tag{B.5.2}$$

where q^{-1} is to be treated as the unit delay operator (not as a complex variable). As the coefficients of the polynomials $A(q)$ and $B(q)$ in $H(q)$ above are the unknowns in our estimation problem, we can rewrite (B.5.2) in the following form to stress the dependence of $\varepsilon(t)$ on θ:

$$\varepsilon(t,\theta) = H(q,\theta)z(t) \tag{B.5.3}$$

Because the polynomials of the ARMA model are monic by assumption, we have:

$$H(q,\theta)|_{q^{-1}=0} = I \qquad \text{(for any } \theta) \tag{B.5.4}$$

This observation, along with the fact that $\varepsilon(t)$ is white and the "whitening filter" $H(q)$ is stable and causal (which follows from the fact that the complex-valued (equivalent) counterpart of (B.5.2), $e(t) = \frac{A(q)}{B(q)}y(t)$, is stable and causal) implies that (B.5.3) is a standard *prediction error* model to which the CRB result of [SÖDERSTRÖM AND STOICA 1989] applies.

Let

$$\Delta(t) = \frac{\partial \varepsilon^T(t,\theta)}{\partial \theta} \tag{B.5.5}$$

($\varepsilon(t,\theta)$ depends on θ via $H(q,\theta)$ only; see (B.5.2)). Then an asymptotically valid expression for the CRB block corresponding to the parameters in θ is given by:

$$\boxed{P_{cr,\theta} = \left(E\left\{\Delta(t)\Lambda^{-1}\Delta^T(t)\right\}\right)^{-1}} \tag{B.5.6}$$

The calculation of the derivative matrix in (B.5.5) is straightforward. The evaluation of the statistical expectation in (B.5.6) can be reduced to ARMA cross–covariance calculations. Since equation (B.5.6) does not require handling matrices of large dimensions (on the order of N), its implementation is much simpler than that of the exact CRB formula.

BIBLIOGRAPHY

Anderson, T. W. (1971). *The Statistical Analysis of Time Series.* New York: Wiley.

Aoki, M. (1987). *State Space Modeling of Time Series.* Berlin: Springer-Verlag.

Bangs, W. J. (1971). *Array Processing with Generalized Beamformers.* Ph. D. thesis, Yale University, New Haven, CT.

Barabell, A. J. (1983). "Improving the resolution performance of eigenstructure-based direction-finding algorithms," in *Proceedings of the International Conference on Acoustics, Speech, and Signal Processing,* Boston, MA, pp. 336–339.

Bartlett, M. S. (1948). "Smoothing periodograms for time series with continuous spectra," *Nature 161,* 686–687. (reprinted in [KESLER 1986]).

Bartlett, M. S. (1950). "Periodogram analysis and continuous spectra," *Biometrika 37,* 1–16.

Beex, A. A. and L. L. Scharf (1981). "Covariance sequence approximation for parametric spectrum modeling," *IEEE Transactions on Acoustics, Speech, and Signal Processing ASSP-29*(5), 1042–1052.

Bhansali, R. J. (1980). "Autoregressive and window estimates of the inverse correlation function," *Biometrika 67,* 551–566.

Bienvenu, G. (1979). "Influence of the spatial coherence of the background noise on high resolution passive methods," in *Proceedings of the International Conference on Acoustics, Speech, and Signal Processing,* Washington, DC, pp. 306–309.

Blackman, R. B. and J. W. Tukey (1959). *The Measurement of Power Spectra from the Point of View of Communication Engineering.* New York: Dover.

Bloomfield, P. (1976). *Fourier Analysis of Time Series — An Introduction.* New York: Wiley.

Böhme, J. F. (1991). "Array processing," in S. Haykin (Ed.), *Advances in Spectrum Analysis and Array Processing*, Volume 2, pp. 1–63. Englewood Cliffs, NJ: Prentice Hall.

Böttcher, A. and B. Silbermann (1983). *Invertibility and Asymptotics of Toeplitz Matrices.* Berlin: Akademie-Verlag.

Bracewell, R. N. (1986). *The Fourier Transform and Its Applications, 2nd Edition.* New York: McGraw-Hill.

Bresler, Y. and A. Macovski (1986). "Exact maximum likelihood parameter estimation of superimposed exponential signals in noise," *IEEE Transactions on Acoustics, Speech, and Signal Processing ASSP-34*(5), 1081–1089.

Brillinger, D. R. (1981). *Time Series — Data Analysis and Theory.* New York: Holt, Rinehart, and Winston.

Brockwell, R. J. and R. A. Davis (1991). *Time Series — Theory and Methods, 2nd Edition.* New York: Springer-Verlag.

Bronez, T. P. (1992). "On the performance advantage of multitaper spectral analysis," *IEEE Transactions on Signal Processing 40*(12), 2941–2946.

Burg, J. P. (1972). "The relationship between maximum entropy spectra and maximum likelihood spectra," *Geophysics 37*, 375–376. (reprinted in [CHILDERS 1978]).

Burg, J. P. (1975). *Maximum Entropy Spectral Analysis.* Ph. D. thesis, Stanford University.

Cadzow, J. A. (1982). "Spectrum estimation: An overdetermined rational model equation approach," *Proceedings of the IEEE 70*(9), 907–939.

Cantoni, A. and P. Butler (1976). "Eigenvalues and eigenvectors of symmetric centrosymmetric matrices," *Linear Algebra and its Applications 13*, 275–288.

Capon, J. (1969). "High-resolution frequency-wavenumber spectrum analysis," *Proceedings of the IEEE 57*(8), 1408–1418. (reprinted in [CHILDERS 1978]).

Childers, D. G. (Ed.) (1978). *Modern Spectrum Analysis.* New York: IEEE Press.

Choi, B. (1992). *ARMA Model Identification.* New York: Springer-Verlag.

Cleveland, W. S. (1972). "The inverse autocorrelations of a time series and their applications," *Technometrics 14*, 277–298.

Cohen, L. (1995). *Time-Frequency Analysis.* Englewood Cliffs, NJ: Prentice Hall.

Cooley, J. W. and J. W. Tukey (1965). "An algorithm for the machine calculation of complex Fourier series," *Math. Computation 19*, 297–301.

Cramér, H. (1946). *Mathematical Methods of Statistics.* Princeton, NJ: Princeton University Press.

Daniell, P. J. (1946). "Discussion of "On the theoretical specification and sampling properties of autocorrelated time-series"," *J. Royal Statistical Society 8*, 88–90.

Delsarte, P. and Y. Genin (1986). "The split Levinson algorithm," *IEEE Transactions on Acoustics, Speech, and Signal Processing ASSP–34*(3), 470–478.

Demeure, C. J. and C. T. Mullis (1989). "The Euclid algorithm and the fast computation of cross–covariance and autocovariance sequences," *IEEE Transactions on Acoustics, Speech, and Signal Processing ASSP–37*(4), 545–552.

Durbin, J. (1959). "Efficient estimation of parameters in moving-average models," *Biometrika 46*, 306–316.

Durbin, J. (1960). "The fitting of time series models," *Rev. Inst. Int. Stat. 28*, 233–244.

Faurre, P. (1976). "Stochastic realization algorithms," in R. K. Mehra and D. G. Lainiotis (Eds.), *System Identification: Advances and Case Studies.* London, England: Academic Press.

Fisher, R. A. (1922). "On the mathematical foundations of theoretical statistics," *Phil. Trans. Roy. Soc. 222*, 309–368.

Friedlander, B., M. Morf, T. Kailath, and L. Ljung (1979). "New inversion formulas for matrices classified in terms of their distance from Toeplitz matrices," *Linear Algebra and its Applications 27*, 31–60.

Fuchs, J. J. (1987). "ARMA order estimation via matrix perturbation theory," *IEEE Transactions on Automatic Control AC–32*(4), 358–361.

Fuchs, J. J. (1988). "Estimating the number of sinusoids in additive white noise," *IEEE Transactions on Acoustics, Speech, and Signal Processing ASSP–36*(12), 1846–1854.

Fuchs, J. J. (1992). "Estimation of the number of signals in the presence of unknown correlated sensor noise," *IEEE Transactions on Signal Processing 40*, 1053–1061.

Fuchs, J. J. (1996). "Rectangular Pisarenko method applied to source localization," *IEEE Transactions on Signal Processing 44*(10), 2377–2383.

Georgiou, T. T. (1987). "Realization of power spectra from partial covariance sequences," *IEEE Transactions on Acoustics, Speech, and Signal Processing ASSP–35*(4), 438–449.

Gersh, W. (1970). "Estimation of the autoregressive parameters of mixed autoregressive moving–average time series," *IEEE Transactions on Automatic Control AC–15*(5), 583–588.

Golub, G. H. and C. F. Van Loan (1989). *Matrix Computations, 2nd Edition.* Baltimore: The Johns Hopkins University Press.

Gray, R. M. (1972). "On the asymptotic eigenvalue distribution of Toeplitz matrices," *IEEE Transactions on Information Theory IT–18*, 725–730.

Hannan, E. and B. Wahlberg (1989). "Convergence rates for inverse Toeplitz matrix forms," *Journal of Multivariate Analysis 31*, 127–135.

Hannan, E. J. and M. Deistler (1988). *The Statistical Theory of Linear Systems.* New York: Wiley.

Harris, F. J. (1978). "On the use of windows for harmonic analysis with the discrete Fourier transform," *Proceedings of the IEEE 66*(1), 51–83. (reprinted in [KESLER 1986]).

Haykin, S. (Ed.) (1991). *Advances in Spectrum Analysis and Array Processing.* Englewood Cliffs, NJ: Volumes 1 and 2. Prentice Hall.

Haykin, S. (Ed.) (1995). *Advances in Spectrum Analysis and Array Processing,* Volume 3. Englewood Cliffs, NJ: Prentice Hall.

Horn, R. A. and C. A. Johnson (1985). *Matrix Analysis.* Cambridge, England: Cambridge University Press.

Horn, R. A. and C. A. Johnson (1989). *Topics in Matrix Analysis.* Cambridge, England: Cambridge University Press.

Hwang, J.-K. and Y.-C. Chen (1993). "A combined detection-estimation algorithm for the harmonic-retrieval problem," *Signal Processing 30*, 177–197.

Iohvidov, I. S. (1982). *Hankel and Toeplitz Matrices and Forms.* Boston, MA: Birkhäuser.

Janssen, P. and P. Stoica (1988). "On the expectation of the product of four matrix-valued Gaussian random variables," *IEEE Transactions on Automatic Control AC–33*(9), 867–870.

Jenkins, G. M. and D. G. Watts (1968). *Spectral Analysis and its Applications.* San Francisco, CA: Holden-Day.

Johnson, D. H. and D. E. Dudgeon (1992). *Array Signal Processing — Concepts and Methods.* Englewood Cliffs, NJ: Prentice Hall.

Kailath, T. (1980). *Linear Systems.* Englewood Cliffs, NJ: Prentice Hall.

Kay, S. M. (1988). *Modern Spectral Estimation, Theory and Application.* Englewood Cliffs, NJ: Prentice Hall.

Kesler, S. B. (Ed.) (1986). *Modern Spectrum Analysis II.* New York: IEEE Press.

Kinkel, J. F., J. Perl, L. Scharf, and A. Stubberud (1979). "A note on covariance–invariant digital filter design and autoregressive–moving average spectral estimation," *IEEE Transactions on Acoustics, Speech, and Signal Processing ASSP–27*(2), 200–202.

Koopmans, L. H. (1974). *The Spectral Analysis of Time Series.* New York: Academic Press.

Kumaresan, R. (1983). "On the zeroes of the linear prediction-error filter for deterministic signals," *IEEE Transactions on Acoustics, Speech, and Signal Processing ASSP–31*(1), 217–220.

Kumaresan, R., L. L. Scharf, and A. K. Shaw (1986). "An algortihm for pole-zero modeling and spectral analysis," *IEEE Transactions on Acoustics, Speech, and Signal Processing ASSP–34*(6), 637–640.

Kumaresan, R. and D. W. Tufts (1983). "Estimating the angles of arrival of multiple plane waves," *IEEE Transactions on Aerospace and Electronic Systems AES–19*, 134–139.

Kung, S. Y., K. S. Arun, and D. V. B. Rao (1983). "State-space and singular-value decomposition-based approximation methods for the harmonic retrieval problem," *J. Optical Soc. Amer. 73*, 1799–1811.

Lacoss, R. T. (1971). "Data adaptive spectral analysis methods," *Geophysics 36*, 134–148. (reprinted in [CHILDERS 1978]).

Lagunas, M., M. Santamaria, A. Gasull, and A. Moreno (1986). "Maximum likelihood filters in spectral estimation problems," *Signal Processing 10*, 19–34.

Levinson, N. (1947). "The Wiener RMS (root mean square) criterion in filter design and prediction," *Journal of Math. and Physics 25*, 261–278.

Li, J. and P. Stoica (1996a). "An adaptive filtering approach to spectral estimation and SAR imaging," *IEEE Transactions on Signal Processing 44*(6), 1469–1484.

Li, J. and P. Stoica (1996b). "Efficient mixed-spectrum estimation with applications to target feature extraction," *IEEE Transactions on Signal Processing 44*(2), 281–295.

Marple, L. (1987). *Digital Spectral Analysis with Applications.* Englewood Cliffs, NJ: Prentice Hall.

Mayne, D. Q. and F. Firoozan (1982). "Linear identification of ARMA processes," *Automatica 18*, 461–466.

Moses, R. L. and A. A. Beex (1986). "A comparison of numerator estimators for ARMA spectra," *IEEE Transactions on Acoustics, Speech, and Signal Processing ASSP–34*(6), 1668–1671.

Moses, R. L., V. Šimonytė, P. Stoica, and T. Söderström (1994). "An efficient linear method for ARMA spectral estimation," *International Journal of Control 59*(2), 337–356.

Mullis, C. T. and L. L. Scharf (1991). "Quadratic estimators of the power spectrum," in S. Haykin (Ed.), *Advances in Spectrum Analysis and Array Processing.* Englewood Cliffs, NJ: Prentice Hall.

Musicus, B. (1985). "Fast MLM power spectrum estimation from uniformly spaced correlations," *IEEE Transactions on Acoustics, Speech, and Signal Processing ASSP-33*(6), 1333–1335.

Naidu, P. S. (1996). *Modern Spectrum Analysis of Time Series.* Boca Raton, FL: CRC Press.

Onn, R. and A. O. Steinhardt (1993). "Multi-window spectrum estimation — a linear algebraic approach," *International Journal on Adaptive Control and Signal Processing 7*, 103–116.

Oppenheim, A. V. and R. W. Schafer (1989). *Discrete-Time Signal Processing.* Englewood Cliffs, NJ: Prentice Hall.

Ottersten, B., M. Viberg, P. Stoica, and A. Nehorai (1993). "Exact and large sample maximum likelihood techniques for parameter estimation and detection in array processing," in S. Haykin, J.Litva, and T. J. Shephard (Eds.), *Radar Array Processing*, pp. 99–151. New York: Springer Verlag.

Papoulis, A. (1977). *Signal Analysis.* New York: McGraw-Hill.

Paulraj, A., R. Roy, and T. Kailath (1986). "A subspace rotation approach to signal parameter estimation," *Proceedings of the IEEE 74*, 1044–1045.

Percival, D. B. and A. T. Walden (1993). *Spectral Analysis for Physical Applications — Multitaper and Conventional Univariate Techniques.* Cambridge, England: Cambridge University Press.

Pillai, S. U. (1989). *Array Signal Processing.* New York: Springer-Verlag.

Pisarenko, V. F. (1973). "The retrieval of harmonics from a covariance function," *Geophys. J. Roy. Astron. Soc. 33*, 347–366. (reprinted in [Kesler 1986]).

Porat, B. (1994). *Digital Processing of Random Signals — Theory and Methods.* Englewood Cliffs, NJ: Prentice Hall.

Porat, B. (1997). *A Course in Digital Signal Processing.* New York: Wiley.

Priestley, M. B. (1989). *Spectral Analysis and Time Series.* London, England: Academic Press.

Proakis, J. G., C. M. Rader, F. Ling, and C. L. Nikias (1992). *Advanced Digital Signal Processing.* New York: Macmillan.

Rao, B. D. and K. S. Arun (1992). "Model Based Processing of Signals: A State Space Approach," *Proceedings of the IEEE 80*, 283–309.

Rao, B. D. and K. V. S. Hari (1993). "Weighted subspace methods and spatial smoothing: Analysis and comparison," *IEEE Transactions on Signal Processing 41*(2), 788–803.

Rao, C. R. (1945). "Information and accuracy attainable in the estimation of statistical parameters," *Bull. Calcutta Math. Soc. 37*, 81–91.

Riedel, K. and A. Sidorenko (1995). "Minimum bias multiple taper spectral estimation," *IEEE Transactions on Signal Processing 43*(1), 188–195.

Roy, R. and T. Kailath (1989). "ESPRIT—Estimation of signal parameters via rotational invariance techniques," *IEEE Transactions on Acoustics, Speech, and Signal Processing ASSP-37*(7), 984–995.

Scharf, L. L. (1991). *Statistical Signal Processing — Detection, Estimation, and Time Series Analysis.* Reading, MA: Addison-Wesley.

Schmidt, R. O. (1979). "Multiple emitter location and signal parameter estimation," in *Proc. RADC, Spectral Estimation Workshop*, Rome, NY, pp. 243–258. (reprinted in [KESLER 1986]).

Schuster, A. (1898). "On the investigation of hidden periodicities with application to a supposed twenty-six-day period of meteorological phenomena," *Teor. Mag. 3*(1), 13–41.

Schuster, A. (1900). "The periodogram of magnetic declination as obtained from the records of the Greenwich Observatory during the years 1871–1895," *Trans. Cambridge Philos. Soc 18*, 107–135.

Slepian, D. (1954). "Estimation of signal parameters in the presence of noise," *Trans. IRE Prof. Group Inform. Theory PG IT-3*, 68–89.

Slepian, D. (1964). "Prolate spheroidal wave functions, Fourier analysis and uncertainty — IV," *Bell System Technical Journal 43*, 3009–3057. (see also *Bell System Technical Journal*, vol. 40, pp. 43–64, 1961; vol. 44, pp. 1745–1759, 1965; and vol. 57, pp. 1371–1429, 1978).

Söderström, T. and P. Stoica (1989). *System Identification.* London, England: Prentice Hall International.

Stewart, G. W. (1973). *Introduction to Matrix Computations.* New York: Academic Press.

Stoica, P., B. Friedlander, and T. Söderström (1987a). "Approximate maximum-likelihood approach to ARMA spectral estimation," *International Journal of Control 45*(4), 1281–1310.

Stoica, P., B. Friedlander, and T. Söderström (1987b). "Instrumental variable methods for ARMA models," in C. T. Leondes (Ed.), *Control and Dynamic Systems — Advances in Theory and Applications*, Volume 25, pp. 79–150. New York: Academic Press.

Stoica, P. and R. L. Moses (1990). "On biased estimators and the unbiased Cramér-Rao lower bound," *Signal Processing 21*, 349–350.

Stoica, P., R. L. Moses, B. Friedlander, and T. Söderström (1989). "Maximum likelihood estimation of the parameters of multiple sinusoids from noisy measurements," *IEEE Transactions on Acoustics, Speech, and Signal Processing ASSP-37*(3), 378–392.

Stoica, P., R. L. Moses, T. Söderström, and J. Li (1991). "Optimal high-order Yule-Walker estimation of sinusoidal frequencies," *IEEE Transactions on Signal Processing 39*(6), 1360–1368.

Stoica, P. and A. Nehorai (1986). "An asymptotically efficient ARMA estimator based on sample covariances," *IEEE Transactions on Automatic Control AC-31*(11), 1068–1071.

Stoica, P. and A. Nehorai (1987). "On stability and root location of linear prediction models," *IEEE Transactions on Acoustics, Speech, and Signal Processing ASSP-35*, 582–584.

Stoica, P. and A. Nehorai (1989a). "MUSIC, maximum likelihood, and Cramér-Rao Bound," *IEEE Transactions on Acoustics, Speech, and Signal Processing ASSP-37*(5), 720–741.

Stoica, P. and A. Nehorai (1989b). "Statistical analysis of two nonlinear least-squares estimators of sine-wave parameters in the colored-noise case," *Circuits, Systems, and Signal Processing 8*, 3–15.

Stoica, P. and A. Nehorai (1990). "Performance study of conditional and unconditional direction-of-arrival estimation," *IEEE Transactions on Signal Processing SP-38*(10), 1783–1795.

Stoica, P. and A. Nehorai (1991). "Performance comparison of subspace rotation and MUSIC methods for direction estimation," *IEEE Transactions on Signal Processing 39*(2), 446–453.

Stoica, P. and B. Ottersten (1996). "The evil of superefficiency," *Signal Processing 55*(1), 133–136.

Stoica, P. and K. C. Sharman (1990). "Maximum likelihood methods for direction-of-arrival estimation," *IEEE Transactions on Acoustics, Speech, and Signal Processing ASSP-38*(7), 1132–1143.

Stoica, P., T. Söderström, and F. Ti (1989). "Asymptotic properties of the high-order Yule-Walker estimates of sinusoidal frequencies," *IEEE Transactions on Acoustics, Speech, and Signal Processing ASSP-37*(11), 1721–1734.

Stoica, P. and T. Söderström (1991). "Statistical analysis of MUSIC and subspace rotation estimates of sinusoidal frequencies," *IEEE Transactions on Signal Processing SP-39*(8), 1836–1847.

Strang, G. (1988). *Linear Algebra and its Applications.* Orlando, FL: Harcourt Brace Jovanovich, Inc.

Therrien, C. W. (1992). *Discrete Random Signals and Statistical Signal Processing.* Englewood Cliffs, NJ: Prentice Hall.

Thomson, D. J. (1982). "Spectrum estimation and harmonic analysis," *Proceedings of the IEEE 72*(9), 1055–1096.

Tufts, D. W. and R. Kumaresan (1982). "Estimation of frequencies of multiple sinusoids: Making linear prediction perform like maximum likelihood," *Proceedings of the IEEE 70*(9), 975–989.

Umesh, S. and D. W. Tufts (1996). "Estimation of parameters of exponentially damped sinusoids using fast maximum likelihood estimation with application to NMR spectroscopy data," *IEEE Transactions on Signal Processing 44*(9), 2245–2259.

van Huffel, S. and J. Vandewalle (1991). *The Total Least Squares Problem: Computational Aspects and Analysis.* Philadelphia, PA: SIAM.

van Overschee, P. and B. de Moor (1996). *Subspace Identification for Linear Systems: Theory - Implementation - Methods.* Boston, MA: Kluwer Academic.

Van Veen, B. D. and K. M. Buckley (1988). "Beamforming: A versatile approach to spatial filtering," *IEEE ASSP Magazine 5*(2), 4–24.

Viberg, M. (1995). "Subspace–based methods for the identification of linear time–invariant systems," *Automatica 31*, 1835–1851.

Viberg, M. and B. Ottersten (1991). "Sensor array processing based on subspace fitting," *IEEE Transactions on Signal Processing 39*(5), 1110–1121.

Viberg, M., B. Ottersten, and T. Kailath (1991). "Detection and estimation in sensor arrays using weighted subspace fitting," *IEEE Transactions on Acoustics, Speech, and Signal Processing ASSP-34*, 2436–2449.

Viberg, M., P. Stoica, and B. Ottersten (1995). "Array processing in correlated noise fields based on instrumental variables and subspace fitting," *IEEE Transactions on Signal Processing 43*(5), 1187–1195.

Walker, G. (1931). "On periodicity in series of related terms," *Proceedings of the Royal Society of London 131*, 518–532.

Wax, M. and T. Kailath (1985). "Detection of signals by information theoretic criteria," *IEEE Transactions on Acoustics, Speech, and Signal Processing ASSP-33*(2), 387–392.

Wei, W. (1990). *Time Series Analysis.* New York: Addison-Wesley.

Welch, P. D. (1967). "The use of fast Fourier transform for the estimation of power spectra: A method based on time averaging over short, modified periodograms," *IEEE Transactions on Audio and Electroacoustics AU–15*(2), 70–76. (reprinted in [KESLER 1986]).

Ying, C. J., L. C. Potter, and R. L. Moses (1994). "On model order determination for complex exponential signals: Performance of an FFT-initialized ML algorithm," in *Proceedings of IEEE Seventh SP Workshop on Statistical Signal and Array Processing*, Quebec City, Quebec, pp. 43–46.

Yule, G. U. (1927). "On a method of investigating periodicities in disturbed series, with special reference to Wolfer's sunspot numbers," *Philos. Trans. R. Soc. London 226*, 267–298. (reprinted in [KESLER 1986]).

Ziskind, I. and M. Wax (1988). "Maximum likelihood localization of multiple sources by alternating projection," *IEEE Transactions on Acoustics, Speech, and Signal Processing ASSP–36*(10), 1553–1560.

REFERENCES GROUPED BY SUBJECT

Books on Spectral Analysis

[Bloomfield 1976; Bracewell 1986; Childers 1978; Cohen 1995; Haykin 1991; Haykin 1995; Kay 1988; Kesler 1986; Koopmans 1974; Marple 1987; Priestley 1989; Percival and Walden 1993]

Books about Spectral Analysis and Allied Topics

[Aoki 1987; Porat 1994; Proakis, Rader, Ling, and Nikias 1992; Scharf 1991; Söderström and Stoica 1989; Therrien 1992; van Overschee and de Moor 1996]

Books on Linear Systems and Signals

[Hannan and Deistler 1988; Kailath 1980; Oppenheim and Schafer 1989; Porat 1997]

Works on Time Series, Estimation Theory, and Statistics

[Anderson 1971; Brockwell and Davis 1991; Bhansali 1980; Brillinger 1981; Cleveland 1972; Cramér 1946; Fisher 1922; Janssen and Stoica 1988; Rao 1945; Slepian 1954; Stoica and Moses 1990; Stoica and Ottersten 1996; Viberg 1995; Wei 1990]

Works on Matrix Analysis and Linear Algebra

[Böttcher and Silbermann 1983; Cantoni and Butler 1976; Gray 1972; Golub and Van Loan 1989; Horn and Johnson 1985; Horn and Johnson 1989; van Huffel and Vandewalle 1991; Iohvidov 1982; Stewart 1973; Strang 1988]

Works on Nonparametric Temporal Spectral Analysis

1. *Historical*
   [Bartlett 1948; Bartlett 1950; Daniell 1946; Schuster 1898; Schuster 1900]

2. *Classical*
   [Blackman and Tukey 1959; Burg 1972; Cooley and Tukey 1965; Harris 1978; Jenkins and Watts 1968; Lacoss 1971; Slepian 1964; Thomson 1982; Welch 1967]

3. *More Recent*
   [Bronez 1992; Lagunas, Santamaria, Gasull, and Moreno 1986; Li and Stoica 1996a; Mullis and Scharf 1991; Musicus 1985; Onn and Steinhardt 1993; Riedel and Sidorenko 1995]

Works on Parametric Temporal Rational Spectral Analysis

1. *Historical*
   [Yule 1927; Walker 1931]

2. *Classical*
   [Burg 1975; Cadzow 1982; Durbin 1959; Durbin 1960; Gersh 1970; Levinson 1947]

3. *More Recent*
   [Choi 1992; Delsarte and Genin 1986; Fuchs 1987; Kinkel, Perl, Scharf, and Stubberud 1979; Moses and Beex 1986; Mayne and Firoozan 1982; Moses, Šimonytė, Stoica, and Söderström 1994; Stoica, Friedlander, and Söderström 1987a; Stoica, Friedlander, and Söderström 1987b; Stoica and Nehorai 1986; Stoica and Nehorai 1987]

Works on Parametric Temporal Line Spectral Analysis

1. *Classical*
   [Pisarenko 1973; Kumaresan 1983; Kung, Arun, and Rao 1983; Paulraj, Roy, and Kailath 1986; Tufts and Kumaresan 1982]

2. *More Recent*
[Bresler and Macovski 1986; Fuchs 1988; Kumaresan, Scharf, and Shaw 1986; Li and Stoica 1996b; Stoica, Moses, Friedlander, and Söderström 1989; Stoica, Moses, Söderström, and Li 1991; Stoica and Nehorai 1989b; Stoica and Söderström 1991; Stoica, Söderström, and Ti 1989; Umesh and Tufts 1996]

Works on Nonparametric Spatial Spectral Analysis

1. *Classical*
[Capon 1969]

2. *More Recent*
[Bangs 1971; Johnson and Dudgeon 1992; Van Veen and Buckley 1988]

Works on Parametric Spatial Spectral Analysis

1. *Classical*
[Barabell 1983; Bienvenu 1979; Kumaresan and Tufts 1983; Roy and Kailath 1989; Schmidt 1979; Wax and Kailath 1985]

2. *More Recent*

[Böhme 1991; Fuchs 1992; Fuchs 1996; Ottersten, Viberg, Stoica, and Nehorai 1993; Pillai 1989; Rao and Hari 1993; Stoica and Nehorai 1989a; Stoica and Nehorai 1990; Stoica and Nehorai 1991; Stoica and Sharman 1990; Viberg and Ottersten 1991; Viberg, Ottersten, and Kailath 1991; Viberg, Stoica, and Ottersten 1995; Ziskind and Wax 1988]

SUBJECT INDEX

C

D

E

F

G

H

I

N

O

P

Q

R

S

T

U

V

W

Y

Z